The motto of the Royal Society is '*nullius in verba*' because in science words alone are empty. Scientists are interested in verbal statements only to the extent that they represent hypotheses to be tested and questioned, to be criticized. Because science grows by first recognizing its faults through self-criticism, and then moving to correct those faults, existing conceptual and theoretical constructs must be criticized.

This book offers a critique of contemporary ecology. It accepts that science is a device to provide information about nature but argues that much of ecology cannot be science because it provides none; much of the rest provides information of such poor quality that it can only be soft science. Although these deficiencies have often been identified, their pervasiveness has not been fully acknowledged, nor have the many similarities among problems in different areas been appreciated. If ecology and environmental science are to meet the needs of the present decade and next millenium, ecological scientists will need far more acute critical abilities than they have yet demonstrated.

Ecologists have minimized the importance of predictive power in assessing scientific quality. Instead, they offer logical rationalization, historical explanation and mechanistic understanding. Given this context, ecologists fall prey to a number of minor failings that complicate and confound any assessment of the science. Even when predictions are possible, they are often vague, inaccurate, qualitative, subjective and inconsequential. Modern ecology is too often only scholastic puzzle-solving.

Ecology can be effective. Informative and predictive it is already a reality in autecology, community ecology, limnology and ecotoxicology. Ecology can become a useful practical science, providing the tools we need to defend the earth and protect our own future, but first we must recognize present inadequacies. This book was written to promote such a development. It is suitable for advanced undergraduate and graduate courses in ecology and the environmental sciences. It should interest professionals in both areas, as well as geographers, landscape architects and others who now try to extract useful information from contemporary ecology.

A critique for ecology

A critique for ecology

ROBERT HENRY PETERS

*Department of Biology, McGill University,
Montreal, Quebec, Canada*

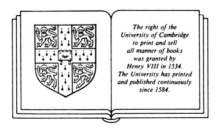

The right of the
University of Cambridge
to print and sell
all manner of books
was granted by
Henry VIII in 1534.
The University has printed
and published continuously
since 1584.

CAMBRIDGE UNIVERSITY PRESS
Cambridge
New York   Port Chester
Melbourne   Sydney

Published by the Press Syndicate of the University of Cambridge
The Pitt Building, Trumpington Street, Cambridge CB2 1RP
40 West 20th Street, New York NY 10011-4211, USA
10 Stamford Road, Oakleigh, Melbourne 3166, Australia

First published 1991

Printed in Great Britain at the University Press, Cambridge

British Library cataloguing in publication data

Peters, Robert Henry
A critique for ecology.
1. Ecology
I. Title
574.5

Library of Congress cataloguing in publication data

Peters, Robert Henry
A critique for ecology / Robert Henry Peters.
    p.   cm.
Includes bibliographical references and index.
ISBN 0-521-40017-1. – ISBN 0-521-39588-7 (pbk.)
1. Ecology.   I. Title.
QH541.P438   1991
574.5'01–dc20   90-21850 CIP

ISBN 0 521 40017 1 hardback
ISBN 0 521 39588 7 paperback

To
A. C. Gregory
J. H. A. Peters
R. Robinson

Contents

Preface   xi

1 Crisis in ecology   1
  Some preliminary disclaimers   2
  Ecologists against ecology   4
  Sociological evidence against ecology   6
  Evidence from the deepening environmental crisis   10
  Academic ecology poses unanswerable questions   13
  Summary – Scientific growth depends on scientific criticism   14

2 Criteria   17
  By definition and example: logic, science and theory   18
  Hypothetico-deductive science   21
  Criteria for judging scientific theories   26
  Summary – A hierarchy of scientific criteria   36

3 Tautology   38
  Tautologies and deductive tools   38
  The principle of evolution by natural selection   60
  Summary – Two tools for two jobs   73

4 Operationalization of terms and concepts   74
  Operationalization of concepts   76
  Typologies and classifications   80
  Conceptual variables – stability and diversity   92
  Non-operational relationships   96
  Atheoretical concepts   97
  Concepts in ecology: the effects of poor examples   100
  Summary – The costs of non-operational concepts for ecology   104

5 Explanatory science: reduction, cause and mechanism   105
  Prediction and explanation: alternate goals for science   106
  Reductionism: an unattainable goal   110
  Causality   128
  Instrumentalist research   136
  Summary – The twin perils of mechanistic and causal explanations   146

6 Historical explanation and understanding — 147

Scientific explanation and understanding — 147
Historical explanations and ecology — 154
Legitimate roles for historical understanding in ecology — 170
Summary – Explanations in ecology — 176

7 Weak predictions — 178

Relevance — 178
Accuracy — 189
Imprecise and qualitative predictions — 196
Generality and specificity — 211
Economy — 216
Appeal — 218
Summary – Practicality and appeal — 219

8 Checklist of problems — 220

The Introduction — 221
Methods — 229
Results — 235
Discussion — 239
Extensions and hypotheses — 250
Summary – The challenge of good science — 254

9 Putting it together – competition — 256

The prevalence of competition — 257
Operationalization — 259
Tautology — 263
Historical explanation — 266
Mechanisms of competition — 268
The theoretical status of 'competition theory' — 270
Summary – The muddles of ecology — 273

10 Predictive ecology — 274

Eight classes of model theories in predictive ecology — 274
The attractions of predictive ecology — 290
Summary – A scientific alternative for ecology — 304

References — 305
Index of names and first authors — 345
Subject index — 352

Preface

The subject of ecology is the relation of living organisms including human beings to their environment. This makes ecology the most important and all-embracing of the sciences. Nevertheless, academic ecology is no more powerful than many other disciplines, and it is weaker than some. Such unflattering comparisons can be attributed to the scope, youth and complexity of ecology. However, the challenge for ecological scientists is not how to excuse modest success, but rather how to succeed more fully by answering the pressing ecological questions of the time. One step in this direction is a critical evaluation of contemporary ecology so that future work can be more effective. This book offers such an evaluation, not because ecology is the weakest of sciences – others are considerably more so – but because ecology is the most important.

The first chapter of this book provides evidence that part of ecology is weak, that many aspects of its weakness are now widely recognized, and therefore that a comprehensive critique of the field may be timely. The second chapter advocates predictive power as the major criterion for such a critique. Thus this book is intended not as an overview of contemporary ecological controversies, although these often illustrate the text, but as a coherent examination of ecology's inability to predict. Chapter 2 acknowledges that science uses both theories (defined as constructs that predict) and non-theories (defined as constructs that do not). Each class is important in science, but each performs different functions. When they are confused, neither performs its role well.

The next four chapters consider four kinds of non-theories in ecology. Chapter 3 deals with tautologies, constructs that identify the range of possibility but do not distinguish the more probable events within that range. Chapter 4 examines non-operational concepts, those ideas that are so vaguely defined that different scientists associate different phenomena with the same term and concept. Chapter 5 and 6 treat the scholarly goal of explanation through various mechanistic, causal and historical models, contending that explanations have become alternatives to prediction in ecology and thereby subverted the science as a source of predictive knowledge. All four chapters seek to show that non-theoretical constructs are common in ecology where they have obscured the need for prediction because they inappropriately replaced theory, thereby exceeding their proper roles in rationalization and inspiration of research.

Next, the book addresses ecological constructs that make predictions and are therefore theories under the criterion of Chapter 2, to show that conformity with this basic requirement does not ensure powerful or interesting theories. Chapter 7 describes aspects of ecological theory that affect the degree of predictive power and therefore serve to distinguish powerful theories from weaker ones. Chapter 8 examines primary research papers in ecology to illustrate that a host of large and small problems debilitate these building blocks of the science; that review can serve as a starting point for critical examination of research papers. Chapter 9 collates various elements of scientific criticism into a critical overview of competition theory, thus demonstrating that theoretical short-comings are not encountered singly but in combination, and that this interaction compounds and reinforces the difficulties of criticism.

In art, criticism has a grim finality because a work of art is fixed forever. In science, criticism is not such a hopeless exercise. It is simply the first step to improvement. Thus, the growing wave of critical comment, of which this book is part, provides no reason for despair, but only for hope. Once the problems of a science are identified, the science can grow around them. To foster this hope, the final chapter describes modern alternatives to traditional ecology which already offer more predictive power and applicability. These achievements serve as models for a future, more predictive, and more scientific ecology.

If parts of ecology are weak, other parts are not. The problems facing us are great, but we have resolved such problems before. We can also meet the challenges of this generation, but to do so we must make the very best use of our capacities and tools. For scientists, this necessity requires that we hone our theories and research until they are as keen and sharp as they may be. The whetstone against which we sharpen our tools is criticism. Criticism, the evaluation of current scientific constructs and the identification of their short-comings, is therefore essential to scientific strength.

Because scientists are proud of their achievements, scientific criticism can hurt, and criticism of a construct is easily confused with criticism of the construct's originators and proponents. This is understandable, but unfortunate. I regret any pain this critique may cause. It is not my intention to offend any ecologist. I have enjoyed meeting many of those whose work I criticize, I admire all, and I am pleased to consider some my friends. To blunt the appearance of malice, I have made a conscientious attempt to avoid splitting the discipline into those who do 'true science' and those who do not. Researchers whose work is criticized at some point in the book will often find other work or other aspects of the same publication commended elsewhere; some of the research by noted scientific critics receives criticism when it is merited. Moreover, most of the examples I criticize were selected to show that even the most respected ecologists are sometimes open to criticism so the criticisms are levelled because the work is well known and because the authors are intellectual leaders of the discipline. In that sense, to be criticized in these pages is intended as a form of compliment, however it is received. Certainly, those who are criticized will find themselves in the very best company.

One of the advantages of science over art is that critics are also expected to create and so they are subject to their own criteria. Would-be critics of my research can rest assured that many problems outlined here I first recognized in the pages of my own publications, and I have made a point of occasionally criticizing my own work in this critique. It would be pleasant to work without error, but since that is an impossibility, it is at least more honorable to acknowledge past errors than to ignore or repeat them. Scientists should feel no embarrassment in fallibility, because we can all do better, but any who feel unfairly criticized herein can find comfort in the thought that a critic who admits past failures may well have erred again.

The fallibility of all scientists puts a great onus on scientific readers. There is a great danger of distortion when a paper or book is reduced to a paragraph or phrase in critical analyses. Readers of this critique must, therefore, read it critically. When possible, my views should be compared with those the author expressed in the primary publication. If I, as the critic, am guilty of misinterpretations, then I must shoulder all the blame. However, if readers repeat my errors without their own critical evaluation, they have also earned their share of my blame.

This text has benefitted from counsel and critical reviews of Chris Chambers, Yves Prairie, and Bill Shipley in Montreal, from Antoine Morin in Ottawa, from Larry Barnthouse, Lyse Godbout, and R. V. O'Neil at Oak Ridge, from John Grehan then at the University of Vermont, and from John Birks, Steve Cousins, Bob Paine, Alan Crowden and anonymous readers of the Cambridge University Press. I could not adopt all of their suggestions, for unanimity of opinion about all the topics in this book could not be expected, and naturally none of these advisers should be blamed for my views. Their input was essential to let me see my writing with fresh eyes and so avoid many blunders I would otherwise have made. Chapter 6 was heavily influenced by the late Ralph Robinson III and Chapter 8 owes much to conversations with John Downing. I am grateful to Sara Griesbach for her continuing help throughout the preparation of the manuscript, to Guylaine Richer for correcting the references, and to Natalie Richter, Anne Hilton, Lucy Byrne, and Erin Coull, for the preparation of the typescript, and also to the staff of Cambridge University Press. Once again, I am indebted to Riccardo de Bernardi, director of the Istituto Italiano di Idrobiologia in Pallanza, the researchers at the Institute, and the Institute staff for their hospitality as I wrote the first drafts of the book. Finally, I thank my wife, Antonia Cattaneo, and our children, Julian and Elisa, for their help, patience, and understanding through a long and sometimes difficult gestation.

31 July 1990   RHP

1 · Crisis in ecology

It is an elemental proposition that, if we want to get someplace, then we must know where we are and where we want to go. For similar reasons, science progresses more easily when present limits and future goals are known. Such knowledge is achieved through scientific criticism. Scientific criticism shows where we are, it can help decide where we should go, but it does not provide the vehicle to carry us from where we are to where we want to be. That difficult problem must be solved by the creative genius of the scientific community. This critique therefore evaluates the present state of ecology and suggests appropriate goals for the science, but except in its final chapter, it does not deal with how we can achieve those goals.

Criticism is easier where the standards for performance are widely accepted, well developed and consistently applied. Such fields have a high potential for growth because they explicitly recognize the shortcomings of present achievements relative to the goals of the science. In other fields, goals and criteria are poorly enunciated, less accepted, and more sporadically applied. These sciences are less coherent, they contain many constructs of dubious merit, and their growth is lethargic. It is my thesis that much of contemporary, academic ecology belongs with the latter group. The science has languished, whereas public demand and the practical necessity for attractive, powerful, ecological theory has mushroomed.

To many contemporary ecologists, the weakness of ecology is patent and the problem needs little elaboration. Di Castri & Hadley (1986) list three perceived short-comings – lack of scientific rigour, weak predictive capability, failure to harness modern technology – and many indications that ecology is in a difficult period – lack of testable theory, low research budgets (compared to other biological sciences), lack of employment opportunities, proliferation of uncontrolled, uncoordinated studies, inadequate contacts with specialists from other disciplines, a tendency of ecologists to demagogy and polemics, and the rarity of interaction between ecologists and planners. Given that such shortcomings are widely felt, this introductory chapter, whose aim is to provide evidence that ecology is sometimes a weak science in need of harsh critical scrutiny, can be brief. It examines four sets of evidence to show first that many ecologists believe contemporary ecology needs critical re-evaluation, second that patterns of citation suggest that ecology is a soft science, moving slowly under uncertain and subjective criteria, third that, although our environmental problems require a dynamic, effective, practical science to confront some of the greatest dangers that humanity has ever faced, the extent of the environmental problem shows that we have just the opposite, and fourth, that the traditional questions and goals of ecology work against the forging of a new and effective ecological science.

This introductory chapter shows that the health of ecology is not good, but also that its disease is now widely recognized. Since part of the cure for weak science is unrelenting critical reevaluation of its parts, the recognition of weakness augurs for a return to growth and development, perhaps even for a scientific revolution. Therefore flourishing criticism is not negative or damaging to the science. It is part of rebirth. It is essential to growth.

Some preliminary disclaimers

This is a book of criticisms. Its basic rationale, content and intent have been sketched briefly in its opening paragraphs and somewhat more fully in the preface. This section is instead intended to forestall certain misgivings by describing what the book is not, in terms of its scope, its purpose, its application and its limitations.

Scope

Ecology, broadly defined, includes just about everything involving man and his environment, and that includes just about everything. This is decidedly not the subject of this monograph. Nor is this criticism directed at those broadly ecological or environmental initiatives that would be more properly part of public health, agronomy or land management, like disease control, forest management or eutrophication abatement (Cantlon 1981). Research in those areas has provided a highly sophisticated set of predictive relations to control pests and diseases while nurturing beneficial organisms, even if these are largely limited to humans and their commensuals. These disciplines include some promising developments that could be instructive for other parts of ecology, but they have had little effect on the mainstream of ecology, as can easily be verified by the contents and citations in ecological journals and texts. Subjects like agronomy and epidemiology could be relevant to ecology, but in practice there has been little exchange (Paul & Robertson 1989).

The ecology to which this book is directed is that found in ecology courses offered by biological departments at most of the world's universities. It is the basis of the standard texts by Odum (1971), Ricklefs (1979), McNaughton & Wolf (1979), Krebs (1985), Colinvaux (1986), Begon, Harper & Townsend (1986), Pianka (1988), and others. It is the ecology that fills journals like *American Naturalist, Journal of Ecology, Ecology, Oecologia, Oikos* and the *Annual Review of Ecology and Systematics*. The ecology dealt within this book is therefore ecology *sensu strictu*. It is, incidentally, also the ecology with which I have worked for all of my professional career and which I have, in some small way, helped to build.

Purpose

Because this book is intended to help practicing ecologists with their subject matter, it rarely criticizes other sciences. Nevertheless, I am confident that analogous problems could be found elsewhere, from the allied fields of forestry, fisheries and agriculture to that 'queen of the sciences', physics (c.f. Cartwright 1983). Because my criticisms usually apply broadly in science, I am not concerned that this critique will be misappropriated as a weapon to discredit ecology to funding agencies or the public. Those who wield such a weapon will find criticism a two-edged sword.

For ecology, I hope the book will help establish a stronger science of the environment and I anticipate that such a development would improve our funding. At present, money is pushed to ecology and environmental sciences partly in the hope that we will address our mounting environmental problems. In the future, I hope that a record of strong theory and cogent advice will attract even more money so that a new, more effective science of ecology will grow even more quickly. The purpose of this critique is to foster such a development by recognizing current deficiencies.

Such recognition requires close scrutiny of contemporary ecology to identify its weakness. In doing so, the demands of brevity make it inevitable that the positive aspects of the science be acknowledged only very briefly. This one-sided approach is justified because science grows by correcting its failures, rather than by enjoying complacently its successes. In any case, the ecological literature has so stressed its positive characteristics, that the negativism of a critique only restores a balance.

For the same reason, this critique has little space for the excuses ecologists offer to explain the inadequacies of the science: the complexity of the material, the youth of the field, the inadequacy of funding, etc. These complaints are not germane to this critique, for its purpose is not to understand our failure but to correct it. I doubt that such a correction is possible if we begin by accepting our own excuses.

Perhaps other sciences succeeded better because their tasks were easier. Perhaps, present weaknesses in ecology reflect the youth of ecology, and perhaps similarly unflattering numbers could be assembled for sciences like physics, chemistry and medicine in other centuries. Unfortunately, we cannot wait centuries for ecology to mature. Instead, we must hasten development with our available intellectual and financial resources so that ecology at the turn of the millennium can confront the problems that assail us. We must force ecology to grow and grow quickly. Criticism is the goad for such progress.

Limitations

A recurrent theme in this book is the vagueness of ecological constructs, and this critique cannot explore all possible interpretations. So much of the science is phrased so ambiguously that the meaning of most constructs is open to reinterpretation by both critic and defendant. Indeed, one role of criticism is to force this reinterpretation so that vague constructs become less so. Throughout this book, I have been forced to define and interpret ecological ideas to give them sufficient form and content so that they could be evaluated. If this necessity has led me to misinterpretation, I have erred, but part of the blame must lie in the uncertainty of ecological formulations and part of the solution must be better descriptions.

Ecologists against ecology

Ecology is currently undergoing a re-evaluation. Some leaders in the field have renounced their earlier work (Rigler 1975a; Simberloff 1976a; Dayton 1979; Wiens 1983) and, using the metaphor of the sickness of their science, they have sought a cure by fundamentally changing their approach to their material. The contemporary literature offers scientific criticism in the form of debates (e.g. Wilson 1975 vs. Lewontin 1979;

Gould & Lewontin 1979 vs. Mayr 1983; Connor & Simberloff 1984a,b vs Gilpin & Diamond 1984a, b; Lehman 1986 vs Peters 1986; Stearns & Schmidt-Hempel 1987 vs. Pierce & Ollason 1987; Hall 1988 vs Caswell 1988), iconoclastic views (Hailman 1982; Wiens 1983), critiques (Smith 1952; Watt 1975; Howe 1985) and simple devil's advocacy (Fowler & Lawton 1985). This literature has heightened our awareness of scientific standards.

Lively debate is not new in ecology. Haeckel questioned the quantitative techniques of the planktologist Hensen in 1891 (McIntosh 1985), Ganong (1904) objected to the speculative nature of the science at the turn of the century, W. R. Thompson voiced suspicions about the relevance of mathematical models in the 1930s (Kingsland 1985), Haskell noted the inability of ecology to predict in 1940, the revolt against superorganisms and the debate over population regulation filled the literature of the 1940s and 1950s (McIntosh 1985), and group selection was questioned in the 1960s (Wilson 1983). Major ecological symposia focussed on important questions of the day seem to recur every decade (May 1984): The British Ecological Society symposium on the ecology of closely allied species in 1944, the Cold Spring Harbour Symposium of 1957, and the Brookhaven Symposium of 1969, followed rather closely by the meeting on the Ecology and Evolution of Communities (Cody & Diamond 1975).

Characteristically, these debates did not resolve the issues they addressed. They provided a forum to air views and the combatants gained some appreciation of opposing positions, but the problems remained. Thus Strong (1984, 1986a, b) can still address the debate over density-dependence and independence, Lovelock (1979) can propose the super-organismic 'Gaia hypothesis', and 'decent folk' can again speak of group selection (Wilson 1983). Ecology has not so much lacked for criticism, as for effective criticism whereby we learn from the debate, rather than just debate.

There are signs that the critical literature in ecology is now so extensive that it cannot be ignored. In the last decade, the pace of such symposia and group reviews has increased dramatically: The Seattle conference of 1980 (American Zoologist 1981), Saarinen's (1982) book on ecological concept, The Wakulla conference in 1981 (Strong *et al.* 1984), the Asilomar meeting of 1987 (Roughgarden, May & Levin 1989) and symposia on competition, island biogeography (*Atti Zoologici Fennici* 1982, 1987; *Oikos* 1980), methodology and competition (*American Naturalist* 1983; Salt 1984; *American Zoologist* 1987). The frequency with which symposia are held to assess the state of the science, the number of introspective articles, and the level of emotions in the debate of the last decade seem unprecedented (Lehman 1986; Slobodkin 1986 a,b; McIntosh 1987).

The evidence shows that reputable scientists feel ecology to be less credible and weak or increasingly fractious, critical and introspective. Such subjective evaluations are important, but treacherous. Science is not a democratic institution, and the opinions of the majority, or a substantial minority, are necessarily suspect measures of scientific worth. There are always contrary views and these views may be valid. Moreover, because science doubles every 10 to 15 years (Price 1986), the amount of any material, including ecological criticism, always seems to be exploding. Subjective evidence must therefore be interpreted in the light of other evaluations.

Sociological evidence against ecology

The charge of weakness in ecology can be developed into hypotheses about the characteristics of the science and tested with a number of metrics reflecting the behaviour of ecologists with respect to other scientists. This section examines several such tests, the results of which are consistent with the view that much of ecology is scientifically weak.

Citation frequency

Every scientist expects strong papers to be read and used by other scientists, while weak ones are ignored. As a result, good papers will be cited more frequently than average, and weak papers cited less. Thus one could test the theory that much of ecology is weak by determining whether ecological papers are cited less frequently on average than non-ecological papers. Several lines of evidence support the hypothesis.

Garfield (1972) compared literature use by a number of journals in 1969. His data show that the ratio of the journals cited in the journal, *Ecology*, to the number citing *Ecology* was the highest of all the journals examined. He asked the rhetorical question 'What does this say about *Ecology* or ecology?'. To this, I suggest it implies that ecologists tend to consume scientific knowledge rather than produce it. Ecology takes material from other sciences but is itself less uninteresting to them.

A second line of evidence is presented in Fig 1.1 which compares the frequency of citation from 1961 to 1981 of broadly ecological, beha-

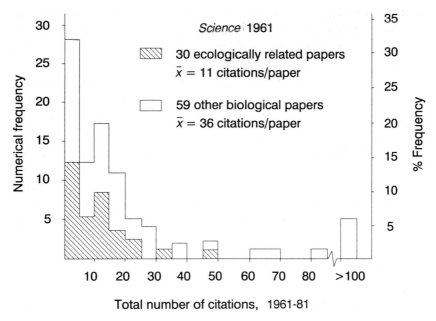

Fig. 1.1. Distribution of citations received over 20 years by biological articles and letters published in the journal *Science* in 1961. Broadly ecological works are less frequently cited.

vioral and evolutionary articles appearing in the journal *Science* in 1961 to the citation frequency of other biological articles in the same issues. The distributions suggest that ecological articles proved only half as interesting as those in other fields. The difference is particularly marked among heavily cited articles. It may be objected that popular articles in other sciences are more likely to achieve high citation rates because other biological sciences cover bigger fields that interest more people. This seems simply another facet of the weakness in ecology. It is a small field because of its frailty, not vice versa.

A third line of evidence ideally would compare average citation rates in ecology with those in the rest of science. This is not possible, because the average rates of citation in the field of ecology are not known. Peters (1989a) found that the average citation rates in the journals *Ecology* and *Limnology and Oceanography* were almost exactly once per paper per year, although a few papers were heavily cited and many were cited more rarely. Because *Science Citation Index* shows these to be among the most cited ecological journals, their citation rate should exceed the average for science, if ecology is a science like the others. Instead, Price (1986) reports that the average for science is also once per paper per year and the definition of a 'citation classic' in any issue of *Current Contents* suggests a similar average. Since papers in the most cited ecological journals are only performing at the average rate for science, most ecological papers must be cited less frequently than average.

Other indices

More evidence for weakness in ecology can be gleaned from the comparative tables published by the *Institute of Scientific Information* based on patterns of citation in the *Science Citation Index* (Table 1.1). Among the listed statistics are the half-life of the references cited by a particular journal (the citing half-life), the half-life of the citations to the particular journal (the cited half-life), the impact factor (the ratio of number of citations the journal received in the two previous years to the number of papers it published), and the immediacy (the number of citations to papers published that year). The few statistics in Table 1.1 indicate the position of ecology relative to other sciences: Ecology uses the older literature more because its half-lives are long. Its papers are cited less frequently than those of other sciences, so its impact is low, and low immediacy suggests that there is little haste to use the most recent work. Indeed, the pace of publication in ecology is so leisurely that it would be difficult to cite the most recent literature without relying on preprints and other rapid means of communication. This is surely another part of the same problem.

Ecology is simply not the competitive, high pressure, interactive discipline that some other sciences are. This has some real advantages for the working environment of ecologists, but perhaps it is not conducive to the development of the best science.

Table 1.2 presents another index of performance of different disciplines, the rates of rejection of submitted manuscripts by learned journals. This table shows that a higher proportion of submitted manuscripts are rejected in ecology than in other sciences. Although high rejection rates might suggest high standards, they are more likely to reveal disparity in the expectations of the ecological contributors and their referees. Ecologists, like historians and philosophers, are more likely to invoke subjective criteria in evaluating their peers. Therefore, ecological scientists may submit papers which are acceptable under their own set of standards, but ecological reviewers may see the same paper as unacceptable under another. As a result, acceptance by an ecological

Table 1.1 *Some comparative statistics from the* Scientific Citation Index *on the use of top-ranked journals in ecology and other sciences in 1986.* The half-life is the median age in years of (A) all references to the journal cited by the literature in a particular year or (B) all references the journal cites in that year. Impact is the ratio of citations the journal receives in the preceding two years to the number of citable items published in the same years, and immediacy is the ratio of citations to items published in 1986 to the number of items published in 1986.

	Half-life			
Journal	A	B	Impact	Immediacy
Ann. Rev. Biochem.	5.9	3.9	31.6	3.1
Cell	3.8	3.5	20.1	3.4
Ann. Rev. Plant Phys.	6.5	4.7	16.4	1.1
Nature	5.7	3.8	15.2	3.2
Science	5.7	4.2	12.4	3.0
Proc. Nat. Acad. U. S. A.	5.4	4.1	9.2	1.5
J. Cell Biol.	5.4	4.5	8.8	1.6
Embo J.	2.4	3.6	8.1	1.5
Ann. Rev. Physiol.	5.1	4.6	7.8	1.2
Phys. Rev. Letter	4.4	4.4	6.5	1.5
★Ann. Rev. Ecol. Syst.	7.6	6.7	6.5	0.1
★Ecol. Monogr.	>10	9.6	5.7	1.1
★Adv. Ecol. Res.	>10	9.8	5.2	1.6
J. Amer. Chem Soc.	9.3	6.9	4.4	0.8
★Limnol. Oceanogr.	8.9	6.9	3.1	0.8
★Amer. Natur.	8.2	7.7	2.9	0.5
★Evolution	7.3	7.7	2.8	0.7
★Ecology	9.7	8.2	2.6	0.5
★Can. J. Fish. Aquat. Sci.	3.8	7.4	1.8	0.6

★ Ecological journals.

journal sometimes feels like winning a lottery and more than one author has complained of the injustices this system can produce (Van Valen & Pitelka 1974; Dayton 1979). By contrast, the standards in 'hard' sciences are known well enough that both author and reviewer work to similar criteria and are more likely to agree about the value of a manuscript. Thus rejection rates in hard sciences are low and those in soft sciences and non-sciences are high (Merton 1973; Peters 1989a).

There is other evidence that ecology is in a period of eclipse.

Table 1.2 *Rejection rates in journals in science and the humanities in 1967 (from Merton, 1973) compared to those in ecological journals. n is the number of journals surveyed.*

Field	%Rejected	n	Field or journal	%Rejected
Arts & soft sciences	82	28		
History	90	3		
Literature	86	5		
Philosophy	85	5		
Political science	84	2		
Sociology	78	13		
Intermediate sciences	61	20	Ecological sciences	54
non-experim. psych.	70	7	*Ecology* 1986/7	65
economics	69	4	*Amer. Natur.* 1966/8	47
experim. psych.	51	2	*Amer. Natur.* 1986/7	62
Maths & Stat's.	50	5	*Limnol, Oceanogr.* 1967	35
Anthropology	48	2	*Limnol. Oceanogr.* 1986	60
Hard sciences	27	33		
Chemistry	31	5		
Geography	30	2		
Biology	29	12		
Physics	24	12		
Geology	22	2		
Linguistics	20	1		

Undergraduate enrollments in ecology courses have fallen, graduate student applications have declined. In many institutions, the number of ecologists on staff is being reduced to allow the growth of molecular biology and genetic engineering. Government funding has been restricted (di Castri & Hadley 1986). All of these indices suggest some failings in ecology.

Evidence from the deepening environmental crisis

The third class of observations showing ecology to be a weak science is the most apparent and the most incontrovertible. It is that the problems that ecology should solve are not being solved. They are worsening, growing more imminent, more monstrous. Every day at least 20 000 hectares of forest are destroyed (Brown & Lugo 1988). Every year, one or more vertebrate species disappears forever (Chiras 1988). The atmosphere is less pure, the waters more foul, and the land more polluted. Deserts spread, top-soil erodes, water tables fall. And all of this occurs against the stark increase in human population, pressing us against the limits of our capacity, consuming our gains, and carrying us closer to the edge of disaster.

There have been spectacular successes in applied ecology and conservation. Some species have been brought back from the brink: the alpine ibex (*Ibex*), the American and European bisons (*Bison*), the whooping crane (*Grus americana*), and the Hawaiian goose or nene (*Branta sandvicensis*). The number of parks and preserves is increasing, the public is more aware of the value of wilderness and wildlife, there have been improvements in some aspects of water and air quality, governments and industry have shown themselves willing to invest in pollution abatement (Cantlon 1981). Even the world press has remembered the ecological crisis, although this needed the global threats of the eroding ozone layer, the greenhouse effect and Chernobyl, and local horrors like the chemical spills at Seveso, Italy, Bhopal, India, and on the Swiss Rhine. Nevertheless, the balance is clearly negative.

This charge will seem unfair. To many, it is unreasonable to fault the science of ecology for its inability to come to grips with global problems of the environment. The social forces of population, politics, and economics that threaten our ecosystem are massive, the ecological complexity of the biosphere is overwhelming, and funding for ecological science is woefully inadequate. All this is undoubtedly so, but if ecologists are to take their message of doom seriously, then the problems are real and the problems are increasing. The defence that ecology is too small and the problems are too big simply excuses inaction. Worse, it reinforces the predicament of man by discouraging scientific solutions. Whether ecological problems are harder than those of other sciences or not, someone must address them. As many have noted before (Southern 1970; McIntosh 1985; Schlesinger 1989; Woodwell 1989), if ecologists do not provide solutions, who will?

Others believe that the problems that confront us are no longer ecological, because ecologists can already provide global answers, like zero population growth, radical conservation, and complete recycling. Yet unattainable solutions are no solutions at all. The current problem for ecologists is to provide feasible solutions that will lead us towards these ultimate answers and clear choices that will make appropriate, but costly, actions desirable. Politicians have not proven deaf to sound ecological advice given in implementable terms. The governments of most developed countries have invested billions of dollars in phosphorus abatement, and before that in sewage control. The government of Sweden has embarked on a costly and extensive program of lake and stream liming to neutralize the worst effects of acid rain (Forsberg 1987). In all cases, the scientific models provided objective bases for rational management whereby the decision maker could weigh the costs of action against the benefits and, with these tools, society has reversed some of the damage it has wrought.

We must find ways to fund and encourage more work and more scientific tools that will provide more effective advice about more ecological topics. One element in this encouragement is a realistic and critical assessment of the political importance of contemporary ecology.

Ecologists could influence policy by entering the public arena with essentially political arguments. This route has been successfully adopted by political scientists, economists, lawyers, and even sociologists who therefore seek to provide, respectively, political, economic, legal and social reasons for decision. There is similarly an ecological rationale (Dryzek 1983) which is exploited by astute ecologists, by 'green' politicians, and by conservation groups to impose more ecologically balanced policy on the basis of public pressure, passionate belief, conservationist ethic and enlightened self-interest. Political involvement likely requires all these elements, but individual ecologists must strive to distinguish their science and research from views they hold as citizens.

Political pressure was the basis for ecological action in the 1960s and 1970s, and it was successful in advancing conservation. However, because of the weakness of ecological theory, this approach could not be scientifically based. It required ecologists to step outside their roles as scientists and speak in unfamiliar territory where they bear little or no authority. The failure of early prognostications (Mellanby 1987) and the zeal of the environmental absolutism which inspired them continues to haunt ecologists who try to give either political or scientific advice. Policy makers have learned that they have the right and the duty to ask those who proffer ostensibly scientific advice for the basis of their suggestions; and many are understandably unimpressed by global doomsayers who are unable to predict within their scientific specialty.

The political role of ecologists has fallen increasingly to non-scientists (Vallentyne 1974). In some ways, this is fortunate. It encourages ecological scientists to separate political beliefs from scientific expertise and perhaps to concentrate their efforts on the development of scientific tools for environmental action.

Experience has shown that useful and meaningful ecological advice is often taken seriously (Cantlon 1981), because it depends on the authority given by scientific expertise. To use this dependency effectively, ecologists must distinguish their beliefs from their science. The former may serve as an ethical basis for political action and there should be no barrier to ecologists also acting as concerned citizens, just as other members of the body politic base their suggestions on legal, economic, social or political rationales. However visceral beliefs in the good of conservation or the danger of pollution are incidental to scientific capabilities. Indeed, we often succeeded as conservationists or concerned natural historians, but as ecological scientists we often failed, letting good intentions replace sound, scientifically based, advice in the public arena. It is time that we recognize that failure, and repair past damage by reasserting ourselves as scientists.

Academic ecology poses unanswerable questions

Because the ecological crisis looms so large, one might expect to find a new sense of urgency and application in the scientific questions of the day. However, the mainstream of ecology has instead maintained its traditional material and questions.

Science has been called 'the art of the soluble' (Medawar 1967) because science succeeds by answering questions, not simply by posing them. Unfortunately, ecology has often asked intractable questions (Rigler 1982a). This is particularly apparent in introductory textbooks, because the writers of these texts strive to present the central issues of the contemporary science for their readers and may be less concerned with the scientific status of these questions. Table 1.3 lists a series of questions that have been presented for consideration by university students.

For reasons that form the basis of this book and are developed more fully in subsequent chapters, the questions quoted in Table 1.3 are weak. Unlike scientific hypothises, they tell us little about the world around us. Some refer to entities which we cannot identify unambiguously, others call only for definitions, or are phrased in terms that can only be answered by circumlocutions and discursive statements of personal opinion. Still other questions encourage infinite research programs or scholastic debates, because no specifiable observation would constitute an answer. As a result, ecologists can continue to ask the same questions, to elaborate their answers, to refine their approaches, to collect data and to express their opinions without any danger of finishing their research.

If ecology is to succeed we must shun questions like these and learn to pose testable hypotheses. Such hypotheses often involve questions that begin with 'how much', 'how many', 'when', and 'where'. Intractable questions frequently begin with 'why' or 'how come'.

The use of introductory texts to assemble Table 1.3 underlines one of the perils of unanswerable questions. They are perpetual conundra, as puzzling to one generation as the next. Because of this longevity, they accumulate in the literature, crowding out resolvable problems, and becoming paradigmatic ecological questions. As a result, one generation mis-trains the next, and ecologists become progressively less able to confront real problems because they have less and less experience with real answers (Hall 1988).

The problem does not end with the questions taught burgeoning ecologists. Many leading ecological thinkers seem more fascinated by unanswerable questions than testable hypotheses. For example, Cody (1974) once held that the 'the goal of ecology is to provide explanations that account for the occurrence of natural patterns as products of natural selection'; the goal of supporting natural selection inverts the widely accepted view that the role of science is to test theory and to point out its failures, so that the theory can be improved. Trail-blazing monographs by other ecologists echo the intractable questions of Table 1.3: Fretwell (1972) poses the general question, 'What are the factors regulating the distribution and abundance of species populations'; and Hassell (1978), writing of important species interactions, states that 'the principal aim in studying the dynamics of these interactions is to explain the distribution and abundance of animal populations'; In 1982, Tilman believed 'the most fundamental question that an ecologist can ask' to be 'Why are there so many kinds of animals?' These ideas, like those of many other ecologists, concern the search for 'explanation', 'cause', 'mechanism', and 'understanding'. As I will argue in Chapter 5 and 6, these elusive and ethereal goals have made ecology a discipline of insoluble questions rather than a source of information and hypotheses about our environment. The confusion of questions in our introductory texts is only one result.

Summary – Scientific growth depends on scientific criticism

At any time, science, that body of knowledge that describes our universe, is imperfect. The descriptions it offers are the best we have, but they are incomplete, imprecise, awkward, and limited. Sometimes, they

Table 1.3 *Some intractable or problematic questions posed in introductory texts in ecology*. To the extent that these texts accurately represent the field, such questions indicate the inability of ecology to pose soluble scientific questions.

From Whittaker (1975a):
'Given a landscape of various communities, how is one to recognize the climax?'
'How are we to interpret the relative stability of populations in natural communities?'

From McNaughton & Wolf (1979):
'In density dependent populations, what are the possible factors that can define carrying capacity?'
'What are the relative roles of genetic changes and acclimation to increasing fitness of individuals to variations in the environment? To what periodicity of environmental changes do organisms respond by acclimation or by genetic changes? How do organisms adapt to unpredictable environmental fluctuations?'

From Ricklefs (1979):
'What are the major attributes of the community?'
'How does competition operate? How can the principle of competitive exclusion, verified by laboratory experiments, be reconciled with observations of similar species coexisting in nature?'

From Krebs (1985):
'What is a community?'
'What are the mechanisms that permit a large number of species to thrive in tropical habitats?'

From Begon *et al.* (1986):
'Are coexisting species necessarily different?'
'Why are there allometric relationships?'

From Colinvaux (1986):
'Why do individuals do the things they do?'
'Are complex communities more stable than simple communities of few species?'

From Pianka (1988):
'What are the effects of indirect interactions among populations and/or guild structure on the assembly, structure, stability and diversity of communities?'
'Can guild structure evolve, even when resources are continuously distributed, as a means of reducing diffuse competition?'

are wrong (Weisskopf 1984). The recognition of these shortcomings is essential to the growth of science, because they show us where improvement is necessary. Epistomologists have long recognized this by dividing their studies of science into two phases dealing with creation or 'the context of discovery' and criticism or 'the context of justification' (Caws 1969). Popper (1985) made the same distinction in describing science as a sequence of bold conjectures and critical refutations.

Kuhn (1970) reserved a special place for criticism in his influential book, *The Structure of Scientific Revolutions*. He believes that science cycles through periods of normality when scientists refine the details of the over-arching paradigms of the field, ignoring problematic observations and criticisms to continue the traditions of the field. In time, as the appeal and fertility of the traditional subject areas wane, the importance of certain incongruities becomes more obvious, established techniques and tools seem inadequate for the questions of the day, and scientists try to reformulate the bases of the science to accommodate these changing views. Kuhn terms this period of questioning and re-evaluation a 'crisis' in science, and holds that such a crisis precedes the reformulation that constitutes a scientific revolution.

It is not too far-fetched to hope that ecology has entered a period of crisis on the way to revolutionary change. Contemporary ecology cannot answer many of the relevant environmental questions of the day. The traditional topics, constructs and theories have come under harsh scrutiny, and the traditional tools of the science, like simulation (Watt 1975), mathematical theory (Strong 1986a; Hall 1988), large research teams (like those in the International Biological Program; McIntosh 1985), and laboratory experiment (Diamond 1986) seem inadequate. Many ecologists are actively seeking alternatives (Price, Slobodchikoff & Gaud 1984).

It is exciting to be part of a science when the limits posed by traditional thought are lifting and new ideas are encouraged. It is also a difficult period for individual scientists as long accepted premises are re-examined, criticized, and changed. Inevitably, this new critical attitude erodes the sense of common purpose and camaraderie that once characterized the science. Some researchers will be left behind with ideas that no longer seem relevant, others will explore new tangents and questions which eventually prove peripheral. Progress in science demands that the future overwhelm the past, correcting its misconceptions, improving its descriptions, and discarding its mistakes. Criticism lays the ground work for scientific advance.

2 · Criteria

Scientific criticism compares a body of the literature to a set of standards or criteria. This book judges the constructs of contemporary ecology against the criterion that scientific theories must describe some aspect of the universe. Proposed theories are therefore judged according to their capacity to provide information about what we will and will not encounter. This simply reformulates the widely accepted view that scientific theories must make testable predictions about the phenomena of nature. One purpose of this chapter is to advocate that view, but also to recognize that as an activity, science is more than a collection of theories. Nevertheless, because this book is a critique of ecological constructs as theories, discussion of the role of non-predictive devices is limited, and largely limited to this chapter.

If scientific theories are characterized by predictive ability, the branches of science are distinguished by the objects of prediction. Ecology seeks to predict the abundances, distributions and other characteristics of organisms in nature. Ecologists may deal with other problems, like global warming or lake phosphorus concentrations, but these biogeochemical topics are ecologically significant only because they impinge upon organisms. This book contends that much of contemporary ecology predicts neither the characteristics of organisms nor much of anything else. Therefore it represents neither ecological nor more general scientific knowledge.

A number of criteria besides predictive power can be invoked to judge scientific theories. Some of these, like accuracy, scope, testability and simplicity (Kuhn 1977; Fagerström 1987), are elements of predictive power. Others, like consistency with existing views, inspirational or heuristic effect (Kuhn 1977), beauty (Fagerström 1987) and understanding (Lehman 1986) involve questions of individual reaction and personal taste. Regardless of the importance given these subjective standards, one criterion whereby a theory is judged must remain its capacity to inform us about the external world; this is done by making 'predictions' or 'falsifiable' statements about what we are likely to encounter. If scientific theories are to provide objective information about the external universe, subjective criteria should complement, not confound, predictive power. A second purpose of this chapter is therefore to describe the place of criteria that do not depend on predictive power, and therefore that may either enhance or confound criticisms based on that defining characteristic.

This chapter begins by establishing the importance of predictive power for scientific theory, while acknowledging the importance of non-predictive elements, like analogy and logic, in contemporary scientific research and hypothetico-deductive science. It then examines the implications of the proposition that predictive power is a defining characteristic of scientific knowledge. Because these implications follow from the definition of scientific knowledge, they are essential character-istics of scientific theories and will be used to assess contemporary ecological constructs in subsequent chapters.

By definition and example: logic, science and theory

The logically possible relations among selected aspects of the external world could be determined by reason alone, without further input from that world. However, that input is crucial to any knowledge of the probabilities of the different possible relations. This limitation of reason is the key to separating two fundamentally different aspects of human knowledge. Identification of the possible is an intellectual, abstract activity called 'logic'; identification of the probable is an empirical, applied activity termed 'science'. As a body of knowledge, logic allows us to list everything that might happen given certain premises, whereas science allows us to determine which events are more likely.

In this book, all constructs that make potentially falsifiable predictions are called 'theories'. Under this definition, theories isolate probable phenomena from the larger set of logically possible phenomena, and therefore provide information by telling us what probably will be encountered and what probably will not. Theories are falsified when unlikely events occur more often than expected, and the vast web of all unfalsified scientific theories constitutes scientific knowledge.

Predictive constructs can be most easily illustrated by the use of an average or a regression to make a prediction, because such humble theories do not entail the problems of the grander theories of ecology that fill most of the pages of this book, they are closer to the daily working experience of most ecologists, and their consideration as theories demystifies the term.

For example, the geometric mean for the abundance of raccoons

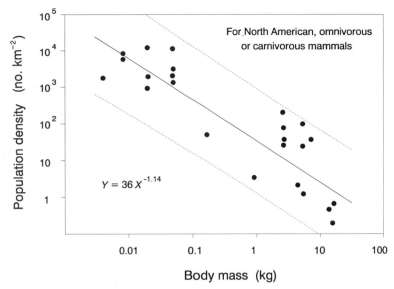

Fig. 2.1. A simple theory to predict the density of raccoons and other mammals (Peters & Raelson 1984). If the preconditions are met (i.e. that the animal is a mammalian omnivore or carnivore living in North America), then the density of the animal may be estimated as $Y \pm$ error where Y is the median abundance estimated by substituting the weight of the animal for X in the regression equation.

(*Procyon lotor*) is 10 animals km^{-2} (95% confidence limits: 1 to 100; Lotze & Anderson 1979). These statistics may be taken only as a summary of past observations, but if it is supposed that these statistics also describe as yet unobserved abundances of raccoons, then the statistics also serve as an empirically derived theory. Such a simple theory specifies its goal (the estimation of mean abundance), implies the domain of application (raccoons, the user must derive any further specification from the raccoon literature), states the expected value (10 km^{-2}), and the likely error around this range (approximately ± 2 S.E.). In this case, the theory specifies a probable range of raccoon abundance which is a much smaller subset of the logically possible range of zero to infinity.

This theory is informative because it limits what we can expect. It is falsifiable because other observations of raccoon abundance could lie outside the predicted range more than 5% of the time. However, until this falsification occurs, the theory may be the best available tool to predict raccoon abundance.

A slightly more complex ecological theory (Fig. 2.1) predicts the abundance of temperate North American carnivorous and omnivorous mammals (Y) by means of a relation that instructs us to measure body mass in kg (X) and to substitute this value into an equation. Simple statistics then allow calculation of the confidence limits around this average. For raccoons (mean weight = 6.4 kg; Lotze & Anderson 1979), the median abundance is predicted to be 4 km^{-2} (95% confidence limits: 0.2 to 100 km^{-2}). Like the theoretical use of average density, this regression can also be used as a theory, for it informs by predicting that the probable values will lie within certain limits and would be falsified if other observations occur with significant frequency.

Like most bits of scientific information, these examples apply to a narrow range of phenomena. Individual scientists might dream of making basic changes in their field through fundamental new insights, but realistically, most make their contributions at the level of the average and the regression. This is part of what Kuhn (1970) terms 'normal science' and this is the level where science is applied to the problems of humanity. This lesson is too little appreciated in ecology where synthetic and conceptual works often take pride of place over less wholly intellectual, but more useful and informative, special theories.

In this text, 'law', 'hypothesis', and 'theory' are synonyms for predictive constructs, but the terms may be used in many other ways in the literature. Ecologists commonly use 'theory' to refer to a body of heterogeneous ideas and constructs, as in 'niche theory' (Pianka 1981), 'optimality theory' (Oster & Wilson 1978), and so on. Sattler (1986) differentiates theory, law, hypothesis, and model by their degree of confirmation. Loehle (1983) distinguishes between theories as explanatory models and as calculating tools. Since definitions are only a matter of convention, the problems of alternative definitions should be ephemeral, provided the terms are defined and used consistently.

Nothing is gained by arguing definitions, but differences in the meanings behind the words can be important. The distinctions between logic and science, or between theory and non-theory, are not the arbitrary impositions of meddlesome philosophers; and predictive power is not just an easily recognized feature of scientific knowledge, like the field mark in bird-watcher's guide. Prediction is the fundamental element in a whole class of knowledge about the external world.

Non-predictive constructs in science

Explicit ecological predictions are increasingly common (Fretwell 1975). Indeed, the criterion of predictive power and falsifiability has been so widely embraced in ecology that critics accuse its supporters of 'Popperphilia' (Diamond 1986) and zealotry (Slobodkin 1986a), presumably threatening intellectual freedom (Lehman 1986; Roughgarden *et al.* 1989) with a morbid acceptance (Diamond & Gilpin 1982) of Karl Popper (1968, 1979) and an obsession for some doctrinaire method of science (May & Seger 1986).

These misgivings distort the position of Popper's ecological supporters. Popper receives support because he has proven an eloquent and thoughtful spokesman for a sizable fraction of the scientific community, but the lament that ecology is only weakly predictive predates that philosopher's influence (Haskell 1940; Kingsland 1985). Moreover, neither Popper nor his supporters suggest any restriction of our freedom to create or that science consists of predictive statements alone, only that theories be testable and tested after they have been created.

Science consists of both knowledge and the process by which this knowledge is created, research. Although research succeeds by building and testing better theories, the process involves constructs, concepts, and activities that are not themselves predictive. Non-predictive constructs serve science as logical devices, memory aids, inspirational prods, incentives to thought, political opinions, personal ideals, half-formed notions, odd beliefs, and unexpressed ideas. These elements are not 'bad' or unscientific. They form a prescientific soup from which each scientist draws inspiration and from which the disciplined human mind has constructed modern science. No demonstration can show that a part of this mix is inappropriate and any such attempt would be futile. However, when we know which constructs are theories, and what non-theoretical constructs can do, we will make better use of both.

Hypothetico–deductive science

This section describes the differences and interdependence of theory and non-theory in science. It begins with a general model of scientific development and then describes how two different types of non-predictive constructs, analogy and logic, fit into this model.

The hypothetico–deductive model depicts research as an alternation between creation and criticism (Fretwell 1975; Hutchinson 1978; Southwood 1980; Romesburg 1981; Sattler 1986; Mentis 1988; Fig. 2.2). Science begins in private reflection and thought as a researcher focusses on a problem worthy of solution and compares various hypothetical solutions with the available evidence and theory. This private phase involves the individual scientist in an alternation between unbridled conjecture to create new hypotheses and critical analysis of the newly created hypothesis. When the private phase is over, the most promising hypotheses are ready to be tested against new data, aired publicly in scientific exposition and evaluated by other researchers.

The private and public phases of science have been associated with hypothesis and deduction, conjecture and refutation (Popper 1968), or synthesis and analysis (Caws 1969), but the same dichotomies occur in both phases. Effective evaluations of the hypothesis often compare several competing hypotheses (Chamberlain 1890; Platt 1964), but minimally, the theory's predictions are compared with measurements of the predicted variable. Since experimental artefact, technical malfunction and sampling bias are always possible, even measurements are fallible representations of the external world. In such cases, theory helps evaluate the data (MacArthur 1972a). This is echoed in the view that there are no hard facts in ecology (Fagerström 1987). Measurements are therefore treated as highly specific hypotheses about particular cases and events.

Synthesis and analysis

The creation of new hypotheses is one of the most obscure and demanding aspects of science. It is captured in Darwin's notebooks 'in which different processes tumble over each other in untidy sequences – theorizing, experimenting, casual observing, cagey questioning, reading, etc.' (Gruber & Barrett 1974 as cited in Roughgarden et al. 1989). This is the synthetic phase where non-predictive elements are most important and where analytical criticism plays a secondary role.

Picturesque examples of the scientific imagination at work show that inspiration is where you find it (Morris 1967). Kekulé is said to have intuited the ring structure of benzene while day-dreaming of snake-molecules seizing their own tails (Franklin 1986). Watson (1968) was drawn to the double helix because of the ubiquity of paired structures in biology, Galileo was inspired to a theory of gravitation by a swinging chandelier, and Newton was hit by his apple. No one gives greater credence to the hypotheses because of their inspirations, but similarly no one should be unwilling to examine them because they were inspired. The source of the creative spark is subjective and incidental to the theory.

Most scientific hypotheses have no such charming ontogenies, but are based on a mixture of common sense, commonplace and common

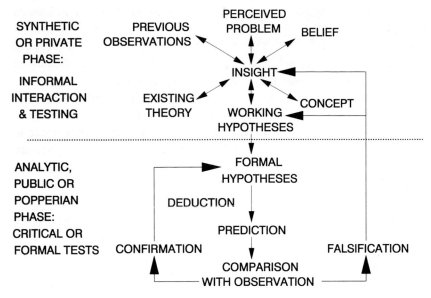

Fig. 2.2. A schematic diagram of the hypothetico-deductive method, indicating the separation of private and public phases of theory building.

practice, and most are less revolutionary. For the working scientist, the richest sources of inspiration are the traditions of science, personal scientific experience and the larger culture in which these are imbedded (Medawar 1967; Caws 1969).

The study of scientific creation and creativity has been largely the domain of visionaries, sociologists, and historians. Visionaries, like Koestler (1969), never quite plumb the topic, perhaps because they believe that the creation is unfathomable. Instead, they obscure the process by introducing unhelpful, even mystical, concepts like 'genius', 'imagination', and 'insight' (Medawar 1967; Caws 1969). Sociologists, like Merton (1973) and Zuckerman (1977), treat the subject as a phenomenon. For example, they show that the most creative scientists (e.g. American Nobel Laureates) typically went to the best schools and studied with the best professors, but not how these experiences made their subjects creative. Phenomenology also appears in the writings of historians of science, like Kuhn (1970), who can suggest the general characteristics of sciences ripe for revolution or, like Cohen (1985), who recounts the particulars of past revolutions. None of these approaches has been particularly valuable in promoting new theories or new revolutions, and my brief overview of the synthetic phase is no better. It seeks only to acknowledge the importance of non-predictive elements in science.

Analogy and metaphor

Many ideas begin as metaphors or analogies, taking lessons from one set of experiences and applying them to another. Information theory has been applied to problems of species diversity (Margalef 1968); game theory offers insights into competition and evolution, and catastrophe theory into problems of instability; analogues to Ohm's law for the relation between electric current and resistance have been drawn in hydrology as 'Darcy's law' for groundwater flow (Shaw 1983) and in ecology for population growth (Chapman 1931); Vollenweider (1968) predicted phosphorus concentration in standing waters by comparing lakes to continuously stirred, instantaneously mixed, single reactors. The success of these analogies varies. Information theory appears sterile for ecology (Hurlburt 1971; Peet 1974); game theory cannot be applied in its entirety or without adjustment to biological problems (Maynard Smith 1972), and catastrophe theory now seems uninteresting (May 1981a). An analogue to Ohm's law applies to groundwater, if calibrated for each application, but not to population growth (Kingsland 1985). However, Vollenweider's work is fundamental for eutrophication control and his analogy eventually won him the recognition of the prestigious Tyler Prize in Ecology.

There is neither merit nor blame in analogy. Analogy is a step in the growth of our ideas, but its validity can only be judged by testing the theory that sprang from analogy against experience. In science, the proof of the pudding is in the eating, not the making.

The insufficiency of synthesis

For the creative process, in science as elsewhere, anything goes. There is no obvious way to cull non-predictive constructs when they appear barren or exhausted, because they are subject only to the oblique criticism that they have not led to new creative insights yet or recently (Goodman 1975; Gray 1987; Pierce & Ollason 1987), and they are always defensible by the contention that future events will prove their worth (Stearns & Schmidt-Hempel 1987). If some constructs, like the climatic climax eventually fade away (Whittaker 1953), others, like the logistic and Lotka-Volterra equations (Hall 1988) or traditional models of succession (Drury & Nisbet 1971) accumulate, clogging the mind of the researcher and cluttering the texts of the science. The presence and nature of these non-predictive constructs should be more widely appreciated, not just to prevent their being confused with theory, but also to appreciate what their roles might be and so to allow them their appropriate use.

Roughgarden *et al.* (1989) suggest the anarchism of Feyerabend (1975) is an appropriate guide for pluralistic ecology. Feyerabend is a radical because he does not separate synthesis from analysis. Therefore, when he demonstrates that the creative phase is essentially structureless and acritical, he can conclude that there are no criteria at all. From this, it follows that anything goes, not just in creation, but in criticism as well. This extreme position would free science from the requirements that it must describe the external world or indeed that science should do anything in particular. Many contemporary scientists are surely unwilling to take this step into chaos.

The refusal to apply criteria to science is summed up in the definition, 'science is what scientists do'. This expansive circularity does not allow scientists to act non-scientifically, it characterizes as 'science' some behaviours which are not peculiar to scientists and others which are very peculiar to particular scientists, but incidental to their research. This view receives tacit support from many thinkers who are not themselves scientists and therefore feel that it would be presumptuous to correct researchers. They see their role as *descriptive* of what scientists do. This approach should be contrasted with *prescriptive* approaches (Thompson 1981), like that of Popper, that provide strong criteria for scientific knowledge and accept that much of what appears in scientific journals is not likely to pass these tests.

Practicing scientists are drawn to a prescriptive approach, because they have little to learn from descriptions of the daily lives of their peers and accept that recognition of current shortcomings is one step towards the improvement of their science. They recognize that science is more than a set of rules and the application of a standard method, but also that existing standards, rules, and criteria still have their place in the analytical phase of science.

Analysis and logic

Compared to the creative phase of the hypothetico–deductive method, the public or deductive phase (Fig. 2.2) in which the hypothesis is tested and refined is well known. Its components have been dissected and described by many writers, the criteria used in evaluation are well known, and the steps required to modify the existing theoretical structures are standard. One first deduces what one can from the hypothesis and compares this deduction to observation. If they agree, one tests the theory further. If observations do not confirm the theory, some minor adjustment of the hypothesis, based on some minor inspiration, may restore agreement. Otherwise, a major revision may be needed. In any case, the theory is always hypothetical and therefore open to further tests.

The deductive element in analysis involves mathematics or logic. Neither predicts, but both identify the relations that follow necessarily from fundamental axioms of the hypotheses. These relations do not depend on the reality of the hypothesized entities. Thus $1 + 1 = 2$, whether we are discussing unicorns or rabbits, and the applicability of this simple relation provides no evidence for the existence of either species. Similarly, the propensity of rabbits to multiply does not disprove the proposition that $1 + 1 = 2$, but only shows that this equality may sometimes be inappropriate for some systems.

Mathematics is essential in every theory that combines a general relation with specific measurements of the independent variables to deduce their logical consequences. For example, if the density of temperate mammalian omnivores is given by the equation $Y = 36W^{-1.14}$ and if raccoons have a mean weight of 6.4 kg and are temperate mammalian omnivores, then it follows logically that the median density of the animal is 4 km^{-2}. If the observed density differs, then at least one premise does not hold. Because faulty deductions can be discovered without observation, observations are not required to determine whether a deduction is fallacious, but are essential to show whether the premises hold. Although mathematics does not make predictions, it is an extremely useful, varied, and widely known system of logic and notation. Mathematics makes logical quantitative deductions easy and straightforward. If it did not exist, it would have to be recreated each time a quantitative theory was used.

The importance of logic, mathematics, and analogy demonstrates that science does not consist solely of predictive constructs. Although non-predictive devices are not central to this critique of the predictive power of contemporary ecology, such a focus does not deny the value of non-predictive constructs in science.

Criteria for judging scientific theories

This book is concerned with the public or analytical phase of the science of ecology, with tests and refutations of existing theoretical structures. It asks what can be deduced from ecological theories, whether these deductions have been or can be tested against experience, and what significance these deductions have. It does not address the basis of conjecture, the source of ecological hypothesis or the nature of scientific intuition. It accepts that constructs can fail the test of predictive power and still be scientifically valuable for their logical strength, heuristic content or inspirational effect. It rejects any suggestion that these concepts substitute for prediction and affirms that past failure to separate concepts and predictions has slowed the development of ecology. Conversely, it maintains that recognition of the capacities and limitations of both predictive and non–predictive constructs will accelerate that development.

Figure 2.3 provides a general model of a scientific theory to serve as a basis for discussion. The plane of the graph represents the set of all logically possible combinations of the predictor and response variables, X and Y respectively. The theory states that under a set of explicit preconditions (C_1, C_2, . . . C_n), a calculation ($Y = a + bX + \text{error}$) will yield the likely range of Y; this appears as the confidence band in Fig. 2.3. This model can be extended to encompass purely qualitative variables or more predictor variables, but Fig. 2.3 suffices to illustrate most of the important properties of a scientific theory.

If simple, limited, generalizations like this are accorded the status of theory, further criteria are needed to select better theories. The most powerful of these criteria derive directly from predictive power, but others are independent. Both types are summarized in this section and applied throughout this book.

Goal definition

The hardest part of any study is the first step: definition of the study's goal. A good choice directs the entire research program by providing a tangible endpoint so that any part of the study may be judged by its relevance to the goal. A poor choice leaves the research direction so vague that both study and student are confused. In the model of Fig. 2.3, the goal can only be the response variable, Y. In the example of the average population density, the goal is raccoon abundance. This limits interest in the theory to those with specific questions about that phenomenon.

Given our present level of ignorance and the pressing problems of the age, it should be easy to define goals for ecology. We need only specify what aspects of which organisms to predict and then build theories for

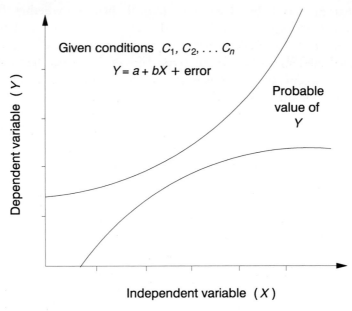

Fig. 2.3. A schematic representation of a scientific theory that predicts, under certain boundary conditions ($C_1, C_2, \ldots, C_n$), the probable range of values ($\pm$ error) of a dependent variable or response (Y), given the value of the independent or predictor variable (X) and appropriate calculus ($Y = a + bX \pm$ error).

that purpose. Unfortunately, the process of goal definition has been obscured by poor examples (Downing in press). Authors claim that their goal is 'to examine' some phenomenon, 'to investigate' some process or 'to illuminate' some observation. Often we teach that the role of research is 'to understand' nature or 'to explain why' the world is as it is. Such goals are nebulous (Woodford 1968). They have the allure of grandeur, but fail to distinguish science from philosophy, theology, history, fine arts, and many other disciplines which justifiably offer the same virtues. Science differs because it is not simply a discourse, but a tool to deal with nature. In this difference lies the vigour that makes science so powerful.

If goals are to be defined for scientific research, they must be phrased in terms of the distinctive characteristic of science, predictive power. Research goals are properly defined only by a statement of what variable the study was intended to predict. Without this, one can never know if a study succeeded.

Relevance

Because the number of things to predict is so vast, simply identifying a particular variable as the research goal is insufficient. Thoughtful scientists choose to study the most important phenomena within their competence. They consider the purpose and utility of their goal by considering what they will do with the theory once constructed and with its predictions once made. If we need to know the abundance of raccoons, then an average value may be relevant. If not, limited scientific resources have been wasted in pursuing that goal.

This advice is banal, but many find it easier to hide in the vaguer promises of understanding or explanation than to state explicitly what one wants to know and how this knowledge can be used. In general terms, the relevance of the response variable in any theory is measured by the role this variable plays in other theories, predicting still more variables, and providing still more information. In the raccoon example, the prediction could be used in regulating hunting or trapping, in assessing the need for control procedures, or in testing the validity of more general models, like that in Fig. 2.1. In any case, the selection of a relevant goal is an important part of the wisdom of a scientist.

Immediacy

Once a relevant goal has been identified, the theories of first choice are those where this goal is the response variable. If this is impossible, because appropriate theories do not exist and because one cannot take the time to build one, then a less–direct solution may be required. For example, it may be necessary to measure an intermediate variable and then to use this to predict a second intermediate which can be combined with a third to yield the desired prediction. This is cumbersome and statistically undesirable (because the uncertainty of the prediction will increase with each manipulation), but may be countenanced if there is no alternative. In any case, the theory involving fewest intermediates is probably preferable.

Relations that predict raccoon density are immediate only for the small range of problems that ask 'what is the abundance of raccoons?' They are intermediate in questions about annual permissible kill of raccoons, or wild faunal reservoirs for rabies, or threats to waterfowl nests, but may be essential to addressing those problems. What must be avoided are lines of research which are not tightly connected to the variable of ultimate interest. If one is unsure exactly how the theory one is using will lead to the prediction of the variable claimed as the goal of the research, one must either accept some other goal or look for some other theory.

Operationalism

Science lies at the interface between the abstract constructions of our mind and the phenomena of the external world. This interface is possible only if the observed phenomena can be associated with the abstract terms of the theory. To apply the theory in Fig. 2.1 to raccoons, we must know what constitutes a raccoon, a mammal, an omnivore, North America, body mass, and population density. If no such association is possible, the theory can never be applied to reality and can never inform us about any part of the world.

Every term in the theory must be sufficiently realistic or 'operational' that other users of the theory will associate the same phenomena with the same theoretical terms. Since it is impossible to specify all characteristics of any entity, the association is never perfect. Nevertheless, the terms in good theories are usually sufficiently well defined or generally known that most practitioners agree about which theoretical variables represent which phenomena. Imperfect agreement is sufficient for science (Hull 1968), but without it, no theory could have the objectivity that allows us to teach and acquire scientific knowledge.

Accuracy

The extent that theories inform us about the external world is measured, in part, by the similarity of the predicted and measured values of response variable, in other words, by the accuracy of the theory. If previous measurements are known to fall within the limits predicted by the theory (Fig. 2.3), we can claim that the theory has been accurate in the past. Nevertheless, because past accuracy may have been achieved by chance, philosophers from Hume (1739, 1740, 1748) to Popper (Popper & Miller 1983) have argued that a history of success never guarantees future accuracy. The impossibility of such a guarantee means that all scientific knowledge is uncertain and hypothetical.

Although a successful past does not assure future accuracy, past failure certainly discourages further use. Scientists therefore prefer theories with a record of accuracy, and most papers provide this record in the form of results consistent with the theory.

An individual may choose to ignore results which would otherwise falsify a theory, because the theory is believed to be more accurate than the data in hand (MacArthur 1972a; Fagerström 1987). Researchers do this routinely in rejecting 'spurious' analyses and this attitude has sometimes been spectacularly successful. Mendel's data are so good that Edwards (1986) surmised that Mendel or his assistants rejected some dubious values thereby skewing their statistics in favour of what came to be Mendelian ratios. The oil-drop experiments that revealed the unit charge of the electron were rendered somewhat more convincing because Millikan rejected some non-confirmatory data as unreliable (Franklin 1986). A more scrupulous opponent, Ehrenhaft, could not confirm the American experiments with German apparatus (Holton 1978), but subsequent development has shown that Millikan made the better choice.

Despite these counter-examples, consistent disregard for observation is a hallmark of fanaticism and quackery, so most scientists are very careful when they ignore established inaccuracies in a theory. When they do, they essentially treat the data as a competing theory of very limited scope. For example, the specific theory for raccoon abundance and the very general theory for omnivore abundance (Fig. 2.1) can be compared with the still more specific theory represented by data from one locale: the maximum density of raccoons (407 km^{-2}) was observed in Missouri one January (Twichell & Dill 1949). The difference between this value and that expected from more general theories likely represents an anomalously dense population of raccoons, but it could reflect some artefact in one study. In the first case, one would claim that the general theory is inaccurate in that season or locality (and perhaps in others like it); in the second, one would suggest that the data on which the highly specific theory is based are fallacious and the specific theory inaccurate. In practice, one could only choose between these alternatives on the basis of more information.

Generality

Because even small points of fact can be interpreted as theories, contemporary sciences must control the number of theories or risk being overwhelmed by details. One way to achieve this control is to rationalize specifics under more general theories. For teaching, for approximation, for broad-scale application, and non-specialized uses, these general theories often suffice. As a result, the scientific community often places a premium on generality.

In terms of Fig. 2.3, general theories are those that have fewer or less-restrictive preconditions, and those that apply across a greater range of the predictor variable. Because such theories predict likely characteristics over a greater range of external conditions, they are more informative. For example, the regression in Fig. 2.1 is more general than the theory that the average abundance of raccoons lies between 1 and 100. Both of those theories are more specific than a regression describing the allometry of animal density for all mammals (Peters & Raelson 1984) or all animals (Peters & Wassenberg 1983).

Precision

The precision of a scientific theory is a measure of the uncertainty associated with prediction. More precise theories have narrower confidence bands and indicate that a larger range of the possible combinations of predictor and response variables is improbable. Precision is desirable because more precise theories alleviate more of our doubts about the configuration of the external world and provide a sounder basis for future action.

Generality and precision are often opposing virtues. Levins (1968) claimed that they cannot be simultaneously maximized in realistic (i.e. operational) theories. The difference in the confidence limits of the specific theory for raccoon abundance (1 to 100) and the general theory for omnivore abundance (Fig. 2.1) applied to raccoons (0.2 to 100) shows that in this comparison, the more general theory is slightly less precise. However, both examples might be made more precise, without losing generality, by simultaneously considering additional predictor variables, like mortality, site productivity, etc. There is no necessary conflict between precision and generality: the confidence bands in Fig. 2.3 can be narrowed or broadened without changing the domain over which they apply. Generality and precision are complementary because they both reduce uncertainty by excluding more possibilities as unlikely. Since the two qualities are independent, we can seek theories that are both general and precise. Since they are both desirable, we should.

Quantification

The form of Fig. 2.3 is intended to suggest a quantitative theory and, although the features discussed in this chapter also apply to qualitative theories, they apply better to quantitative ones. When the theory makes quantitative predictions, it is easier to decide if a theory has been accurate, to determine if the preconditions are met, to make the calculations required for prediction and to state the precision of the prediction. Qualitative theories must invoke subjective evaluations based on 'reasonable agreement' and 'satisfactory comparisons' (Romesburg 1981). For example, it is easy to recognize that the high raccoon densities reported from Missouri do not confirm either of the more general theories about animal abundance. In contrast, qualitative predictions, predictions about whether raccoons are rare or many, are almost invariably vague. They should only be used where there are no alternatives.

This preference for number is common scientific practice. It is summed up in a phrase attributed to Lord Kelvin: 'If you cannot measure, your knowledge is meager and unsatisfactory'. As a result, quantitative theories prevail over qualitative competitors (Kuhn 1977).

Economy of effort

Scientific theories reduce uncertainty and doubt, but their application is not always easy. Many theories require considerable education, instrumentation, and man-power to be used at all. This effort is warranted by the significance of the question. However, when theories which make similar predictions are compared, that which yields the most information for the least effort should be preferred.

One measure of the advantage offered by a theory is the difference between the costs required to measure the predicted variable directly and that required to measure the predictor variable and preconditions, and then to apply the theory. If a response variable is easier to measure than to predict, wise researchers ignore the theory. Conversely, if the theory makes the response variable much easier to predict than to measure researchers will find the theory more useful.

Economy is one of the attractions of allometric relations, because body weight is almost always easier to estimate than the predicted variable. For example, raccoon densities might be estimated by intensive or extensive field programs, by complex models of population dynamics, or by the allometric relation in Fig. 2.1. The last is clearly most economic, but the choice of approach depends on whether the high uncertainty of the regression offsets its economy in addressing the problem at hand.

Social elements of scientific criticism

Theory selection also has a social dimension. Contemporary scientists are extremely busy. They are writing papers, analyzing data, proposing new directions for their personal research, teaching, discussing, advising, going to scientific conferences, serving on editorial boards, and administering the bureaucracies of their disciplines and institutions. Consequently, they have limited time to learn new techniques, concepts, and approaches. This is especially so as the theories addressed become less central to the scientist's primary specialization, so the importance of appeal increases with the generality of the theory. Some obvious properties (e.g. practicability, simplicity, consistency and heuristic power) render some theories more attactive and easier to use than others.

The characteristics of appeal differ from those of predictive power (i.e. from operationalism, relevance, generality, precision, quantification, and economy). The latter fall logically from the proposition that the information content in a scientific theory is reflected in predictive power, that is in the number of possible states that the theory excludes as improbable. The former reflect a number of other personal and social precepts. Although the scientific value of our work depends on predictive power, no theory has any value unless it is read and used by the scientific community. Attention to social criteria may help a theory capture that attention.

Practicability

Theories should be easy to put into practice. They should use variables that are familiar to the majority of the audience, and they should require as little new instrumentation and method as possible. Such conservatism in proposing theories attracts the interest of at least the scientific community already studying the response variable. The new theory also attracts researchers using the same predictor variables because the theory shows how these measurements can yield more information. In addition, theories based on established techniques and common equipment allow many other researchers to use these theories without redirecting their resources to re-tooling.

Simplicity

Theories should be conceptually simple. They should not ask for more effort than is necessary to use the theory and they should expound the theory succinctly and clearly. One should therefore avoid extensive preambles, divarications in the line of argument, and elaborate dissections of the points made. Some of this material may be provided after the fundamentals of the theory have been expounded, but detail can confuse the reader and obscure the main point. Simplicity can also be achieved by minimizing the mathematical treatment of the problem, limiting the notation used to those which most working ecologists are capable of using, and by maximizing the range of the phenomenon addressed by the theory.

Consistency

The requirement of simplicity does not imply that new theories must be phrased in words of one syllable. Contemporary scientists have mastered a range of sophisticated techniques and concepts, and new theories should certainly build on this foundation. Whenever possible, a wise writer should use the norms of the science to make his theories familiar and therefore easily acceptable to the audience. For example, new systems of classification and unfamiliar notations are justified only where these are essential to the theory. To do otherwise is to confuse the theory with the mode of presentation. This confusion may make the whole package too difficult to follow and leave both theory and presentation in obscurity. This caveat is simply a restatement of Ockham's razor: 'Do not proliferate entities unnecessarily'.

None of this denies the need for or desirability of new state variables, new notations or new ideas in ecology or in the rest of science. When these are required to express a theory, they should and must be used, as Newton used the calculus, the concept of action at a distance, and the distinction between mass and weight in developing his physics. However, when theories compete, those demanding less from the reader, while predicting substantially the same thing, are more likely to persist.

Heuristic power

Kuhn (1977) makes an important counterpoint to this conservative perspective on new modes in science. He argues that, although the intellectual coherence of a science depends upon its conservatism, new directions depend on bold new initiatives. He sees the balance between these opposing forces as an 'essential tension' which both invigorates and protects the scientific enterprise.

A scientific construct may succeed despite its failure to meet the above criteria for predictive power, if it is so fertile or 'heuristic' that other scientists build upon this then fundamental work. For example, the papers of G.E. Hutchinson have had their greatest impact by inspiring others to explore new directions in ecology. Since the strength of such contributions does not lie in the predictions they make, the standard criteria of predictive power do not apply. Regrettably, there are few Hutchinsons and appeals to heuristic power set a dangerous precedent for scientific judgement since they can become a defence of last resort for bankrupt theory (Gray 1987; Pierce & Ollason 1987).

Summary – A hierarchy of scientific criteria

Science consists of both theories and the range of non-predictive devices that help in the construction and application of these theories. Because this book is a critical overview of the theoretical content of ecology, it deals only briefly with non-predictive elements in the science. This limitation is justified because the distinguishing characteristic of scientific knowledge is predictive power. Scientific theories identify some configurations of the external world as more probable than others by instructing us to measure certain predictor variables under certain conditions, and to apply these values in a calculus that estimates the likely range for the theory's predicted variable. To be valuable, the theory must be directly relevant to the explicit goals of the researcher, these goals and the other terms in the theory must be phrased in operational terms, and the theory must predict that at least one of the logically possible configurations is improbable.

Many scientific constructs are theories under this broad definition. Some are trivial, so further criteria are needed to select the more powerful theories. These criteria identify theories which provide greater certainty because, relative to other theories with the same response variable, they identify a smaller set of probable conditions from the set of all logical possibilities. Such theories are more general, more precise, more quantitative and technically easier to use.

To these two sets of criteria, a third must be added that identifies which theories are more likely to be put into practice. These subjective criteria reflect the proposed audience for the theory. Relative to this public, the theory should be conceptually simple and consistent with other concepts held. If the theory ignores the state of its audience, the audience may simply ignore the theory and continue to measure the dependent variable.

With the exception of predictive power, all of these criteria are relative. With that exception, there is no absolute criterion which will unambiguously separate theory from non-theory. Instead one must compare available constructs in the light of these criteria to select that which is most appropriate for a given application. Since these characteristics will vary independently among different theories, a theory which is best in some contexts will be less desirable in others. Researchers therefore need to be wise critics to identify the constructs that are appropriate for their intended applications.

3 · Tautology

This chapter examines constructs that describe the logically necessary relations among the assumed or measured properties of ecological entities. The chapter begins by defining purely logical arguments as tautologies, acknowledging that tautologies can prove extremely useful when dealing with complex logical systems. However, the main point of the chapter is that logic alone can only identify the possible and so must not confound the capacity of theory to identify probabilities.

Ecological tautologies exist at many levels of complexity. This chapter illustrates that range with a series of examples of increasing complexity, from simple definitions and mathematical truisms to the logistic equation and theoretical ecology. This survey should alert the reader to the varied character of ecological tautologies, to some of the ways that tautologies complement scientific theory, and to the difficulties that arise when tautologies substitute for theory. The last third of the chapter deals with an example of special concern to ecology, the principle of evolution by natural selection. Because this paragon for theoretical ecology has long been accused of tautology, the ensuing debate offers many examples of relevant counter-arguments and replies. Natural selection therefore serves as an important and familiar example of the characteristics, limitations and issues associated with the charge of tautology in ecology.

Tautologies and deductive tools

Logical constructs are transparent to the rational mind, so in principle, we can recreate any logical argument just by thinking it through. In practice, few scientists are so rational that they do not appreciate a little logical help. For example, arithmetic is a logical system (Nagel & Newman 1958), but it is so complex that many theorems, proofs, and sample problems are indispensable for its use. Similarly, practice with the rationalizations of other logical systems allows scientists to complete deductions more rapidly and surely, without the effort of thinking through the problem afresh at each new application.

Ecology, like most sciences, contains many constructs that are model solutions to common logical problems. Experience with these models prepares ecologists to make deductions that might otherwise prove difficult and the teaching of these deductive aids forms a part of most introductions to the field.

Logical models complicate the development of ecological theory because the limitations of logic may be unclear. The elegance and security of logic may be so seductive that deductive certainty becomes an acceptable substitute for theory and prediction. A particular model can become so familiar that alternative premises and deductions appear strange and unacceptable. Because models are used to avoid the work of logical thought, they can erode the logical faculties, leaving scientists less able to confront novel problems. Indeed, some deductive models are so much a part of the drill of ecological teaching, that the models become essential for, rather than illustrations of, deductive thought. Student efforts are then wasted memorizing theorems, rather than learning to think logically about problems. In short, the abundance of logical models obscures the poverty of ecological theory and, for many students, these models have become essential to solve problems that are trivial to scholars with more developed logical acuity.

Logical models need a term under which they can be discussed. They have been called theorems, logical models, deductive systems, rationalizations, syllogisms, or tautologies. I prefer 'tautology' because this distances logical argument from theories, but minimizes other terminological confusions. 'Deduction' and 'logic' play essential roles in the application of a theory to particular cases. 'Theorem' and 'syllogism' have well-known, special uses in mathematics and logic respectively. 'Rationalization' has a perjorative connotation in everyday conversation. Tautologies are purely logical constructs that describe the implications of given premises and never reveal more than those premises contain.

Tautology can appear in many different contexts (Sober 1984). In everyday conversation, it is saying the same thing in different words – a circular argument that ends up where it began. For the philosopher Reichenbach (1951), it consists of any purely logical argument. According to Sober (1984), its meaning in logic is 'P or not-P', i.e. a statement and its converse: either it is raining or it is not. In the context of knowledge as defined in Chapter 2, a tautology is something that identifies the range of possibility. Given two unrelated variables, X and

Y, their combination defines the range of all possibilities as the plane of Fig. 2.3. All usages agree that a tautology is empirically empty because it tells us nothing new about the world around us.

A tautology is to logic as a theory is to science. Both are important tools, but they do different jobs. A tautology identifies the range of possibilities under certain premises. A theory identifies the smaller set of probabilities within this range. This distinction is consistent with Sober's definition of tautology, but expands it to allow more detailed specification of the possibilities than the dichotomy, P or not-P. This definition is also more contiguous with scientific practice and therefore more helpful to scientists. It is also consistent with the view that simple tautologies, like definitions, are circular, since the only possible deduction allowed by a definition is restatement, and with the view that all purely logical arguments are tautologies because, although logic only reveals the implications of the premises, it can reveal all of them. Because a tautology identifies everything which can happen under certain conditions, it always corresponds with what is observed: It is always either raining or not raining. Thus, in a sense, a tautology is certain whereas a theory is hypothetical, risky and dubious.

The same thing in different words

Sciences generate jargon. This specialized terminology is necessary to express complex and novel ideas clearly and succinctly, but terminology often becomes a linguistic barrier around the science, protecting its practitioners from too close a scrutiny by the public or other scientists and providing the camaraderie offered by a common, but private, language.

In weak sciences, jargon substitutes for ideas and theories so that even a hopelessly uninformative field can share come trappings of success. For example, in its early days (Clements 1905), ecology offered neologisms for a spade ('geotome'), for a myriad of seral stages (e.g., actium, a rocky seashore formation, and agrium, a beach formation) and for the plants in those stages (e.g. actad, a plant of the rocky seashore, and agad, a beach plant; etc.). One definition described ecology as 'the science given over entirely to terminology' (McIntosh 1976) and another as 'the science which says what everyone knows in language that no one understands' (Elton 1927). This is not just an historical problem (McIntosh 1985). A recent dictionary of ecological terms (Lincoln, Boxshall, & Clark 1982) lists 10 000 entries, yet includes only 60% of the terms in the indices of ecological textbooks (Downing 1985). This implies that there are over 16 000 ecological terms. No doubt many are needed, but some surely just say the same things in different words.

Other simple verbal circularities

The problem extends beyond the proliferation of terms. Ecologists write vast amounts of prose and more than one has fallen into circularities. For example, D. C. Culver (1976), a respected authority on the ecology of cave organisms, has written, in one of the most respected ecological journals, that:

the advantage of cave communities is that the length of isolation in caves is marked by the degree of regressive evolution of some easily visible characteristics, assuming that the degree of regressive evolution in caves indicates relative age of isolation . . .

This circularity may only be a verbal slip but, both in context and out, it uses a circularity to justify evolutionary and ecological study of cave organisms. This may direct the attention of the scientific public away from the necessary investigation of the assumption itself (Barr & Holsinger 1985).

Material included in the BioCO-TIE program provides a second example. This program was conceived by the University of Colorado and the Institute for Ecology as an introductory course for colleges without staff who are competent to teach ecosystem ecology. In the section on biogeography, a smiling cartoon map explains,

Every organism is everywhere, except where it:
. couldn't get
. got to but couldn't stay
. got to and changed.   (BioCO-TIE 1975).

This neat circularity ought not be excused as a teaching device. Popular writings and teaching materials often reflect scientific precepts far more faithfully than the guarded prose used to communicate with scientific peers. In any case, the argument is insidious because it is directed to an audience that is less able to discern its hollowness.

Spurious self-correlation

Spurious self-correlation represents a special case of saying the same thing in different words. Self-correlation results when a variable is plotted or regressed against some function of itself, as in the simple case

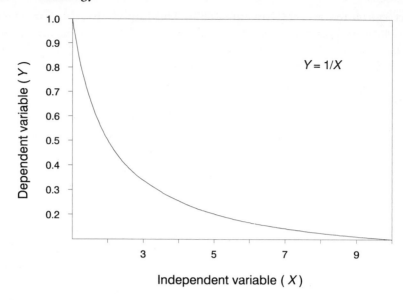

Fig. 3.1. A self-correlation generated by plotting a series of numbers against their inverses. Logical necessity allows only one possible combination of each pair $(X, 1/X)$ and so the possible range of values is described by a line, not by the plane of the graph. The self-correlation is tautological because it describes only the range of possibilities; it is spurious and uninteresting because the deductions involved are trivial.

when a variable is plotted against its inverse (Fig. 3.1). In such cases, the plane of a graph no longer represents the range of the possible. Instead, the only possible relationship is that described by the curve and the graph shows nothing that could not have been easily calculated. Such a relation may be called spurious when pretentious presentation or interpretation asserts that the logically necessary relationship represents additional scientific information. Ecological examples of self-correlations include plots or regressions of the biological half-life of a contaminant against the rate constant (k) of clearance of that contaminant from the body, because half-life is defined (Riggs 1970) as ln $0.5/k$. Another is the inverse relationship between population growth rate (r) and generation time (T) implicit in the definition of the variables ($T = \ln 2/r$; Stearns & Schmidt-Hempel 1987).

   When self-correlations are combined with other, more independent, variables, the necessary relations are less evident, but still impose limits on the realm of possibility. For example, the relation between the volume and area of a solid can only vary within limits set by geometry:

The minimum area (A_{min}) that a solid of given volume (V) can have is the area of a sphere with that volume ($A_{min} = 5.0V^{2/3}$), but the maximum area is not restricted by geometry. Reynolds (1984) uses this restriction in his discussion of the area : volume ratio of algal cells to demonstrate that the surface areas of algae range upwards from the minimum. His point is not that the ratio of area : volume is never less than $5\ V^{-1/3}$ but rather that algal cells press on this logical lower limit. In contrast, the upper limit is a matter for empirical determination.

Self-correlation does not necessarily involve tautology. For example, allometric relations that describe the connection between individual weight and some weight-specific properties are self-correlated, but they are tautological only when they are redundant (Fig. 3.2). The relation between growth rate (G, in g fresh weight d^{-1}) and body weight (W, in g fresh weight) for a broad range of organisms is

$$G = 0.018\ W^{0.78} \tag{3.1}$$

This equation and its accompanying plot (Fig. 3.2A) clearly indicate that the growth rates increase with body size and occupy only a portion of the possible combinations.

Equation (3.1) also implies that the weight specific growth rate (G/W, in g g^{-1} d^{-1}) declines as body size increases, but dividing both sides of the equation by W shows this more clearly:

$$G/W = 0.018\ W^{-0.22} \tag{3.2}$$

Because Equation (3.2) can be obtained analytically (i.e. deductively) from Equation (3.1), it provides no additional information. Because error is multiplicative in this example, neither plots nor regressions of specific growth rate against weight (Fig. 3.2B) provide information not inherent in the analyses represented by Fig. 3.2A.

The regression statistics for both plots contain identical information, the regression equations make predictions that are entirely consistent, the uncertainty around the regression is unaffected by the self-correlation, either plot and regression can be extracted by analyzing the other, and, in both cases, the plane of the graph represents possible combinations of weight and growth rate. Since the two formats are simply logical reworkings of the same information, the choice of formulation depends on the application (Prairie & Bird 1989). If one wishes to show that specific growth rate decreases with individual size, then the self-correlated version (Fig. 3.2B) is preferable. If one wishes to show that larger animals grow at faster rates than small ones, Fig. 3.2A is better.

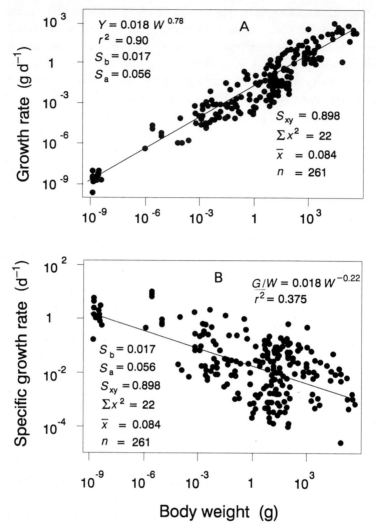

Fig. 3.2. Tautology can result when two or more plots of regressions differ only because one of the two incorporates a self-correlation. For example, the relation between (A) individual growth rate of a wide range of organisms and average size contains the same information as (B) that between specific growth rate and size. There is no gain in presenting both, since one is implied by the other (Peters 1988).

Tautology could result if the same information were used twice, but not because one plot is intrinsically wrong. Tautologies neither add nor remove information, but lengthy elaborations of both logical variants are unnecessary and undesirable.

The self-thinning rule, which describes the relations among the biomass (B), density (N) and individual size (W) of monospecific plant stands (Gorham 1979; White 1981; Westoby 1984), provides an analogous example. Weller (1987) argues that formulations of the self-thinning rule which regress W on N are spurious because W is often calculated as B/N. He illustrates the problems this creates by examples in which regressions of W on B can mislead 'the unwary'. In some stands, W and N are inversely related, even though B and N bear no statistically significant relationship to one another. This relationship reflects the calculation of W as B/N and therefore the self-correlation of plotting B/N ($= W$) against N. In a second example, Weller (1986) shows that an obvious curvature in plots of B against N is less obvious when W is plotted against B. In the first example, Weller worries that the unwary may think that biomass changes as density declines and in the second that the data are linear. Both misapprehensions would support the hypothesis that there is a general self-thinning line for plant populations. Weller (1987) opposes this view because his interest lies in the implications of departures from a single curve.

Weller's specific reservations are well taken, but the general thrust of his argument is misplaced. None of the formulations of the self-thinning rule is incorrect. If an author wishes to discuss the inverse relationship between individual size and density, this is shown most effectively in plots of W against N. Similarly, the curvature in the relation between B and N is not removed in plots of W against N, it is only less striking and presumably less relevant to the intent of the original author. Notwithstanding Weller's objections, there is an inverse relation between W and N and the data can be taken to support a very general, but imprecise, self-thinning rule.

Weller's analysis shows that the general self-thinning rule is imprecise and probabilistic. The utility of this relationship will therefore be limited and the utility of the plot for particular researchers will depend on the immediate context of the research. The unwary will be misled only if they ignore the quantitative implications of the statistics or fail to scrutinize the residuals. The solution is not to remake science for the unwary but to educate scientists so that necessary, simple, logical relations are self-evident. If this is not now the case, it is a condemnation of our educational system.

A number of schools in ecology use similarly self-correlated variables to stress certain properties of interest. For example, hydrologists often calculate the total amount of material transported by a stream as the product of material concentration and the rate of flow of water (Kenney 1982). Regressions of material transport against flow rate usually yield strong positive correlations and high correlation coefficients. The latter are largely artefacts of the self-correlation, but both graph and plots correctly represent the strong effect of flow on transport and the uncertainty in our calculations.

A complex example

Self-correlations are becoming more common and more complex in ecology, probably because computer assisted analysis has made data manipulation so easy. Complexity renders the identification of spurious self-correlation more difficult, but commonness makes this ability more necessary.

For example, food-web analysis makes wide use of a complex variable called 'connectance' (C) which expresses the number of trophic links (L) in a food web of S species as a proportion of all possible trophic links: $C = 2L/S(S-1)$ (Paine 1988). The maximum value of connectance, when everything is connected to everything else is 1. Since each member of a food web must be connected to at least one other member, the minimum value of L is $S-1$ and the minimum value of C must be $2/S$. Plots of C against S (Fig. 3.3) will necessarily be constrained between these values (Auerbach 1984; Peters 1988b) and the plane of the graph of C against S overestimates the range of possible combinations. Briand (1983) is exceptional in making the lower limit explicit.

Another variable used by food web analysts, $(SC)^{-1/2}$, is considered important because it bears on questions of community stability (May 1974). Because this variable is an inverse function of C, it takes a maximum value of $(S2/S)^{-1/2}$, which reduces to $2^{-1/2} = 0.71$, when C is minimal. The minimum values vary as $S^{-1/2}$ and are achieved when $C = 1$. As a result of these limitations, the minimum values of C and $(SC)^{-1/2}$ decline hyperbolically as S increases, whereas the maximum values remain fixed. Because the minima decline hyperbolically as larger numbers of species are considered and because the ranges are limited to high values when the number of species is small, plots of either variable

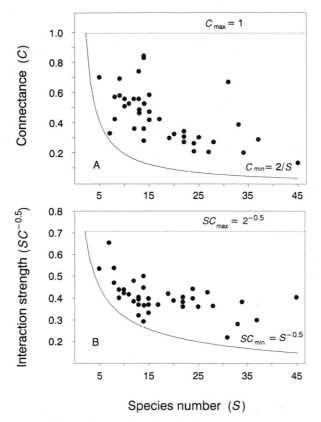

Fig. 3.3. Two self-correlations from food web analysis. (A) The possible combinations of species number (S) and connectance (C) in a food web do not occupy the whole plane of the graph, because the definition of connectance constrains the variable between its maximum possible value, 1, and its minimum, $2/S$. (B) The possible range of values for $(SC^{-1/2})$ of food webs as a function of the number of species in web is similarly constrained between a minimum of $(S^{-1/2})$ and a maximum of $2^{-1/2}$. Data from Briand (1983).

against S may give a false impression of the decline in C and $(SC)^{-1/2}$ in more species-rich communities. Auerbach (1984) shows that other constraints used in drawing up the food webs enhance this effect by reducing maximum values still more when S is large. He urges that any biological discussion of food-web variables be first viewed in terms of the artefactual mathematical properties introduced by definitions and calculations. Biological properties can only be meaningfully examined when the possible range of values is known. As models become more complex, we must take special care to specify the range of possibilities in our writing and to discover this range in our reading. The presence of self-correlated variables is a good warning sign that such analyses are needed.

Identifying the possible

Tautologies are practical tools for organizing scientific knowledge. They allow us to see the complete implications of our premises and can ensure that every eventuality has been considered and, if necessary, named. Conversely, any system of classification which has a place for every possibility is, by itself, tautological. The classification becomes a system of theories when the classes are associated with characteristics beyond those which define the classes, but the approach is over-used if classifications become ends in themselves.

A very simple case is provided by the division of all organisms into two mutually exclusive classes. This was a favorite stratagem of Robert MacArthur who used it in developing theoretical approaches to generalist vs specialist consumers, fine-vs coarse-grained habitats, scramble vs interference competition, density-dependence and independence, and r- and K-selection (Schoener 1972; Fretwell 1975; Kingsland 1985). As bare classifications, these terms simply name alternate possibilities and are therefore logically exhaustive. This is indicated by redundancy. For example, anything which is not a specialist must be a generalist, and vice versa.

Eventually, these comparisons should lead to theories based on this classification or be abandoned as a false lead. At the simplest level, such a theory might quantify the frequency of occurrence of the different classes (Hutchinson 1978). More informative theories would also point to the consistent differences in the characteristics of each class. The theory might then take the form: Any organism which can be called a specialist on the basis of its patterns of consumption will also have this or that other characteristic. The critical response to any classification is first to ask what theory has developed from it and then to judge this theory on predictive power.

If no prediction is made, the construct should explore some logically necessary relation which depends on the classification. The critical reader should then focus on the classification to determine if it is likely to lead to theory or at least if it can apply to ecological phenomena. For example, in the generalist-specialist case, one should ask if all consumers can be unambiguously classed as generalist or specialist. Often dichotomous divisions reflect the poles of a continuum, and discussions based on these extremes may be irrelevant to a majority of animals lying in the excluded middle. When a continuum is suggested, the behaviours of real organisms should be considered to determine if they correspond to the linear ranking.

An equally familiar example of a logically exhaustive classification is the description of population distributions in terms of mean (X) and associated variance (S^2). These distributions are called 'random' (when the mean is not significantly different from the variance), 'clumped' (when the mean is less than the variance) or 'even' (when the mean is greater than the variance). These three classes represent every possible relation between mean and variance. This very simple classification is unfortunately complicated by alternative terms for even (regular, uniform, overdispersed) and clumped (aggregated, contagious, underdispersed). Such terminological complexity is gratuitous and confusing. It hides a more fundamental question which asks if the subdivision of a quantitative continuum into mutually exclusive, qualitative, classes is a useful exercise. This classification may separate similar populations simply because the variance of one differs significantly from the mean, whereas the variance of the other does not. On the other hand, populations with variances which are only slightly, but significantly, greater or smaller than the mean may be lumped with those whose variances differ very much. Thus this division of a continuum into arbitrary, heterogeneous classes is likely less effective than alternatives, like Taylor's power law $(S^2 = aX^b$; Taylor 1961), that treat the continuum as a continuum.

Most classifications are not so simple. The BioCO-TIE program identifies the possible 'coactions', a term which represents the interactions between two organisms, one of which is more powerful and one less (Fig. 3.4). A coaction may be positive, negative, or neutral for each organism. Since there are three possible results for each of two interacting organisms, there are a total of (3^2) nine possible combinations or coactions. Each of these has its own term and all of these together exhaust the possible interactions under the premise that organisms interact in pairs (Burkholder 1952).

This device has several imperfections, some more important than others. It is a rather long elaboration of a simple problem in probability (how many combinations of two things are possible if each occurs in three states?) and may obscure the more general approach. Because not all combinations are actually of any interest – some, like allolimy, are

Strong organism

		+	0	-
		+ + Mutualism (symbiosis)	+ 0 Commensalism	+ - Parasitism
Weak organism	+			
	0	0 + Allotrophy	0 0 Neutralism	0 - Allolimy
	-	- + Predation	- 0 Amensalism	- - Competition (Synnecrosis)

Fig. 3.4. The coactions: the table defines terms to describe the possible interactions between a strong organism and a weak one, under the assumption that the interactions can be classed as positive ($+$), negative ($-$), or neutral (0). Terms are from BioCO-TIE (1975) with alternatives (Burkholder 1952) in parentheses.

never studied and may not even occur – the categorization introduces unnecessary concepts into the literature. Worse, it introduces unnecessary terms for these unnecessary concepts into a literature already replete with jargon. The table of coactions is largely a mnemonic device for recalling these terms. The predictions now associated with these combinations are few and trivial – one might be that allolimy is rare – and the need for this mnemonic is in serious doubt.

One of the attractions of such logical arguments is that by listing all possibilities, the tautology focusses future thought and discussion. However, these possibilities are exhaustive only under the particular premises of the argument and other premises would permit other deductions.

In the case of the coactions, the nine possibilities can easily be deduced when necessary and the predictions associated with these possibilities are uninteresting, so one can ignore the coactions as self-evident and instead focus on the premises of the classification. One must ask if the sum of all interactions between pairs of organisms can be classed as wholly and consistently positive, negative, or neutral and if organisms can be usefully considered in isolated pairs, without reference to the rest of their environment. If one is not willing to grant these assumptions, there is no

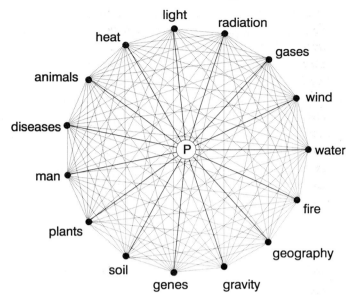

Fig. 3.5. The possible interactions between a plant (P) and its environment. The number of factors is intended only to be indicative of a large and varied set (from Billings 1952).

need to pursue the coactions further. In practice, work on two-way interactions has been limited to competition, predation, symbiosis, and parasitism, but even these distinctions are often hard to draw.

One of the most famous identifications of the possible is Billings' (1952) representation of the interactions in the environment of a plant. Figure 3.5 shows that all elements of the environment act on the plant and on each other. Every possibility is implied. This figure expresses a central issue in ecology: the possible extent of ecological interaction and complexity is immense. Given such logical possibilities, the critical questions become 'Are the premises tenable? Is everything connected to everything else? Must we consider everything in order to do anything?' These are empirical questions, resolved by showing that some predictions can be made even if some logically possible interactions are not considered.

Tautologies in 'theoretical ecology'

As tautologies become more complicated, they become more interesting, because more elaborate analyses are required to reveal their implications. These complicated, logical constructs form a major component of what contemporary ecologists call 'theoretical ecology'.

Theoretical ecology is necessarily a loose term. Its contents are open to debate (McIntosh 1985) and little would be gained by arbitrarily restricting membership in the field. Nevertheless, the term invariably includes a group of acknowledged intellectual leaders. John Harper (1977) writes of a Princeton-Harvard-Imperial College of London axis and any list of major figures would include Robert MacArthur and his students (Cody & Diamond 1975; Fretwell 1975) and G. E. Hutchinson and his (Kohn 1971; Edmondson 1971). Many key contributions appear or are reviewed in the Princeton Monographs, the two editions of *Theoretical Ecology* prepared by Robert May (1976a, 1981c), and the Asilomar Conference (Roughgarden *et al.* 1989). May (1977) identified work with the logistic curve, with models of competition, and with predator–prey relationships as characteristic areas of interest for theoretical ecology.

The need for tautological constructs in theoretical ecology is widely recognized. Hutchinson (1978) gives 'logico-deductive constructs' a central role in theoretical population biology. Wangersky (1978) wrote that such devices 'serve to deduce the form of possible solutions' and Lewontin (1968) felt they describe 'the limits of possibility'. For May (1977), theoretical ecology is the study of mathematical models, and for Levin (1981a), theoretical ecology looks at the logical consequences of assumptions. The latter statements are provocative exaggerations since theoretical ecologists often use empiricisms and make predictions. Nevertheless, the statements show that leading theoretical ecologists see the elaboration of tautologies as an important part of their subject.

Use of the term 'theoretical' to indicate tautologies is at variance with the definition of theory as a predictive construct which was advanced in Chapter 1. This disparity should present no difficulty in comparing concepts underlying the two approaches. In this critique, 'theoretical ecology' will be used in its contemporary sense and not to indicate the body predictive constructs in ecology. The latter will be called ecological theory, but intended meanings should be clear from the context.

The distinction between the tautologies of theoretical ecology and the deductive phase of hypothetico-deductive theories may seem subtle, but it is fundamental. All sides agree that deduction by itself is tautological and that the conclusions from deductions are only reworkings of the logical implications of the premises. The claim for predictive power therefore does not reside in the deduction itself but in the assumption that the premises (and therefore their logically necessary implications) apply to nature. Hypothetico-deductive theories can therefore be distinguished from tautologies by asking if the premises or conclusions could be proven false by observation.

For example, Slobodkin (1986) confuses logic with theory when he argues that arithmetic models are theories. Slobodkin correctly holds that a statement to the effect that one rabbit plus another rabbit gives two rabbits is a theory and that this theory would be disproven if one of the rabbits ran away, or died, or reproduced. However he errs in suggesting that this would test the proposition that $1 + 1 = 2$. Whatever number of rabbits we counted, we would not doubt the equality, for if we did we would reject the system that allows us to do the counts. Instead, we would doubt the application of the equality to this situation, perhaps because we doubt our capacity to census rabbits accurately. In any case, deductions from a theory about rabbits can be tested against experience, deductions within the logical framework of arithmetic cannot.

In theoretical ecology, the logic of the models is not disproven when they fail to describe an ecological phenomenon, for failure only indicates that the premises of the model did not fit the ecological system to which they were applied. Mismatches of model and observation are not counted as disproofs, but as indicators of limitations which may eventually describe the domain of a yet unarticulated theory. Models become theories when we have enough experience to predict when they will apply and when they will not. Until we gain this experience, the models are safe from disproof. Many constructs in theoretical ecology may therefore be steps towards theory, but not yet theories because their domain of application is unknown. Until then, they are only systems of logic. This distinction is essential to any assessment of the status of contemporary ecology.

Because tautologies are not corrected by confrontation of observation and prediction, theoretical ecologists stress other criteria in their work. Strong (1986) holds that this luxury has encouraged a 'prim' science, concerned with elegance of logical or mathematical argument rather than reality, applicability and prediction. Theoretical ecology is therefore free to isolate, simplify, homogenize and mathematize ecological phenomena to fit existing logical frameworks. This allows easier and more graceful mathematical manipulations, but not necessarily more effective description or prediction. This division lies behind many of the conventions that characterize theoretical ecology and encourage tautologies to replace theories (Table 3.1).

Table 3.1 *A comparison of the 'prim' constructs of theoretical ecology that allow mathematical caricatures of ecological phenomena and more realistic approaches that may permit prediction (Strong 1986 a, b).*

	Theoretical ecology	Predictive ecology
Mathematics	The object	One of several subjects
Trophic links	Tight	Loose
Population control	Constant, absolute	Variable, contingent
Population limits	Constant, K	Hazy, variable
Negative feedback	Consistent, strong	Inconsistent, vague
Rates	Instantaneous, r	Finite, R_0
Factors	Single to few	Several to many
Key variables	K, density, kin selection	Environmental mosaic, weather, phenology, migration
Determinism	Absolute	Moderate
Stochasticity	Ignored or trivial	Moderate to great
System	Closed, one-celled	Variously open, many celled
Environment	Homogeneous or cyclical	Heterogeneous, fluctuating

The logistic

Because the constructs of theoretical ecology are necessarily complex and because their tautological nature is not necessarily in question, this discussion can be limted to a part of one line in theoretical ecology, the development of the logistic equation and some of its descendants (Wangersky 1978; Hall 1988). The logistic equation is one of the oldest constructs in ecology (Verhulst 1838, 1845), although it was better developed about half a century ago (Kingsland 1982). The logistic equation describes the growth of a population increasing at a rate determined by the intrinsic rate of increase (r) and population size (N) to an environmentally set maximum (K):

$$dN/dt = Nr(K-N)/K \tag{3.3}$$

This equation describes a sigmoid curve asymptotically approaching K. Although an infinitely large set of equations could describe such growth, the logistic has been disproportionately popular, possibly because its terms can be given biological meaning rather easily and because it has been championed by leading biologists (Kingsland 1982).

The logistic performs a useful function as a simple linearizing transformation when population data follow a sigmoid growth pattern, but the continuing appeal of the logistic does not reflect its utility as a transformation, much less its success in describing population growth. Most populations are not increasing, but fluctuating around some long-term mean, and even growing populations need not fit this curve particularly well (Kingsland 1982; Fagerström 1987; Hall 1988). Neither does the appeal of the curve lie in its effective representation of biological processes. Its limitations in that regard are well known: Since all individuals in a population are not identical fractions of a homogeneous N, but differ in their reproductive potential and their environments, r and K vary with time and space, and among individuals. Moreover, growth of real populations may involve time lags, stochasticities, and higher-order interactions which are not represented in this simple equation (Dorschner *et al.* 1987).

The success of the logistic reflects a double standard that allows ecologists to count only successful applications of the model. When the data follow a roughly sigmoid growth, the logistic can be used, by definition, and its application is supported by the successful fit. However, when the data do not follow such a pattern, they are irrelevant to the logistic, and cannot be used to judge it. Thus the two results of comparing the logistic curve to data are either that the data are not appropriate or that the data fit the curve. Adherence to the logistic models of growth therefore involves an implicit tautology because all possibilities are permitted. This indulgence freed the logistic from critical scrutiny and ensured it a long life in ecology.

Kingsland (1982) suggests that the appeal of the logistic lies in its use as a logical ideal. This model shows how a population might grow if certain premises are met. By looking at deviations of real populations from this norm, one gets a better idea of the factors affecting growth in the real population and what adjustments are needed if the model is to describe this growth. The logistic curve serves as a 'positive heuristic' (i.e. an aid to fruitful growth of the field) by directing empirical studies of additional factors and by encouraging revisions that make the basic model more adequate. An analogous argument has been applied to the models of population genetics in general (Michod 1981) and to the Hardy-Weinberg model in particular (Emlen 1973; Sober 1984). The strength of such logically true, but often empirically false or inapplicable, models lies in their utility as null models, and in the ease with which one can tinker with their basic equations to fit new data.

The proposed benefits of the logistic as logical comparators and positive heuristics are difficult to accept. The failure of a logical model to describe the behaviour of the population only indicates that at least one of the premises of the model does not apply. The failure is negative. It tells us what does not obtain and can never indicate what does. Thus the only increase in understanding which the logistic's failure can provide is to tell us that some unspecified alternative is needed.

When an hypothesis is falsified, a new one must be erected (Hutchinson 1978; Southwood 1980). The easiest of an infinitely large number of alternatives is some modest revision of the original model and, in that sense, the failed model provides a positive heuristic, albeit a weak one. This adjustment requires little originality and offers no challenge to the intellectual college already committed to the basic model. Such tinkering is symptomatic of normal science and is justified if it augments scientific knowledge by increasing our predictive powers. If it does not, it will lead to an infinite spiral of falsification and revision, until the science grows bored with the approach or the problem. The heuristic offered by a failed model can thus become negative and deleterious by discouraging the investigation of alternative working hypotheses (Chamberlain 1890).

Lotka-Volterra equations

Despite the weaknesses of the logistic, theoretical ecology did not abandon the approach, but developed it. The best known of these developments is the Lotka-Volterra model for competition between two populations, N_1 and N_2, growing towards carrying capacities, K_1 and K_2, at rates, dN_1/dt and dN_2/dt, such that

$$dN_1/dt = r_1 N_1 (K_1 - N_1 - aN_2)/K_1 \qquad (3.4)$$

$$dN_2/dt = r_2 N_2 (K_2 - N_2 - bN_1)/K_2 \qquad (3.5)$$

where r_1 and r_2 are the intrinsic rates of increase of the two populations, and a and b are the competition coefficients that determine the effect of population 2 on population 1 (a) and of population 1 on 2 (b).

Since there are eight unknowns, these two equations are flexible enough to allow any combination of N_1, N_2, and t (Smith 1952). If the constants or variables are further specified, then the ranges of possible values are more circumscribed and the implications of the models more interesting. For example, the combinations that yield a steady state may be obtained by simplifying these equations for zero growth when $dN_1/dt = dN_2/dt = 0$, so that

$$K_1 - N_1 - aN_2 = 0 \text{ or } N_1 = K_1 - aN_2 \tag{3.6}$$

$$K_2 - N_2 - bN_1 = 0 \text{ or } N_2 = K_2 - bN_1 \tag{3.7}$$

The steady state will then depend on the relative values of each variable in these equations for the 'zero growth isoclines'. A typical didactic application of this model is to set one population (e.g. N_1) to its zero growth isocline (Equation 3.6) and then to determine the effects of various assumed values for N_2 on the trajectories of development of both populations. For example, when $N_2 < K_2 - bN_1$, $dN_2/dt > 0$; therefore N_2 will increase and N_1 will decline along its zero growth isocline until it is extinct or N_2 stabilizes. This exercise can be repeated with various assumed values until the logical possibilities of the equations are exhausted. Then it can be repeated holding N_2 on its zero growth isocline and varying the characteristics of N_1. Figure 3.6 sketches a popular graphical treatment of this logical argument which identifies these possibilities (e.g. MacArthur 1972a, b; Hutchinson 1978; Begon *et al.* 1986).

The limitations of this model are those of the logistic, adjusted for added complexity. Since the various parameters of the model can rarely be specified a priori, they are obtained by fitting the data to the model empirically. When populations act as the model suggests, values of the parameters can be specified to generate a quantitative description that constitutes a predictive theory for other instances of that particular system. The equations of Gause (1935) describing competition in cultures of micro-organisms are a case in point. Unfortunately, real populations are rarely so obliging. Other aspects of the biotic and abiotic environment impinge on the two competitors changing their rates of increase, carrying capacities and competition coefficients, so the specified model no longer applies. These situations are irrelevant to the model which is only expected to apply where the parameters are fixed. Regrettably, this cannot be known without first making the observations we hoped to predict.

The further elaborations of the logistic seem legion (see reviews in MacArthur 1972a, b; May 1976a; Christiansen & Fenchel 1977; Hutchinson 1978) yet they share the same limitations. None applies everywhere and we lack rules which indicate which model to apply to which populations. Therefore we approach the world accepting that the model will either fit or not. In the former case, the model applies and in the latter, some other model is required. Our expectations are always

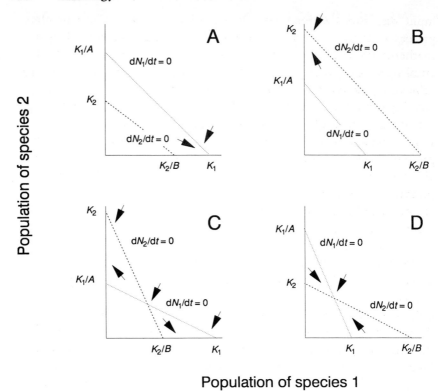

Population of species 2 (vertical axis label)

Population of species 1 (horizontal axis label)

Fig. 3.6. The Lotka-Volterra model of competitive exclusion showing various solutions for equations (3.6) and (3.7) describing the zero growth isoclines of populations 1 and 2. Regardless of initial population size, each population will move, as indicated by the arrows, towards its isocline. If the coefficients are such that one isocline is always above the other, as in panels A and B, the species with the higher isocline will always exclude that with the lower. If the two isoclines cross, the outcome of competition may depend on initial population sizes (panel C), or there may be a stable equilibrium to which both populations tend (panel D).

satisfied. This is a basic characteristic of tautology and results in the untestability of many constructs in theoretical ecology.

Identifying tautologies

Although the identification of a tautology sometimes requires consider-able thought (Sober 1984), several characteristics can help a preliminary identification. Because a logical argument has no need of empirical input, discussions without data often involve tautology. These are often presented as mathematical theorems or graphical analyses. Given a mathematical argument, one must ask if the coefficients are restricted empirically or theoretically. If they are, a deductive argument will achieve predictive power by elaborating these restrictions and critical scrutiny should focus on them. If the axioms have no empirical content (i.e. if they admit of their converse, as the logistic theory accepts that some populations fit the model while others do not), the argument will probably be tautological. In graphical analyses, unscaled axes (Mitchell 1974) often indicate a tautology, because they reflect unrestricted premises. Finally, one should ask what predictions are made by the construct and what the author might conclude if contrary observations were made. Theories would be falsified by non-corroboration, but tautologies, like the logistic, carry on.

Because the mental effort in analyzing a tautology is usually directed at the body of the argument, an effective strategy when dealing with a possible tautology is first to examine the premises or conclusions of the argument. For each premise, one must ask if the converse would falsify the argument. If not, then the premise does not restrict the theory. If all the premises admit of their converse, the construct will be a tautology. Since premises are often difficult to identify, one should also ask what logical possibilities are permitted under the premises. If the conclusions of the argument exclude some possibility, then the argument is likely not tautological and the source of this exclusion should then be scrutinized.

The logic of theoretical ecology is not always easy and most feel a sense of accomplishment when we have studied the rationale well enough to reproduce the deductions and so claim an understanding of the argument. This effort teaches the range of inter-relations which the premises of the equations allow, and it assures us that any biological system meeting these premises will act according to one of the possible outcomes. It does not indicate if such a biological system exists nor can it assure us that systems behaving in accordance with one of the deductions of the model also obey its premises. In other words, we gain no new information about nature by working through the model and, as a result, effort expended in theoretical ecology is sometimes poorly repaid.

To the extent that theoretical ecology refers to ecology that can be derived without reference to experience, it is an ecology based on the logical implications of certain assumptions. Its value lies in identification of the limits of possibility, in the preliminary probing of unstudied areas to see if intensive study might yield interesting results, in the evaluation of the implications of available information, even if this is relatively sketchy, and in the provision of logical models which may be made empirically relevant by specification of constants and unknowns measured for particular adaptations. These are valuable contributions. They should form part of university education, because they help students to think, and part of every research proposal, because this is the only way of demonstrating the value of data which have not yet been collected and theories which are not yet built. However, the constructs of theoretical ecology must always be seen as preliminary to science. They are not ends in themselves.

The production of new models and more publications to describe them and more lecturers to teach them are indices of the health and fertility in research. If self-perpetuation and expansion are considered sufficient ends for a science (Grene 1983; Lakatos 1978), then this fruitfulness justifies the research. Under that criterion, theoretical ecology in general and the logistic in particular have succeeded. However under the more stringent criterion that valuable science tells us something about the natural world, both have failed because they only tell us what we already know: either population data fit the theoretical curves or they do not. Tautologies accommodate every possibility.

The principle of evolution by natural selection

The fondness of ecologists for logical argument no doubt has many sources. In part, it may be a reaction against a dull empiricism promoted by some field biologists (McIntosh 1985). In part, it reflects the difficulty in finding appropriate data to answer the relevant questions of the day, and it may reflect the youth of a discipline wherein practitioners wisely explore potential new routes for research logically, before expending resources to examine them empirically. In addition, the role of logical argument is enhanced in ecology because of the pre-eminence given to one tautology, the principle of evolution by natural selection.

Because it is seen as a paradigm for ecological achievement (Ehrlich 1989), many ecologists have emulated the principle of evolution by natural selection. This regrettably leads ecology away from predictive power and into purely logico–deductive argument. It is one of the major influences making ecology a new scholasticism and its analysis provides an example of how one might approach ecological tautologies.

The principle of evolution by natural selection explores the logical implications of three axioms:

(1) Variation – Organisms differ phenotypically.
(2) Heredity – Some differences pass from parent to offspring.
(3) Selection – Only some organisms survive and reproduce.

This is one of many possible formulations. Bradie & Gromko (1981) listed 17 others and did not exhaust the field. Almost all formulations concur in identifying heredity, variation, and selection as essential for evolution by natural selection. Given these premises, evolution (i.e. descent with modification), its opposite (an evolutionarily stable state), adaptation (correlation between variants and survivorship), and extinction are possible.

The tautological nature of the arguments derives from three intersecting sets of tautologies. The first set, which is analogous to the tautology of the logistic, results because each of the premises of the argument admits its converse, and is therefore tautological in itself. The principle of natural selection requires variation among organisms, but it is not disproved by the absence of variation. Instead the existence of genetically identical clones of organisms is universally accepted. Similarly, the premises are not intended to deny the existence of uninheritable traits; such traits are simply irrelevant to the deductive argument. Finally, the theory does not deny the possibility of similar survivorship, it only claims that differential survivorship (in some inclusive sense that may reflect individual longevity, fertility, fecundity, but could also include the evolutionary fate of the progeny) is necessary for evolution. The converse of each premise is allowed but irrelevant to the logic of evolution, since evolution occurs only if all preconditions represented by the premises are met. Thus, each of the premises must be seen as entailing its converse and as a result, each is tautologous.

Figure 3.7 illustrates the logically possible conclusions that can be drawn from the premises of variation, heredity, and selection. This also illustrates a second tier of tautology in evolutionary theory, for all possible results of the combination of the three premises are acceptable under the theory and therefore any observation would confirm it. The population may survive or not, and if it survives, it may change or not. Thus the change of the peppermoth (*Biston betularia*) is as consistent with evolution as the constancy of the cockroach (*Periplaneta*). Similarly, there is a place in evolutionary theory for the extinction of the dodo and the speciation of the sandgrouse (Pteroclididae). Since all logical possibilities are permitted, the principle of evolution by natural selection must be a tautology.

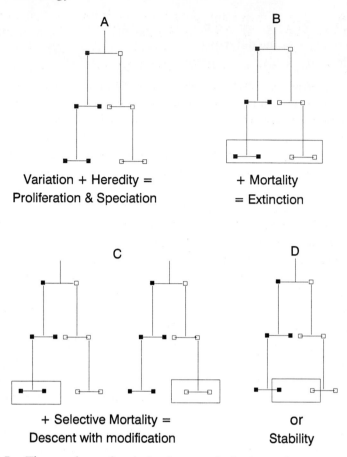

Fig. 3.7. The tautology of evolution by natural selection. When the members of a population differ in their inherited characteristics (A), lines of descent are created that differ in their characteristics. If only some of these descendants survive and reproduce, then there may be survival or extinction of some lines of descent, with the result that the population goes extinct (B), changes (C), or remains the same (D).

Scientific theories often begin by stating a series of preconditions that must be met before the theory can be applied. These usually restrict the generality of the theory in some way and, conceivably, the premises stated above may be preconditions which must be met before we consider evolutionary arguments. If we are to accept the three premises as preconditions for a scientific theory, we must now look at the prediction which is made under these preconditions. Regrettably there is very little left to predict, since extinction, preservation, and descent with modification are already logically entailed.

The remaining elements are adaptation and fitness, and it is to these that the evolutionary theorists turn for the claim that the principle of evolution by natural selection has predictive power. The theory can be stated as: Given variation, heritability and selection, the fittest or best adapted will survive. This statement raises the problem that, in practice, both fitness and the fittest are identified *a posteriori* by survivorship, again where survivorship is taken in very broad sense. A typical evolutionary argument identifies the idiosyncracies of the survivors and claims that these are the adaptations that conferred fitness (Bethell 1976). Unfortunately, this procedure still relies on the empty claim that the fittest, identified by their better survivorship, survive better. This is the most widely acknowledged tautology in evolution.

At this point, some statements of the principle introduce a new factor, adaptation, and claim that the principle of evolution can be delivered from the charge of tautology if we can identify the causal connection between survivorship and the trait or traits which the survivors share. This requires a fourth premise:

(4) Adaptation. Some of the heritable variations in phenotype sometimes enhance survivorship. Ghiselin (1969) and Endler (1986) achieved the same result by rephrasing the premise of selection so that it incorporates adaptation. In either case, it will then follow logically that the frequency of these adaptive variations will sometimes increase among the survivors. This last premise introduces both new and old difficulties.

One familiar difficulty is that the premise does not require that all traits be adaptations or that a particular trait be an adaptation at all times or places, only that some trait sometimes be adaptive in some places. This may be given a statistical flavour by claiming that a trait will be favoured even if it helps only one time in 100 or once in 100 000 times. Such rare cases are well below the statistical limits of detection in most studies and are likely to result in 'type I' statistical errors in which biologically important events are thought to be insignificant. Thus the fourth premise, like the first three, is tautological because it admits of its converse, and every observation with respect to the adaptiveness of a trait, is acceptable. Indeed, processes that may result in the survival of a non-adaptive trait (i.e. in 'non-Darwinian evolution'; King & Jukes 1969) are well known. These include genetic drift, founder effects, pleiotropisms, stochasticities in genetics, survival and adaptation, linkage disequilibria, allometric limitations, and other architectural and biological constraints (Gould & Lewontin 1979; Harper 1982; Mayr 1983; van der Steen 1983b). These non-selective processes are incidental to all statements of the principle of evolution by natural selection, except extreme pan-selectionist formulations wherein all traits are required to be selectively advantageous. Despite the success of adaptationism (Gould 1982, 1983), very few evolutionary biologists endorse so extreme a position. As a result, the absence of apparent selective advantage for a successful trait is not relevant, so this absence does not falsify the theory.

A new problem is that the fourth premise is required only if it permits deductions in addition to those that were made from the first three. Since these were already sufficient to deduce the possibilities of the correlation of heritable traits with extinction, evolution, and survivorship, further correlation of heritable, adaptive traits with these same responses are redundant. Consequently, adaptive and non-adaptive traits are hard to separate because they have the same characteristics. The premise attempts to escape this by invoking causal connection between the trait and survivorship. But, causal connection is extremely hard to demonstrate (Chapter 5), and failure to demonstrate causality is easily explained within the general context of evolutionary theory by reference to non-adaptive processes, low selection coefficients, the discontinuity of selective advantages in space and time, subtle or multiple causation, and the inappropriate identification of the purpose of a given trait. Once again, the charge of tautology can be fairly made because the theory is consistent with any logically possible observation.

Any attack on the status of evolutionary theory is open to misinterpretation. For example, scientific criticisms of evolution are sometimes appropriated by Christian fundamentalists, and Ruse (1982) intimates that the two may have similar motivations. Denial of theoretical status to evolution by natural selection need not endorse theism, but insistence that the principle is a tautology means no contrary position is logically tenable (Manser 1965; Barker 1969). Lamarckism, at least in its simple forms, has been tested and rejected as an alternative or addition to natural selection: the children of weight-lifters or amputees do not reflect the parental anomalies; complex forms of Lamarckism requiring that the change be both desired and desirable may be untestable because they require both existence and knowledge of internal motivational states. In any case, Lamarckism is presently moribund. Creationism holds no interest for those who wish to make predictions about this world because all possibilities lie within the deity's powers. My criticisms are not directed at the idea of common descent nor do they deny that predictions may be based on the presumption of common descent (Williams 1973, 1985). What is denied is the charge that evolution by natural selection can be false and that there is any need for a fundamentalist alternative (Mayer 1984). Instead, tautological status relieves evolution by natural selection of the charge that it is 'just a theory'. To the extent that the principle of natural selection is a tautology, it can claim to be true. Regrettably, to the same extent, it cannot claim to be scientific information.

Ruse's (1982) suspicions about the motivation of scientific critics of evolutionary theory reflect his long experience, protecting the rear and flanks of science from the attacks of religious bigots. Scientists should be grateful for his labours, but his bitter experience is irrelevant for those who accept the logic of selectionist arguments, but recognize that purely logical arguments produce no predictions.

Assaults and defenses

The charge that evolution by natural selection is untestable and tautological is a common one (Manser 1965; Barker 1969; Lee 1969; MacBeth 1971; Popper 1974a; Harris 1975; Bethel 1976; Peters 1976, 1978; Thompson 1981; Brady 1979, 1982) and evolutionists have vigorously maintained that the charge is false (Ghiselin 1969; Lewontin 1969, 1972; Ferguson 1976; Gould 1976; Caplan 1977, 1978; Castrodeza 1977; Stebbins 1977; Hairston 1979; Ruse 1982; Van der Steen 1983b; Naylor & Handford 1985). The defenders' arguments frequently differ in emphasis, but several main lines of defense can be discerned.

Explanatory power

The most common defense is to point to the success of evolutionary theory in explaining particular observations and to ask how a tautology could fit so well. This misses the point. A tautology fits any observation well. This is clearly shown by considering whether the antithesis of any successful evolutionary explanation would have disproved the general model. For example, if peppermoths had remained peppered or gone extinct, instead of evolving into melanistic forms where industrial pollution killed light-coloured lichens on dark tree barks (Kettlewell 1973), we would not discard the general premises of evolutionary theory, but claim that selective factors other than visual predation determined its evolution. Indeed, the details of industrial melanism fit the explanation less easily (Bishop & Cook 1980), but this has not disproven the principle.

Natural selection is often much easier to invoke than to demonstrate

(Endler 1986), and some evolutionary interpretations may be less strong than their description suggests. For example, Ferguson (1976) and Stearns (1982) point to the evolution of eusocial behaviour among wasps and bees as a successful prediction from evolutionary theory based on kin selection (Hamilton 1972; Trivers & Hare 1976). The argument is complex, but the deduction is that because haplodiploid sex determination, a characteristic of Hymenoptera, enhances the genetic similarity of full sisters to the point that a female bee or wasp is more closely related to a full sister than to its own offspring, female Hymenoptera may forego reproduction to help their mothers produce more, reproductive, sisters. This success must be weighed against the failure of eusociality to develop among other haplodiploid arthropods (including non-social bees, some beetles and mites), by its absence among parthenogenetic vertebrates and arthropods, and its appearance among non-haplodiploid vertebrates and termites. Equally damning is the failure of eusocial bees and wasps to meet some of the requirements of Hamilton's model. For example, because queens are not monogamous (Wilson 1971), workers are not as closely related as the model assumes (Thompson 1981). Stearns (1982) lists a series of other failures of the theory of kin selection. None of these need shake our faith in evolutionary logic, because they are simply non-instances, not falsifications. Since negative evidence does not count in evolutionary biology, evolutionary biologists quite rightly ignore failures and celebrate successes. Of course, this is not science.

The ubiquity of adaption

A related defense notes the prevalence of successful adaptations. Logic is sufficient to identify the possibility that if adaptations arise they will be favoured, but not to show that selection of adaptations has been a dominant force in evolution. We might be unimpressed if the apparatus of natural selection applied in only 2 cases in 100, but if it applies in 98 of these cases, Lewontin (1972) feels we would be justified in claiming to have tested the principle of evolution by natural selection. A similar claim could be advanced if we were to predict the probable frequency of adaptation among particular types of traits, organisms, or environments (Ghiselin 1969).

The question of how often natural selection occurs (Hairston 1979) is difficult to resolve, because the evidence for particular cases in nature is not clear cut and because random sampling of all traits, organisms, and environments is likely impossible (Gould 1983). Thus, although Endler (1986) provides strong evidence for natural selection of 314 traits in 141 species, we cannot know if this represents 2% or 98% of the total number of cases and so Lewontin's (1972) criterion for testing the principle of natural selection cannot be applied. If anything, the relatively small number of good examples of selection and the absence of discernible patterns among them after 125 years of research (Endler 1986) must represent a disappointment for selectionists.

Popper (1978) offered a more rigorous variant of Lewontin's (1972) test. Popper (1974a) had previously denied the status of scientific theory to evolution by natural selection, because no conceivable observation would falsify the theory and suggested that it should instead be seen as a metaphysical research program. In 1978, he took the seemingly concilia-tory position that natural selection would be a predictive theory if it asserted all traits to be adaptations. In this case, premise (4) would exclude its converse and the theory would escape the charge of tautology. Thus evolution by natural selection could achieve the status of theory, but only by jettisoning the concept of non-adaptive traits (Lewontin 1978). This is so hollow a victory that Ruse (1982) doubted the sincerity of the retraction. No selectionist claims that all evolutionary change is adaptive, just as none claim that all traits are heritable or that all organisms differ in all traits. Even the pan-selectionist A. J. Cain (1977) advocates this position only as a safeguard against facile 'explanations' based on a supposed selective neutrality.

Specific cases

A third defense argues that the general statement of the theory makes no predictions until it is specified for particular cases, just as the regression for animal density in Chapter 2 requires the average weight of the raccoon to calculate population density. This analogy is false for the regression's parameters are already specified. Nevertheless, there is no question that predictions can be made about evolutionary change in particular cases. What is dubious is the relevance of the success or failure of these predictions to the status of the general theory.

For example, Lewontin (1978) proposes a specific evolutionary theory whereby it is observed that zebras with longer legs more easily out-distance their predators and predicted that longer-legged zebras will increase in abundance. Ferguson (1976) suggests that, if diabetes is an inheritable condition and if more diabetics reproduce since the discovery of insulin, then the proportion of diabetics should be on the increase. Both of these suggestions are acceptable theories and both are necessarily consistent with evolutionary logic. However, if these theories are falsified, both the falsifying observation and a new theory that predicts this observation will still be consistent with the general logic of evolution. Perhaps predation is so minor a factor for zebras that escaping predation is irrelevant, or perhaps such zebras sometimes break their longer legs at high speeds or have difficulty in parturition, so longer legs are not favoured by selection. Perhaps because diabetes was expressed late in life, the symptomatic treatment of diabetes with insulin has had no measurable effect on the reproductive success, and so diabetics are not increasing in the population. The general principle of evolution can accommodate any of these logically possible outcomes and more. Therefore if a specific explanation which is consistent with the general statement is falsified, another one which is just as consistent can be found.

Empirical content of the premises
Some authors (Ghiselin 1969; Lewontin 1969; Caplan 1978; Slobodkin 1986) dismiss the charge of tautology because the premises of the principle of evolution by natural selection can be related to observation. Thus, Caplan (1978) feels that the charge of tautology should be dismissed because the principle contains some minimal existential claims to the effect that organisms display variation, heredity, and differential survival. Kaplan is correct that if these premises are hypotheses about the biological world, then conclusions from these premises would not be tautological. However, if the premises are hypotheses, they must identify some things as being improbable. Instead, common exceptions, like clonal populations, non–hereditary traits, and similar survivorship, do not falsify the premises, and therefore, the premises are not treated as hypotheses. In practice, we do not consider an evolutionary argument unless we were reasonably sure the premises hold. If the premises apply, then some evolutionary pattern is inevitable and if the expected result does not occur, we change the premises to fit the result or discard the example. No general theory is tested in any case. Kaplan is correct that the premises are associated with biological phenomena, but they have no more empirical content than the statement 'it is raining or not raining'. Both the premises of the principle of natural selection and of this obvious tautology about rain, refer to empirical phenomena but are not, thereby, given theoretical status.

Defining fitness
Evolution by natural selection could claim to predict if there were some general characteristic which would identify the traits which are more

likely to flourish. This would constitute a definition of fitness which is independent of survivorship (Dunbar 1982). As a result, many defenders of natural selection have attempted to identify some property (X) which would indicate fitness but be logically independent of survivorship. They hold, correctly, that a true prediction would result if we would rephrase the principle to take the form 'when the premises are met, then the most X organisms will survive'. Unfortunately, this property X has proven impossible to specify and therefore the principle remains mired in tautology.

The reason that this task has proven difficult is that what constitutes fit is defined relative to the environmental context. Sober (1984) claims that fitness is a supervenient property, by which he means it is a non-physical property such that two physically different objects may both have it, but it will necessarily be shared by two objects which are physically identical. Such obscure traits make poor scientific variables.

Several authors have offered alternative, operational definitions of fitness. Stearns & Schmidt-Hempel (1987) illustrate the difficulty in suggesting that different fitness measures be taken as tokens of an unmeasurable general fitness. Thus an appropriate fitness token might be the population geneticists' w, defined as 'that parameter best representing differential reproductive success in such a way that one can predict changes in gene frequencies'; for life-history evolution the token could be 'that parameter which incorporates age-specific effects on differential reproductive success and trade-offs in such a way that optimal integration of life-history traits can be predicted'. Both tokens are defined in terms of their success in predicting survivorship and neither can be measured *a priori*. Therefore neither delivers the principle of natural selection from tautology.

Gould (1976) suggested that principles of engineering design be used to identify the best adapted trait and that the prediction be rephrased as survival of the best designed. This has been formalized in optimality theory. Regrettably, it is often difficult to decide what a trait is designed to do and therefore to decide if the design is appropriate. Ollason (1987) argues that design is imposed by human interpretation following criteria which are necessarily tautological. Moreover, design by natural selection often involves jerryrigging from available parts and is unlikely to win any engineering awards (Gould 1978). One could restrict adaptationist arguments to instances where these decisions can be made easily (Gould & Lewontin 1979; Thompson 1981), but this would so restrict the application of selectionist arguments that would be unacceptable to the

majority of evolutionary biologists. In practice, non–tautological definitions of fitness have always proven too restrictive or would lead to
falsification of the theory of natural selection if they were accepted. As a
result, evolutionary biologists are thrown back on tautological
formulations.

Evolutionary predictions

Another defense attempts to out-flank the charge of tautology by
identifying some prediction of evolutionary theory, on the correct
grounds that even one prediction gives lie to the charge of tautology.
Unfortunately, because these are flanking manuevers, the authors do not
show how the prediction is deduced from the general hypotheses.
Darwin himself felt that his theory would be placed in serious difficulty if
an organ was found that existed solely for the benefit of another species
(Ghiselin 1969), but when put to the test, he accommodated such
instances as the human appendix by supposing that they were rudimen-
tary organs on their way to oblivion. Ruse (1982) believes that the
presence of systematically difficult species and semi-species, and the
absence of paleozoic mammals should count as predictions from evolutionary theory. If this is so, they are not predictions from the simple
premises outlined above but from some larger and undefined body of
science which is incidental to the testability of the principle of evolution
by natural selection.

The hard core of the adaptationist program

An epistomoligical defense claims that the principle of evolution by
natural selection should not be seen in isolation, where it is tautological
and untestable (Van der Steen 1983; Naylor & Handford 1985; Stearns &
Schmidt-Hempel 1987) but as part of a larger, hierarchical, system of
theories. This system has been called a generic theory (Tuomi &
Haukioja 1979), a hyper-theory (Wassermann 1981), and a meta-theory
(Tuomi 1981). One of the drawbacks of this view is that the meta-theory
is dynamic and therefore cannot actually be stated (Ruse 1982). It is
accordingly hard to confront.

    This line of attack derives from the conception of science as a group of
research programs (Lakatos 1978) that characterize and define the
normal science in each discipline. In Lakatos' view, all research programs
contain a 'hard core' which is rendered invulnerable to scientific test by a
soft core of ad hoc hypotheses that protect, define and interpret the hard
core for particular applications. For example, the theories of thermo-

dynamics form part of the hard core of ecological energetics, and ecological observations about energy fluxes and caloric contents form a soft core premised on these principles. Thus the law of conservation of energy, part of the hard core, is not falsified when an energy budget of a bird or a *Daphnia* fails to balance. Instead, we attempt to correct our measurements of flux, a part of the soft core.

In evolutionary theory, the hard core minimally includes a statement of evolution by natural selection like that stated above, and Mendelian genetics. Falsification of an hypothesis which results from the combination of the hard and soft cores only results in a modification of the soft core. Thus, defenders of evolutionary biology claim that maintenance of an untestable central core is accepted scientific practice in all developed sciences and that evolutionary theory is no worse.

This interesting defense of evolutionary theory, that it is no worse than other sciences, can be countered in two ways. First, natural selection does not seem to fit the model, and second the model should not be seen as prescriptive.

Lakatos' model supposes that if the research program were stripped of the protection of ad hoc hypotheses in the soft core, the hard core would be readily falsifiable. This would indeed be the case of Mendelian genetics. Few interesting traits are controlled by single loci exhibiting simple dominance relations; nevertheless the theory is saved from falsification by invoking ad hoc hypotheses about linkage, crossovers, incomplete penetrance, pleiotropisms, multiple alleles, environmental determination of expression, and so forth. This is not the case for natural selection. When its protective coat of ad hoc hypotheses is stripped away, the hard core of evolution by natural selection remains immune to test, because it is tautological.

In a larger context, this defense is inadequate because it makes a description of what science is into a prescription of what it should be. Science may have its sacred cows, but they should not become role models.

The attraction of evolution by natural selection

This extended criticism begs the question, 'If evolution by natural selection is so scientifically weak, why does it fascinate so many scientists?' or perhaps 'Since so many scientists are fascinated with natural selection, what is missing from this critique?'

Natural selection has several strong points in its favour (O'Grady

1984). It is an argument from natural history (Peters 1980) which addresses questions that have interested naturalists for centuries. Since most ecologists are also naturalists, they often ask similar questions. Moreover, natural selection provides a non-theological answer to these questions and so seems the only scientifically acceptable alternative to creationism (O'Grady 1984). Furthermore, because the principle of natural selection entrains tautologies at several levels, it is not open to falsification and it is not easy to criticize. Instead, it can accommodate any observation and therefore is a powerful tool to explain and rationalize observations once they have been made.

The agreement of the principle of natural selection with observation is one of its chief attractions. Because the theory is always right, it is relatively immutable (O'Grady 1984). Material may be added to it, but little need be removed. As a result, teachers can use last year's notes, textbooks can be revised once a decade, and students can learn ideas which will be relatively stable throughout their careers. All seem to benefit because the fundaments of the science are fixed. Moreover, these basics can be taught to large numbers of students in the lecture hall, making use of a range of powerful technical aids and building on a base of natural history and evolutionary arguments which is diffuse in our culture. Given this strength, it has been easy to overlook the fact that natural selection is a logical truth, not a scientific theory.

Because it is a logical truth, the principle of natural selection applies well beyond biological evolution. Indeed wherever variation and selection occur, essentially selectionist arguments apply (Sattler 1986). The approach has therefore been applied to competition among theories in epistemology (Popper 1979), to cultural development in man (Goudge 1961), to economics and sociology (e.g. Wilson 1975; Alexander 1977) and to linguistics (Cavalli-Sforza et al. 1989). This flexibility hints at the lack of empirical content of the theory, for if it was a strictly biological theory it should encounter more obstacles when applied to non-biological problems.

Logical aids, like the logistic, natural selection, and arithmetic, are needed when the argument is sufficiently complex that one can easily be lost. This is certainly the case for arithmetic and it is likely the case for optimal foraging theory and possibly for the logistic curve. This is less often the case for much of natural selection. The popularity of selectionist arguments on television, among school-children, and in the popular press suggests that most evolutionary arguments are transparent. The need for a logical crutch is a personal one and its utility likely depends on context. Nevertheless, I suspect that the logic of survival and evolution is simple enough that it usually can be reconstructed as needed.

Many authors see the heuristic value of natural selection as the basis for its success (e.g. Dunbar 1982; Michod 1981; Ghiselin 1969; Bradie & Gromko 1981), and certainly, the evolutionary literature is vast. However, if Endler (1986) is correct that all this work has uncovered natural selection in only 141 species, one must wonder if the heuristic has really been fruitful. Is it not possible that the principle of natural selection is heuristic because it is so easy? Is it not possible that natural selection, like the logistic curve, seems heuristic because it offers the path of least resistance, the route which requires least thought and originality? Could these logical aids so close our minds to alternatives that our science becomes less creative, less original, and less sensitive to contemporary demands? I suspect the answer to all these questions is affirmative.

Natural selection has spawned a vast scientific literature, an even greater literature for the lay naturalist, powerful films, and beautiful television series. All are based on facile explanations using the logical models provided by natural selection. As a result, when ecologists are asked to make ecological predictions relevant to our contemporary world and its problems, we can rarely do so. We have long since tied our scientific imagination to a tautology which, in the last analysis, can tell us nothing new.

Summary – Two tools for two jobs

Tautologies are distinguished from theories because the former identify the range of possibility whereas the latter identify the range of probability. Tautologies serve science as logical aids or as and organizing tools, but they must not supplant theories. Ecology incorporates many tautologies. Some are too simple to warrant close attention, but others involve aspects of theoretical and evolutionary ecology that are difficult enough to require the help of a logical model. Unfortunately, ecologists have confused tautologies with theories, to the disadvantage of both. Critical ecologists must therefore be alert to the possibility of tautology and to the limitations that such constructs present.

4 · Operationalization of terms and concepts

Every part of a theory begins as a concept. There are concepts that could eventually describe ecological entities and classes, concepts that may develop into variables that are characteristic of these classes, concepts that may become relationships among these variables, and concepts that will never lead to anything of scientific interest. Because weak concepts are hard to eradicate and because new concepts are constantly being created, the ecological literature will always contain concepts at every stage of elaboration (Table 4.1).

The creation of a new scientific theory from the range of available concepts is no easy task. The vague primordia of our ideas must be given form in our thoughts, this form must be interpreted in light of our experience, and then refined to remove incongruous material while preserving some essence of the original meaning. At the end of this process, the concept will be associated with a set of external phenomena and the association likely legitimized with its own term. Only concepts that have gone through this process of operational definition can be used in a theory. Scientific criticism encourages this operationalization by identifying the present capacities, limitations and roles of existing concepts.

The chapter begins with a brief discussion of what constitutes operational definition, the importance of theory and practice in achieving this definition, and the role of non-operational concepts as precursors to theory. Next, the attributes of concepts that discourage operationalization – complexity, open-endedness, relativism, scale, non-scientific antecendents, nebulous goals – and the various terms that have been used to describe non-operational constructs – concept cluster, conflation, pseudocognate, omnibus term, panchestron, non-concept – are illustrated by ecological examples. The first set of examples represent conceptual typologies or classifications, including the most fundamental classification of all, the dichotomy between organism and environment. Stability and diversity are examined to illustrate the difficulty posed by undefined variables. Next, two conceptual relations, the stability – diversity relation and competitive exclusion principle, are used to exemplify problems that arise when concepts are confused with theories.

Table 4.1 *The most popular 25 ecological concepts as identified by a poll of
the members of the British Ecological Society (Cherrett 1988, 1989).*

Concept	Mean	Variance	Number choosing the concept
The ecosystem	5.18	4.50	447
Succession	2.98	3.38	347
Energy flow	2.56	3.48	274
Conservation of resources	2.29	3.41	257
Competition	2.23	3.10	268
Niche	2.14	3.20	245
Materials cycling	2.13	3.13	243
The community	1.88	3.32	189
Life history strategies	1.88	2.94	238
Ecosystem fragility	1.76	3.13	194
Food webs	1.73	2.98	194
Ecological adaptation	1.71	3.03	199
Environmental heterogeneity	1.57	2.95	170
Species diversity	1.47	2.75	176
Density-dependent regulations	1.43	2.78	172
Limiting factors	1.34	2.67	166
Carrying capacity	1.27	2.52	162
Maximum sustainable yield	1.24	2.55	157
Population cycles	1.17	2.50	141
Predator–prey interactions	1.14	2.40	148
Plant–herbivore interactions	1.10	2.40	148
Island biogeographic theory	0.99	2.27	132
Food chain bioaccumulation	0.84	2.11	113
Coevolution	0.82	2.12	115
Stochastic processes	0.78	2.12	99

Note:
Each of 645 respondents selected 10 concepts from a preliminary list of 50 and
scored these from first (score = 10) to tenth in importance (score = 1). The mean
is the average of these scores and variance in this score indicates the amount of
disagreement in these rankings. The lower 25 concepts were, in descending
order, natural disturbance, habitat restoration, the managed nature reserve,
indicator organisms, competition and species exclusion, trophic level, pattern, r-
and K-selection, plant–animal coevolution, the diversity–stability hypothesis,
socioecology, optimal foraging, parasite–host interactions, species–area relations,
the ecotype, climax, territoriality, allocation theory, intrinsic regulation, the
pyramid of numbers, keystone species, the biome, species packing, the 3/2
thinning law, and the guild.

The final set of illustrations considers a series of non-operational statements from environmentalism. Although these propositions are not strictly parts of ecology, they are included in this critique because the public has confused them with ecological theory, and because their simplicity and popularity may make the shortcomings of non-operational concepts easier to appreciate: some concepts seem to advance authoritative, factual propositions and so to replace theories, but this substitution is actually a deception because the propositions are predictively empty and atheoretical. The chapter concludes by summarizing some of the scientific difficulties associated with the dominance of concept over theory in ecology.

Operationalization of concepts

Operationalization or operational definition is the practical specification of the range of phenomena that a concept or term represents. As a result, theories apply to the world around us only to the extent that they are based on operational terms, because operational definition describes this application.

The role of theory in operationalization

The importance of operational definition does not lie in fixing the conceptual entity itself, but in the degree to which such a definition allows access to information through theoretical relations that use the concept. For example, a prime goal of ecology is to predict the population density of an organism, but an estimate of animal density alone is uninteresting. The value becomes interesting because it might be used to predict other attributes, like cropping rates, disease control, or extinction rates. Operational definitions are adopted only if they lead to effective theories, and the most informative definitions function in many different relationships.

The importance of functional definition has long been recognized. G.E. Hutchinson (1953) expressed it with characteristic grace over a generation ago: 'knowledge appears to consist of known relationships between entities which are apparently unanalysable. And as they are studied these entities become relations between still more abstract entities'. A similar concern is reflected in Popper's (1985) disdain for definitions that try to capture the essence or truth of the entity being defined. Popper prefers terms that are simply convenient labels for the operations required to identify observable entities with theoretical elements. Such operational definitions obtain their warrant because they function in prediction.

Operationalization is not achieved by decree, by proclaiming that a given concept will be represented by a designated term, defined in such-and-such a way. In ecology, such attempts (Edwards & Fowle 1955; Milne 1961; Hurlbert 1981; Whittaker, Levin & Root 1973) have rarely been successful. Instead, operational definition is achieved by relating the concept to a series of defining operations, and then demonstrating that such an entity so defined can play an important role in theory.

If a definition seems necessary, that definition is likely doing the work of theory. For example, one could define mammals as hairy homeotherms that give milk to their young. Like all definitions, this would be sterile, unless we had a theory of the form 'all mammals (i.e. hairy, milk-producing homeotherms) are X'. The definition would do the work of a theory if part of the definition were used to identify an entity as a mammal and the remainder to predict some further characteristic: any animal suckling its young is hairy and homeothermic.

Maintaining operationalization

Standardized operational definitions are essential if different workers are to make similar measurements of similar entities. Since no definition, whether operational or not, is so precise that all doubt about the meaning of a word is removed (Sattler 1986), initially unnoticed ambiguities in the definition may lead to misapplication of the theory. As a result, operationalization is always incomplete and operational definitions are often amended over time. However, if a concept is to prove scientifically useful, both the original definition and its replacements must be sufficiently operational that informed users associate the concept with similar phenomena (Hull 1968). This degree of operationalization is essential to any assessment of predictive power.

Because complete operationalization is impossible, procedures diverge as different workers interpret and apply standard methods in slightly different ways. This is one reason that interlaboratory tests vary more than replicated analyses from the same laboratory. To control this definitional expansion, scientists have developed several checks on analytical consistency. Routine application of the defining operations to known standards can detect methodological deviation as a drift in the standard values over time. Comparisons of concurrent estimates of the same entity with independent techniques should reveal inconsistency in operational definitions as the difference in the estimates. A third important constraint is achieved by comparing the predictions of the theory in which the operational entity is imbedded with measurements of the response variables. Discrepancies between present and past standard determinations, or among independent measurements, or between observation and prediction warn of the possibility that the methods have deviated.

The place of non-operational concepts

Initially, operationalization seems hopeless. The concept is nebulous and unfamiliar, its terms incomprehensible, its place in theory and practice undetermined, and its predictive power unevaluable. Only those who are drawn intuitively to the idea are likely to pursue it, since such pursuit requires an irrational faith in the chance of success. As the problem becomes better defined and its potential solution seems more attainable, it attracts other researchers who in turn help to define the concept. If the program succeeds, the concept, its terminology, the operationally defined variables, the theories that have been developed, and the applications of those theories become entrenched by frequent use and paradigmatic example (Kuhn 1970).

The process of theory creation and growth therefore begins with a vague sensation of personal insight and consequent belief, followed by greater public awareness, definition, and involvement as new adherents are drawn into the program. The process culminates in wide acceptance of the operationally defined theory.

Ecology, like every science, desperately needs the intellectual pioneers who will explore new concepts. Without them, the scientific enterprise would grind to a stop. It is therefore normal practice for good scientists to entertain irrational, vaguely formed, poorly defined, unpredictive concepts and their associated terms. These may be the stepping stones that eventually lead to strong theories. Many influential works in the literature do not contain testable theory, but are only propaganda for developing concepts. Presumably, the originator's ideas had developed to a point where they could be expressed, but not to the point where the concepts could be cast as hypotheses. As a result, the ecological reader should be acutely aware of the differences between concept and theory; otherwise, enthusiasm of belief and vigour of exposition may allow concepts that substitute for prediction (Fig. 4.1).

Because a concept is vague – Rigler (1975a) defined it as 'a general notion' – the success of operationalization is uncertain before the fact. Until concepts find their places in testable theories, their place is the

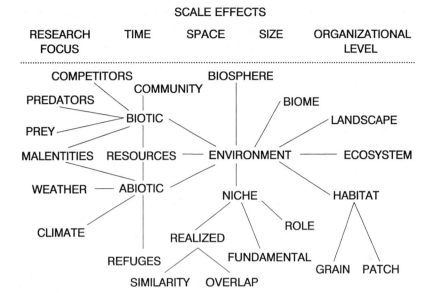

Fig. 4.1. A summary of some interactions among concepts associated with 'the environment'. Such diagrams can only suggest some of the links and associations among concepts, because ecological concepts are so vague and subjective that any rigorous analysis of the concept cluster is sure to fail. Whatever associations are envisaged, they must allow for redefinitions with changes in scale and with changes in the organism upon which the research is focussed.

personal realm that precedes hypothesis (Chapter 2), and they are justifiable only because they might lead to theory. Attempts to operationalize a concept may yield a new theory or a new variable, but they may only produce a succession of non-operational concepts. Although we must use concepts to develop theories, we should be prepared to jettison those that prove unproductive. We must also realize that to teach or promote concepts as theories may be premature; it is to teach the undefined and to promote the unevaluable.

Because concepts are not open to test until they are re-expressed in theories, the early products of conceptual development are difficult to extirpate when their utility fails. The hopeful spawn of past initiatives haunt the science. Some, like the broken stick model (MacArthur 1957, 1966) and the constancy of ecological efficiencies (Slobodkin 1961, 1972), even survive repudiation by their originators (McIntosh 1982). Some ecological questions – e.g. the natures of competition, regulation, stability, the environment – have been the basis for symposia and discussion for over fifty years (May 1984), yet the questions remain. This lack of resolution suggests that the debates dealt with concepts not theory, since, if theories had been developed, the issues should have been resolved on the basis of scientific evidence.

Typologies and classifications

A recurrent problem in operationalization for ecologists is the isolation of distinct classes that are appropriate for further study. Robert MacArthur (1972b) thought that the ecology of the future would see a fundamental two- or three-way classification of organisms and their environments. Given a system of classification, we would seek theories that predict some class characteristics from measurements of other properties of the class or its environment. Before this could be done, the types and classes must be defined.

The typological approach is essential to questions about the behaviour of populations, species, guilds, communities, and ecosystems, in which these concepts are conceived as distinct entities for study (Fig. 4.2). It is also fundamental to the view that nature involves a set of spatial patterns induced by a mosaic of associations, biocoenoses, communities and biomes, undergoing a sequence of temporal steps between different successional stages (Watt 1947). Such a view can only find application when the different spatio-temporal units can be identified and separated. The realization that such identification was difficult in plant community ecology (Cain 1947; Egler 1947; Mason 1947; Whittaker 1953, 1957), precipitated one of the greatest conceptual changes the science has experienced: the shift from a 'superorganismic' conception of the ecosystem to an individualistic or continuum view (McIntosh 1985; Simberloff 1980a). Analogous typological problems arise whenever discussions centre on class characteristics (McIntosh 1970; Wiens 1983; Schoenly & Reid 1987; Shipley & Keddy 1987; Hawkins & MacMahon 1989).

These problems are not great when the class has already been defined and one response to typological problems has been to suggest that ecologists concentrate on recognized classes (Simberloff 1980a). Taxonomic groupings are particularly obvious for biologists, but presumably any recognized classification would do. Whenever the class can be identified by simple, well recognized properties, a description is straightforward and typological problems are small.

If no prior convention about class membership exists, the properties

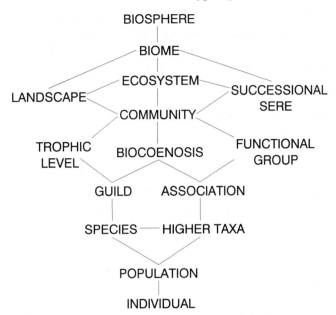

BIOSPHERE

BIOME

ECOSYSTEM

LANDSCAPE

SUCCESSIONAL
SERE

COMMUNITY

TROPHIC
LEVEL

BIOCOENOSIS

FUNCTIONAL
GROUP

GUILD   ASSOCIATION

SPECIES — HIGHER TAXA

POPULATION

INDIVIDUAL

Fig. 4.2. A diagrammatic summary of typological concepts associated with organizational scale in ecology. Although typological analysis depends on the operational definition of the type under study, this is usually only possible with taxonomic entities.

that define the class must be identified as part of the theory. Selecting these properties can be a difficult problem in operationalization. The researcher must sift through the available data applying a series of alternate class definitions, seeking that which allows the most powerful predictions about the relevant characteristics of that class. This requires a sure grasp of what is relevant to the study, great mental flexibility in entertaining multiple working hypotheses (Chamberlain 1890), and a certain ruthlessness in rejecting unsuccessful pet hypotheses. Nevertheless, the step is essential.

In the absence of a clear operational definition, different users of the term may develop independent, even inconsistent, definitions. In this way, the original conception grows by accretion or 'conflation' of meanings (Lewontin 1979), until any single meaning of the concept appears restrictive and inappropriate. By that stage, the term represents a 'non-concept' (Hurlbert 1971) or an 'omnibus term' (Milne 1961) because it carries so many different meanings that one can never be sure which is intended at any one time. Clarity can then only be achieved if the term is defined at each use, but this proliferates definitions and compounds the problem. For example, Hurlbert (1981) was able to find 27 definitions of 'niche', Hawkins & McMahon (1989) found three basic definitions for 'guild', MacFadyen (1957) lists seven for 'community', Milne (1961) 12 for 'competition', Edwards & Fowle (1955) a similar number for 'carrying capacity' (cf. Dhondt 1988), and Bradie & Gromko (1981) 17 for the principle of natural selection. All definitions for the same term differ and in no case do the reviewers presume to have listed all existing variations. As a result, the class of a study object or phenomenon implies only that it has at least one of the many characteristics identified in different definitions. Peet (1974) has called these groups of similar, but not identical, definitions for a single term, 'concept clusters'.

Some typologies and their operational problems

Lake trophic status

The ambiguity of typologies involving concept clusters is clearly illustrated by the case of lake trophic status. Lakes have long been divided into oligotrophic and eutrophic on the basis of their productivity or trophic status (Naumann 1930; Hutchinson 1969; Rigler 1975a). Since this dichotomy was erected, lake trophic status has become one of the most studied and best known areas in limnology (Carlson 1977; Peters 1986; OECD 1982), but there is still no standard definition of eutrophy and oligotrophy. Oligotrophic and eutrophic lakes may differ in nutrient concentrations, nutrient dynamics, distribution and abundance of plants and animals, hypolimnetic oxygen levels, transparency, primary and secondary production, fish yield, lake morphometry, and the economic development of the drainage basin. In principle, the classification could be based on any of these characters and appropriate relations developed to allow translations among typologies based on different properties (Carlson 1977; Vollenweider 1987). In practice, there are few rules and different lakes cannot be consistently classified. Thus if told that one lake is more eutrophic than another, one knows very little because this phrase could mean so many different things. Indeed, when limnologists from different parts of the world compare notes, they may find that what is an oligotrophic lake for one is eutrophic for another. The classification is too subjective.

The terms oligotrophy and eutrophy now do little more than signal a general area of limnology. Thoughtful limnologists try to avoid the terms in precise discourse, using instead quantitative measurements of algal abundance and activity, or nutrient concentrations (F. H. Rigler, personal communication). These measurements lack the stateliness of an all-inclusive term like oligotrophy or eutrophy, but they are more easily used in quantitative predictions (Peters 1986) and less easily misunderstood. Carlson (1984) has suggested that the terms 'oligotrophy' and 'eutrophy' be abandoned as non-concepts.

Trophic levels

The devision of ecosystems into trophic levels (which are completely unrelated to the limnological term, 'trophic status') as outlined in Lindeman's famous essay on trophic dynamics provides a more complex example of typological problems in ecology. Lindeman (1942) suggested that the aggregation of all the organisms in an ecosystem according to their energetic distance from the sun would reveal important generalities in the assembly and regulation of ecosystems. Lindeman sketched the sorts of generalities that might develop, but he saw his essay as a description of a research program rather than a theory. Because available information was insufficient, he hoped to inspire others to make the necessary measurements from which empirical theories could be constructed. His views, supported by the work of other leading ecologists (e.g. Riley 1946; E. Odum 1953; H. T. Odum 1957; Teal 1957), inspired a generation of ecological investigation, including the International Biological Program of 1964 to 1974, and still represent a fundamental concept for ecology. For example, this view is intergral to various treatments of limitations of different trophic levels made popular by the classic work of Hairston, Smith & Slobodkin (1960), and by many subsequent discussions (e.g. Wiegert & Owen 1971; Van Valen 1973; Fretwell 1977, 1987; Pimm 1980; Carpenter, Kitchell & Hodgson 1985). These studies, and many others, presume that the division of ecosystems into discrete trophic levels will reveal patterns in ecosystem structure or function, but ignore the serious operational problems presented by such a division.

In principle, this division involves only the assignment of the component species of the ecosystem to distinct trophic levels based on their diets, but, with the exception of terrestrial plants (Rigler 1975b), this is an extremely difficult task (Cousins 1987). Many organisms are so flexible in their diet that trophic relations and trophic levels change seasonally (Martin *et al.* 1951), ontogenetically (Hardy 1926; Larkin 1978), and geographically (Livingstone 1988; Rigler 1975a). Omnivores

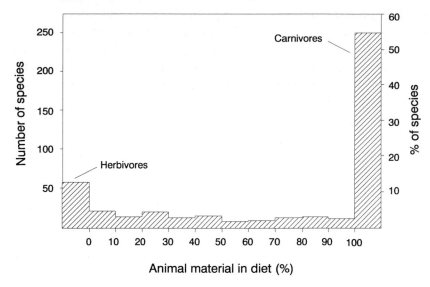

Fig. 4.3. The proportions of plant and animal material identified in gut analyses of species of North American birds and mammals. Only animals studied year round were included in this analysis. From Peters (1977) based on the data of Martin *et al.* (1951).

and detritivores are especially difficult to classify yet the former are common in most ecosystems (Fig. 4.3; Peters 1977) and the latter often dominate energy exchange (Cousins 1985). Finally, the recycling of energy among trophic levels, once considered impossible because of an overinterpretation of the laws of thermodynamics, is impossible to accommodate in Lindeman's scheme (Rigler 1975b), yet it may be quite common in nature (Patten 1985; Cousins 1987).

An example will illustrate some of these difficulties. Herbivores are among the easiest animals to classify to a trophic level (Lindeman 1942). Beavers (*Castor* spp.) eat only plants (Martin *et al.* 1951) and are therefore herbivores, but their position as primary consumers in a trophic dynamic model of their ecosystem is less certain. Not all plant material is equally assimilable (Van Soest 1988), and therefore the beaver's role in establishing the balance between decomposer-based food webs feeding on its feces and herbivore-based webs depending on the beaver will vary seasonally and geographically with the beaver's diet. Even the position of beavers as primary consumers is in doubt; all mammals grow to 10–20% of their adult size while still totally dependent on their mothers' blood supplies or milks for nutrient (Millar 1977), so an age–dependent portion of the beaver's tissue is certainly that of a secondary consumer. In addition, adult mammalian herbivores depend on intestinal flora to digest their food and, in a sense, rodents are secondary consumers of the wastes of this flora. In practice, these observations are seen as physiological technicalities, irrelevant to trophic dynamics, so beavers would be classed as primary consumers.

The problems with this simple solution loom again when we consider free-living herbivorous micro-organisms in the beaver's ecosystem. To be consistent, we should ignore them and class their predators as primary consumers. Actually, we class these free-living micro-organisms as primary consumers, like the beaver, and their predators as secondary consumers. Such a decision assumes that beavers plus their micro-organisms are energetically similar to free-living microbes, even though this seems improbable because of the basic metabolic differences between homeotherms and unicells (Hemmingsen 1960; Peters 1983). Further uncertainty is introduced when we consider the fate of the beaver's feces and carcass. Should the decomposers of the feces be classed with the intestinal micro-organisms because they feed on the plant tissue that was in the gut, or with the beaver because they feed on the digesta of the gut organisms, or with the beaver's predators because some of the fecal contents are derived from the beaver itself? How should any complex trophic interactions between bacteria, protozoa, fungi and invertebrates in the egesta be treated? Are decomposers of the beaver's corpse in the same trophic level as the beaver's predators, wolves and mosquitoes (Kozlowsky 1968), or should they be classed as decomposers, like the coprophages (Ivlev 1943; Wiegert & Owen 1971)? These questions must be answered if we are to carry through Lindeman's research program (Cousins 1987); instead they have been ignored because they are likely insoluble.

When the trophic dynamic classification of an ecosystem is attempted, the same problems occur with each new species. Moreover, the problems are compounded because the trophic position of any higher element is defined relative to the trophic positions of all of its prey. Any uncertainty in the position of lower-order consumers adds to the uncertainty of the positions of its predators. As a result, the more completely we describe the trophic relations within an ecosystem, the less easily we can divide it into levels (Peters 1988b).

Ulanowicz (1986) and Baird & Ulanowicz (1989) have tried to address these difficulties by defining all decomposers to be primary consumers and by allowing different species to assume intermediate trophic levels reflecting use of several producer levels. Thus trophic levels need no longer be integers. This flexibility allows a more realistic representation of different trophic relationships, but does not address the problems of variance around these non-integral trophic levels, nor the increasing uncertainty of the trophic position of the higher trophic elements, nor the arbitrary solution of the problems of decomposers. Ulanowicz has addressed some of the operational difficulties of the trophic level concept, but I am not yet convinced he has solved them.

Unless we can perform the classification, no statement involving trophic levels can be tested since we have no way of recognizing the entities that such a theory would invoke. Nevertheless, because the term trophic level is only a loose concept, it can easily be associated with selected observations of individual plants, herbivores, or predators, even though these are not trophic levels. These associations sometimes appear to support generalizations about trophic levels (Hairston *et al.* 1960; Slobodkin, Smith & Hairston 1967; Fretwell 1977, 1987). They also illustrate the difference between seeking support for a theory and seeking to test it. To test a theory, we must know what evidence would be inconsistent with it (Murdoch 1966); to support a concept, we need only reinterpret selected evidence in terms of that concept. If the concept is sufficiently poorly defined, we can even claim that supporting evidence confirms the generality but that contrary evidence is inappropriate for the test. Rigler (1975b) believed that Slobodkin *et al.* (1967) used just this defense to argue that they were concerned 'with trophic levels as wholes' so that observations about populations were irrelevant, and that their conclusions were supported by observation, even though the practical difficulties of assigning organisms to trophic levels made falsification impossible.

Foods webs

One response to the operational problems of the trophic level concept has been a renewed interest in food webs (Cohen 1978; Pimm & Lawton 1978;Pimm 1982; Yodzis 1980; Briand 1983; Briand & Cohen 1987; Schoener 1989). This research focuses on trophic interaction, not energy flow, and limits itself to those portions of the ecosystem which have proven tractable in analysis by other authors.

Food web analysis has not sucessfully escaped all the operational problems that dogged trophic dynamics (Paine 1988; Peters 1988b). This approach has yet to develop consistent rules for the construction of food webs in different communities and by different authors. As a result, patterns in food webs may represent the conventions and views of particular schools of ecologists, rather than differences among communities. For example, food webs described from studies of the open water tend to be longer and more narrow than those described by terrestrial or inshore benthic workers (Briand 1983; Briand & Cohen 1987). This may reflect a real difference, or only the tendency of workers in the open water to use larger ranges of organism size in building their food webs but to lump more species together at each level. Food web analysis, like trophic dynamics, has tried to generate theory before operationalizing its concepts. As a result, its components are unidentifiable, the validity of its general constructs cannot be assessed, and its status as scientific knowledge is in dispute (Paine 1988).

Dichotomies

Ecology employs many systems of classification that break continua or complex phenomena into a few alternatives. The resultant problematic typologies include r- versus K-selection (Pianka 1970; Southwood 1981; Parry 1981), the stages and endpoints of unidirectional successional change (Whittaker 1953; Odum 1969; Drury & Nisbet 1971), the divisions between specialists and generalists (MacArthur 1972a; Fretwell 1975), density-dependent versus density-independent regulation (Andrewartha & Birch 1954), and semelparous versus iteroparous reproduction (Kirkendall & Stenseth 1985). All these concepts map many, imperfectly correlated characteristics onto a simple pair of alternatives.

To the extent that the different characteristics are independent, simple divisions do not adequately represent the phases of the system under study. In most cases, no weighting factors for different phenomena are specified, so that different authors use the scale subjectively. However, even if objective weightings were available, the collapse of a diverse set of measurements into a few classes involves a drastic loss of information. Different workers must reinterpret the typology in the light of their own experience, so the type means different things to different people and in different ecological contexts. Such constructs proliferate meanings, producing first concept clusters and omnibus terms, then non-concepts.

The organism–environment dichotomy

The concept of a separation between the organism and its environment is a simple classification that seems essential to ecology. The dichotomy is implicit or explicit in many definitions of the ecology, including that of Haeckel (1870, as translated by Allee *et al.* 1949): 'the investigation of the total relations of an animal both to its organic and to its inorganic environments'. This dichotomy has been extended until it now includes any division between the identified object of study – an organism, a population, a community, man – and everything else. It represents a characteristic way in which ecologists see the world (Hughes & Lambert 1984).

This section addresses that classification as an important case of operational difficulties in ecology. In a scientific context, it illustrates common operational difficulties posed by concepts that are open-ended, relative, and interrelated. In an ecological context, the particular example is important because the dichotomy is unquestioned in most ecological analyses.

Alternatives

Some examples will illustrate the limitations imposed by the organism–environment dichotomy. In lakes, phosphorus has long been thought to limit phytoplankton biomass and productivity (Hutchinson 1969), but for many years, no convincing relationship was identified because limnologists compared phytoplanktonic responses to 'free' nutrient levels measured in filtrates (Elster 1958). The failure to find clear results was attributed to 'other factors' and 'system complexity', so that the strong relations that exist between total phosphorus levels and most aggregate measures of biological activity in lakes (Peters 1986) were obscured, and phosphorus abatement programs delayed. Some limnologists still consider relations between total nutrient and biological response to be somehow illegitimate because they do not distinguish between the algae and their environment (Horne 1985); oceanographers continue to measure free nutrient levels and to explain the absence of any relation between nutrient and biological activity by the complexity of the systems.

This is not an isolated example, nor is the problem only relevant to aquatic biology. The primary productivity of terrestrial systems is best predicted by evapotranspiration (Lieth & Box 1972), a process which involves both the plants and their environment. The best general models for animal density (Damuth 1981; Peters & Wassenberg 1983; Peters & Raelson 1984; Fig. 2.2) do not use environmental characteristics at all, but depend on animal body weight. This also holds for a host of other allometric models that are the best available predictors of many autecological traits (Peters 1983; Calder 1984).

The separation of organism and environment is therefore not an essential element in ecology. It represents a perspective that may be useful in some situations, but not all. Typological thinking tends to blind us to alternatives and the organism–environment dichotomy is a prime example.

The operational environment

For the purpose of further discussion, I will presume that the organism can be operationally identified. This is usually the case when the focus of the study is a single organism, but is far from certain when the focus of the study is some aggregate, like a clone, population or community. This uncertainty only compounds the problems discussed below.

Definition of environment

The vagueness of the environment concept has long been recognized by ecologists. At least once in every decade, a new and more rigorous definition is advanced. Haskell (1940) recommended a mathematical reformulation, Mason & Langenheim (1958) analyzed the concept using the theory of signs and recommended that environment be restricted to direct interactions between the organism and its world. Maelzer (1965a, b) and Spomer (1973) reached similar conclusions. More recently, Niven (1980, 1982) provided a 'formal definition'.

As yet, these reanalyses have had little effect (McIntosh 1985). Each definition narrows present usage, so new definitions may seem too restrictive. Some discussions use such unfamiliar philosophical contexts that they are opaque to practicing ecologists. Finally, many ecologists are quite content with a vague term, like environment, and see no need for further definition. However, two leading ecologists, Andrewartha & Birch (1984), consider the work of Maelzer and Niven invaluable in their own search for a functional redefinition of the environment. Perhaps, future developments based on 'the theory of the environment' (Andrewartha & Birch 1984) will require re-evaluation of the utility of these rationalizations of the environment concept. What is certain is that without some restriction and redefinition the concept can never find application in a theory.

In practice, the environment is currently identified by stipulating what it is not: The environment is that which is not the object of investigation. Thus the environment of an entity is everything outside that entity. This sweeping definition of environment introduces a number of operational difficulties.

Open-endedness

The definition of environment refers to an infinite series of properties varying in both space and time. This provides an inclusive concept, but since specification of these entities would be an interminable chore, each investigator must choose those aspects of the environment that are appropriate for study. Thus a single term and concept, environment, is represented by many different operational entities.

Because the concept of the environment refers to so much, it often provides an easy explanation. 'Environmental influences' are said to explain the noise in our relationships; apparent falsification of our theories are ignored because 'environmental variations' are said to have abrogated unstated assumptions that all else is equal, and the study of 'environmental factors' is often proposed to explain new phenomena. As an explanation, environment is a 'panchestron', a word that can be invoked to explain everything and anything (Hardin 1957). Not only are such explanations facile and shallow, they are dangerous because they hide our ignorance and discourage more penetrating analyses.

The concept of the environment can only be employed if reduced to something more manageable. We might limit ourselves to environmental properties that significantly affect the focus of the research, but that approach requires us to identify the properties of the environment which are significant to the object of study. In other words, we must have specific hypotheses about the effects of specific factors. Once such hypotheses have been made, a general concept of the environment no longer has a role. It has been replaced by defined variables in a theory worthy of the name. Thus, many theories relate particular environmental characteristics, like temperature or food level, to specific biological responses, but these characteristics are never claimed to represent more than a tiny piece of the environment.

Relativism

An equally serious problem with this definition is that it defines environment relative to the focus of the study. Any change in focus entails a change in whatever constitutes the environment. Minimally the focal individual of one study becomes part of the environment in another, but usually other elements change too. The environment of a litter mite differs from that of a deer or songbird.

The concept environment must accommodate another relativistic shift whenever the scale of the investigation changes. Conspecifics are an important element in the environment of the individual, but are subsumed into the object of study in population ecology. Much of the environment of the population forms the focus for community ecology and the environment of the community can similarly be the focus for ecosystem studies. In another dimension, environmental properties that are relevant on the scale of a day may be less important on the scale of a year.

The relativity of the environment concept results in necessary discrepancies in the way the term is used as the object of study changes. Minimally, this requires care in specifying where the term applies and considerable mental agility in accommodating the conceptual changes which occur when the focus changes. Failure to do so results in inconsistent use of the term and ineffective communication.

Conceptual interrelations

A number of other ecological concepts which specify a multidimensional location of an organism or group of organisms share analogous problems (Fig. 4.1). The community, as the biotic environment, is clearly as non-operational as 'environment', even without considering the associated concepts of biocoenosis and superorganism (MacFadyen 1957). Any argument which requires specification of community or environment is also likely to be non-operational. For example, Elton (1927) described the 'ecological niche' of an organism as its role in the community, Hutchinson (1957, 1959a, b) as the hypervolume that the organism occupies in n-dimensional environmental space. Both definitions define the niche relative to a focal organism (or population or species), both refer to infinitely large sets of properties and thus neither concept of the niche can be operationalized. Hurlbert (1981) argues that the niche should be defined on the basis of resources, but because there are likely many resources and because resources are defined relatively (Tilman 1982), this solution is too gentle to be effective. Without radical redefinition of the term, no theory involving the niche can be constructed. Both habitat and ecosystem (O'Neill *et al.* 1986), as multidimensional, unlimited, relativistic entities representing the environment are open to similar criticisms. This has not hindered use of 'the environment', 'the niche' or the other related terms as panchestra for ecological explanations.

The interrelations among environment, resource, community, niche, habitat and ecosystem exemplify a further characteristic of concepts in ecology: interdependence or coherence (Whittaker 1957). Ecological concepts intergrade into and build upon one another. This can blunt the force of criticism by diffusing it over a vast area where the limits of the interdependent concepts are hard to identify. However, persistent criticism can be particularly effective by inducing a 'domino effect' in which the identification of one non-operational concept points the way to others.

Some exceptions

This criticism of the concepts of 'environment', 'habitat', 'ecosystem' and 'niche' should not be construed as a condemnation of all uses of the terms and even less of theories which predict characteristics of organisms using properties external to the organisms as independent variables. Because concepts are at best only vaguely defined, they can be used in different, even inconsistent contexts (Egerton 1973). When use is theoretically sound, the concept has been restricted to represent a few dimensions which are considered sufficient for a particular purpose. Thus Box (1981) defined meteorological hypervolumes for different plant life forms and has shown that this approach permits global prediction of life form. Gates (1980) described the 'thermal niche' within which a songbird can maintain homeothermy. The niche of bluegreen algae can be partly defined on the basis of concentrations of nitrogen, phosphorus and light availability (Smith 1983, 1985), and that of stream-dwelling trout on the basis of temperature (Barton & Taylor 1985). Significantly, none of these authors cast their work in terms of niche theory, and all use the operational definition of their particular variables instead of a general concept like 'the environment'.

Conceptual variables – stability and diversity

Operational problems are not limited to systems of classification, but also effect the properties of these systems which we wish to measure and predict. This is nowhere more apparent than in the continuing controversy over 'stability', its meaning and its ecological relevance, so that debate is used to demonstrate the problems that result when insufficiently operationalized concepts are used as theoretical variables.

The origins of this discussion lie in prescientific concepts of the balance of nature which are certainly much older than the science of ecology (Egerton 1973). In recent years, increasingly strident demands for conservation have forced consideration of natural balance and stability in resource management. Since many ecologists are profoundly conservationist and deplore the wholesale despoilation of the earth, they consider the stability of a natural system to be an inherently valuable property that should be protected. This requires the development of theories to predict stability or its converse, the extent of ecological change.

Despite long-standing interest in stability (*Brookhaven Symposium* 1969; May 1974), the term has never been satisfactorily defined. Part of the difficulty in doing so reflects the dynamism of open biological systems. Because biological systems constantly react to both external and internal changes, they cannot be preserved like a work of art. Instead, stability must be defined in dynamic terms. Such contradictions foster misunderstanding.

Many definitions have been proposed (Holling 1973; Westman 1978; May 1974), but all assume that the concept of stability is too complex to be defined and therefore fragment the general concept into smaller ones. The definitions offered by Orians (1975) are indicative. Orians (1975) suggests that stability may mean many different things: the absence of change ('constancy'), the length of survival ('persistance'), resistance to perturbation ('inertia'), speed of return after perturbation ('elasticity'), the displacement from which return is possible ('amplitude'), the degree of oscillation ('cyclic stability'), and the tendency to move towards a similar end point ('trajectory stability'). These independent concepts are sometimes correlated, and sometimes not (Orians 1975); some, like constancy and resilience, may even be inversely related.

If Orians' scheme were accepted, the sort of stability under examination would have to be specified at each use. This has not yet proved necessary because none of these components of stability are sufficiently well defined to represent a variable in a theory. Instead they represent more concepts, fragments of the original vast concept, which still require operationalization. Attempts to solve operational problems by redefinition produces instead hairsplitting scholasticism and infinite regress (Popper 1985). The debate over stability is a case in point.

The operationalization of stability is rendered still more complex because Orians' categorization does not capture every nuance of this multifaceted concept. In comparing the divisions of stability proposed by different authors, Westman (1978) proposes another set based on a primary division of stability into 'stability' (Orians' constancy) and 'resilience' (the ability to restore structure following perturbation). Resilience was further decomposed into resistance to perturbation ('inertia'), time to recovery ('elasticity'), extent of departure from which recovery is possible ('amplitude' and 'brittleness'), degree to which the pattern of recovery differs from the initial pattern of alteration ('hysteresis') and ease of alteration ('malleability'). By considering constancy and several different aspects of non-constancy (amplitude of regular fluctuations, regularity of fluctuations, persistance and structural maintenance) at different levels of organization ('population', 'community', and 'environment'), Whittaker (1975b) was able to arrive at thirteen different meanings for stability, but he did not intend this to be a complete catalogue. May (1974, 1975) suggested an alternative classification consisting of 'global' and 'neighborhood' stabilities (which are related to, but not identical to, considerations of elasticity), 'neutral' stability (Orians' cyclic stability), and 'structural' stability (which refers to the continuity of change and may be related to Orians' persistence). Since different authors use the same term for different concepts and different terms for the same concept, any ordering of the concept of stability requires a clear head and a thorough housecleaning.

The problem of multiple concepts lurking behind a single term is not the only difficulty that stands in the way of operationalizing stability. Since the parts of a system normally change at different rates, any definition of stability must also determine what the object of this definition is. Whittaker (1975b) intimated that stability would have different definitions when interpreted at different levels of organization. Moreover, within each level of organization, stability will vary among components. For example, populations of small animals will seem less stable than those of large ones (Calder 1984); the stability of species composition may not be highly correlated with the stability of their populations; and some processes or structures may be stable while others fluctuate (Schindler 1987). An operational definition of stability must therefore indicate what aspects of stability of which elements at what level are to be considered. The sort of theory that can be developed will depend on these choices.

Stability, like niche or environment, is one component in a web of concepts shading slowly into one another (Fig. 4.4). 'Equilibrium', 'steady state', and 'balance of nature' are related concepts as are the conceptual mechanisms which maintain stability, like 'negative feedback', 'homeostasis', 'regulation', and 'density dependence'. Stability also invokes an antithesis, 'instability' and its conceptual mechanisms, 'perturbation', 'positive feedback', and 'stress'. Instability, which reflects temporal change, has spatial analogues in 'heterogeneity' and, perhaps surprisingly, 'diversity' (Horn 1974).

Many of these concepts have become the subject of recent criticism.

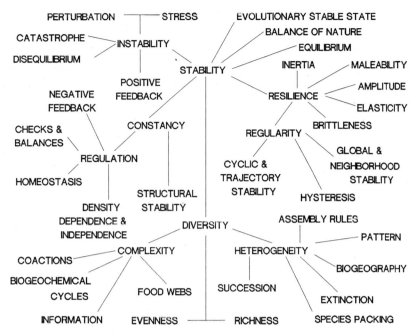

Fig. 4.4. A diagrammatic summary of the relationship among concepts associated with stability. As in Fig. 4.1 and 4.2, these associations are necessarily subjective and incomplete.

There are fundamental problems in operationalizing 'equilibrium' and 'stability' (Murdoch 1979; Caswell 1978; Connell & Sousa 1983; Wiens 1984; DeAngelis & Waterhouse 1986); Egerton (1973) and Ehrlich & Birch (1967) discuss the difficulties associated with the concept of 'the balance of nature'. 'Diversity' has been termed a non-concept (Hurlbert 1971) and a concept cluster (Peet 1974), and its continued utility has been questioned (Hurlbert 1971; MacArthur 1972a). 'Stress' and 'perturbation' have also been attacked as operationally empty phrases because they can only be defined after the fact, in terms of their effects (Peters 1977).

Stability illustrates several common characteristics of troublesome concepts in ecology (Whittaker 1957). Like all the concepts mentioned in this section, it is a 'concept cluster' (Peet 1974) because it 'conflates' (Lewontin 1979) 'multiple meanings' (Hawkins & MacMahon 1989). In addition, it is a 'pseudo-cognate' (Salt 1979) because a meaning for the term is grasped intuitively, without the onerous necessity of operational definition. Regrettably, different scientists intuit different meanings and failure to define this term has ended in a terminological and conceptual morass. Ecologists have had to use subjective judgments to abstract some partial aspect of the concept. Worse, repetition and familiarity have made us so uncritical of the term that it has become accepted as a concrete property of nature. The vagueness of this and related terms has allowed the elaboration of a grand and complex conceptual system that obscures serious scientific shortcomings.

Non-operational relationships

Because ecologists confuse theory and concept, many conceptual relations in ecology are no longer stages in the development of theory, but alternatives to theory. Thus it may be claimed that a 'theory' can link different concepts, even though these are ambiguously defined. Any statement that involves such terms remains impervious to test and will be abandoned only if the interest of the scientific community flags. A first step in this process is an evaluation of existing relationships to establish their theoretical status.

Diversity–stability

The pursuit of the diversity–stability relationship, described as 'the stability–diversity debacle' (Dayton 1979), provides an example. The debacle refers to the fall from grace, if not the demise, of the view that the stability of ecological entities is positively related to the complexity or diversity of those entities or their environment.

The proposed relation dates at least from Thienemann (Hynes 1970). The concept has a great intuitive appeal because it provides a strong rationale for the conservation of species and species richness (Goodman 1975). Elton (1958) and Hutchinson (1959a) lent their support to the relationship, although the vagueness of both terms made unequivocal support (or falsification) impossible. The tide of scientific opinion quickly reversed after May (1975) showed that complexity and stability were inversely related in model 'random communities' and Goodman (1975) was unable to find good supporting evidence in the published literature. Notwithstanding a call for more empirical studies (McNaughton 1977) and the frank admission that the behaviour of randomly assembled, computor communities may be irrelevant for real ones (May 1974, 1981a), the weight of opinion is now against a positive diversity–stability relation (Auerbach 1984).

Despite its difficulties, the diversity–stability relationship cannot be said to be false. The ambiguity of the literature in its regard and the faddish swings of scientific opinion about it reflect the nebulosity of the terms. Unless these can be defined, the relation will never be falsified.

Competitive exclusion

The competitive exclusion principle is another example of a concept masquerading as a theory (Hardin 1961). This old saw claims that at equilibrium, competing species cannot coexist indefinitely in the same niche. The problem is that all terms are so poorly operationalized that the principle cannot be evaluated. As a result, ecologists have long assumed that it applies and interpreted observations as consistent with the principle (Murray 1986). Coexistence is taken as evidence for disequilibrium (Hutchinson 1959a, 1978), niche differences (Gause 1970; Peterson 1975), or the absence of competition (Pianka 1981); differences between species are said to represent niche differentiation (Abrams 1983; Arthur 1988), and exclusion is said to reflect competition (Connell 1980). Other explanations are almost always possible (Underwood & Denley 1984), but the flexibility of undefined terms has allowed the construction of a tautological argument that can accommodate any observation (Peters 1976), so that no other explanation is required. Indeed, competitive exclusion continues to play an important part in community ecology (James & Boecklen 1984) through extensions like limiting similarity (Abrams 1983), diffuse competition, resource partitioning and species packing (Pianka 1981). The absence of predictive power seems only a minor inconvenience.

Atheoretical concepts

The overwhelming majority of concepts discussed in ecology represent notions that could lead to useful theoretical development, if their various problems were resolved. There are however other concepts which will likely never be phrased as a theory and which should be treated with great circumspection. For example, statements about existence, like 'there is a Loch Ness monster', could be confirmed, but never disproved. The operations which would suffice to show the monster's existence are unknown and the failure of any attempt to prove such an existential statement can easily be explained away. Before existential statements can be used, they must be rephrased as formal theories describing what observations made under what conditions would lead us to consider non-existence more probable than existence.

Table 4.2 *Some environmental truisms that appear to be statements of fact, but are really empty concepts, incapable of test or falsification (from T. C. Emmel 1977).*

> Everything is connected to everything else.
> You can't change just one thing.
> Nature knows best
> Nothing goes away.
> Dilution is no solution to pollution.
> There is no free lunch.

Environmental truisms

A number of non-operational concepts associated with environmentalism are also atheoretical. Although their status as part of ecological science is debatable (Evernden 1984) the familiarity of the phrases in Table 4.2 makes them good illustrative material. These environmental concepts (Table 4.2) look like facts or theories, but they are not. Nevertheless, they are often accepted as such by the public and by policy makers, who therefore expect similarly clear advice from science.

The aphorisms in Table 4.2 should not be confused with scientific knowledge. They represent general notions of very limited scientific utility. This is made apparent by trying to reinterpret these general notions in the form of a theory, relating response and predictor variables given certain conditions (Chapter 2). Such an attempt will fail. The aphorisms do not describe what possibilities are probable under specified conditions. Instead, they invoke seemingly universal statements which are entirely untestable because the meaning of their key terms is never stipulated.

At first glance, a statement like 'everything is connected' is so patently false that the statement would normally be dismissed. However, it is protected from falsification by reference to subtle, indirect, or imperceptible connections of an unspecifiable kind. Thus the failure to demonstrate a connection is not taken as evidence against the statement, but only against the adequacy of the definition of connection used.

Well framed theories make it clear what evidence would be taken as contrary to the theory, so that we realize when the theory has been falsified. In this case, a strong theoretical statement would include definition of the nature of the connection and the operations required to reveal its presence. The inadequacy of non-theories is usually easy to

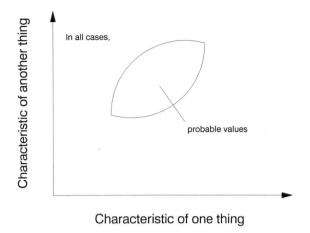

Characteristic of one thing

Fig. 4.5. An attempt to rephrase the statement 'Everything is connected to everything else' as a theory in the form of a Cartesian plot (cf. Fig. 2.1, 2.3). Since the characteristics of the 'things' which are addressed are undefined and the region of expected values is not stated, the statement cannot be applied.

demonstrate by trying to rephrase the construct as a simple Cartesian graph (Fig. 2.3). Such a restatement (Fig. 4.5) would show that there are no preconditions to tell us where the theory should apply, that the axes are essentially undefined because no characteristics have been stipulated, and that the shape of the relation between them is unspecified. The phrases in Table 4.2 are therefore antitheses to science.

The advantage of such formulations is that they are never wrong. This allows environmentalists to claim, with the advantage of hindsight, that their reservations were well founded whenever an environmental project fails. Unfortunately, the problem confronting environmental management is not to criticize acknowledged and obvious failure after the fact, but to predict disastrous consequences beforehand and to provide effective tools so that disasters may be avoided.

A similar analysis can be applied to other statements. This will first show that they appear false: for example, dilution is a solution to some levels and kinds of pollution (like the smoke from a campfire); and some things (like the passenger pigeon or an abandoned apple core) do go away. Second, the analysis will show that there is always an easy escape from falsification because the hypothesis is too poorly phrased to be testable. Finally, it will show that no observation could disprove the statement because the terms are so poorly defined that we really do not know what they are.

Such statements are not an effective basis for scientific testing or for environmental management. They are not intended as pieces of information but as moral strictures for the public and politicians. Proponents of such strictures have made them the basis of ethical and political positions and they may be effective in that role. But they must not become models for scientific theory.

Concepts in ecology: the effects of poor examples

Science is a social activity, so the behaviour of our peers and our leaders has a profound effect on what is acceptable. In ecology, concepts and theory have been confused to the point that the science has been seriously debilitated (Rigler 1975a).

Typological limitations

Birks (1987) identified three distinct problems associated with typological thinking in biogeography, and each has its counterparts in other areas of ecology. According to Birks (1987), typological thinking encourages us to see divisions where none exist, a problem which dogs studies of seral stages (Schoenly & Reid 1987) and of local clinal gradients and mosaics (Shipley & Keddy 1987). Typological thinking also encourages us to identify types which are easily recognized and then to group phenomena which are less easily classified around these types. This appears to be the case for trophic levels, where plants and top carnivores are easier to classify than the bulk of organisms in between. Typological thinking is also easily biased by causal schema, like the organism–environment dichotomy.

Conceptual discussions

Debate and discussion are among the great pleasures of academic life. Regrettably, theories are poor subjects for discussion, because once stated they should be judged on the evidence, not on the quality of exposition. Since concepts cannot be judged on the evidence, they are more appropriate subjects for debate and may be further operationalized in that process.

Prolonged discussions indicate that operationalization is unlikely and that other topics would be more profitable. Participants in conceptual debates should therefore be prepared to conclude that the questions are intractable. Unfortunately, debate makes such a sage decision unlikely because it polarizes opinion and gives all camps some stake in the reality of the question. This mitigates against the realization that the question is insoluble.

Conceptual argument has the advantage that it rarely requires much empirical input. Indeed, discussions can be derailed by authoritative statements of fact. Thus, conceptual arguments are favoured by those who lack the specialist's command of detail: theoreticians, generalists, and novitiates. Writers of textbooks encourage this propensity when they set students to debate questions which do not require any information (Table 1.3). For example, Pianka (1981) suggests that a good 'question for discussion' may be 'Do clusters of functionally similar species exist?' Such a question cannot be answered without operationalizing the terms and setting it in a theory of the form: 'Given that conditions $C_1, C_2, . . , C_n$ occur, then functionally similar clusters will be detected as follows . . . '. McNaughton & Wolf (1979) suggest that a suitable 'hypothesis for discussion' is 'What is the importance of competitive efficiency in niche overlap zones in determining the relative abundance of species?' This leads the student to empty discussion in which boundary conditions ('in niche overlap zones'), variables ('competitive efficiency', 'relative abundance'), and the relationships among them are unknown; the nature of hypothesis is simultaneously misrepresented. Futuyma (1979) proposes the following questions, 'for discussion and thought': 'In a resource limited environment, do organisms of different ages compete as much with each other as with others of their own age? Do different age classes specialize on different resources and so avoid competition?'

The only appropriate response to such questions is to demand operational definitions. This lesson should not require so much reiteration, so one suspects that undergraduate discussions frequently proceed in other directions. Such discussions are fun, but they are not science, so this pedagogical device may be a form of misinformation.

Abstraction and scientific interest

The emphasis on concepts in teaching and writing gives an unwarranted lustre to ineffective definition. Ecologists have repeatedly acted as though an intangible abstraction is more appropriate for theoretical analysis than a clearly defined variable or relationship. As a result, contemporary ecology seems a new scholasticism, interminably debating the fine points of unobservables and formalisms. Inevitably, this unconcern has driven a wedge between the specialists and applied ecologists, on one hand, and generalists and theoreticians on the other.

Each group now fears the other holds it in contempt and communication between them is tenuous (Larkin 1978; Kerr 1980; Werner 1980; Rigler 1982b). Thus the specialist loses the context and direction offered by those with a wider view and the generalist loses contact with scientific realities. Neither benefits from such divisiveness.

Reconceptualization of potential variables

The world of everyday deals with a series of very tangible and very important problems which could effect the wellbeing of every person, and perhaps every being on the planet. Besides the global questions of nuclear war, overpopulation, climate change and the ozone layer, there are crucial questions about fish catch, lumber yield, crop harvest, insect damage, pest and disease levels, water clarity, plant height, the abundance of particular species and a host of other very real problems. These concerns may also be seen as concepts, but they differ from those of the preceding section because they are already relatively narrowly defined. Because of this narrowness, such concepts can serve as readily available variables with considerable economic and social importance.

Many of these have already become the basis of successful programs in environmental management and have led to the development of separate schools and faculties producing professional agriculturalists, foresters, and fisheries managers. Many aspects of these problems are still open for study and in general, ecologists welcome such problems because of the greater definition and purpose provided by societal demands.

There can be a more morbid aspect to the ecological analysis of socially relevant problems. Academic ecologists have been trained to approach a problem by abstracting it. Sometimes this is essential to put the problem in the context of the literature and to allow the application of existing theories which address the same problem using other terminological or conceptual variants. However, this abstraction introduces the danger that the original question may be lost, so that the answer generated by the abstraction may be only peripherally relevant to the question. In extreme cases, the abstraction is so abstract that it is no longer operational. This 'reconceptualization' is more likely if the difference between concept and variable is not distinct. Ecologists are capable of substituting a pallid, intangible concept for a perfectly good, concrete variable without realizing that anything has been lost.

For example, if asked about the factors that effect lumber harvest, an ecologist should make recourse to models describing the effects of various factors on lumber harvest (Downing & Weber 1984). However, if this is not appropriate for some reason, then reference might be made to factors affecting net primary production (Lieth & Box 1972), litter production (Vogt, Grier & Vogt 1986), or carbon dioxide fixation (Reynolds & Acock 1985). The relation between these variables and the original question is increasingly tenuous, and the various models hold different implications for management. A matter of considerable importance for the economy and for conservation of our forest resources may rest on the quality and degree of abstraction. A further step leads beyond the pale to discussions about primary productivity, productivity of a given trophic level, or simply productivity, terms which are subject to a variety of conceptual interpretations.

In the case of fish yields, the process of abstraction has proceeded so far that discussions often revolve around non-operational concepts. A concrete question about the likely size of the fish catch has been abstracted to questions about fish productivity, stocks, recruitment, catchability, maximum sustained yield, and optimum yield (Schaefer 1954; Beverton & Holt 1957; Ricker 1958; Gulland 1974; Healy 1984). Few of these terms can actually be ascertained. Because the size of most fish stocks is unknown, it is impossible to estimate catchability (catch/population) or recruitment (population × eggs/individual), or productivity (individual growth rate × population). Because equilibrium has been left undefined, it is meaningless to speak of a sustainable yield (Larkin 1977; Sissenwine 1978) and the supposed catch–effort relationship underlying this concept is apparently illusory (Godbout 1987). Good estimates of future fish catch can be made on the basis of catches in preceding years (Roff 1983) and successful applications of the sustainable yield concept (Ryder 1965, 1982) owe much more to empirical definitions based on observation than to the concepts to which they pay lip service. There is a desperate need to publicize and analyze successful applications of ecological knowledge in applied areas like forestry, fisheries, agriculture and public health, so these may serve as examples of a useful ecology.

Most ecologists use practical rationales and potential applications in their grant proposals. Most would like to be used by their society because they feel that the preservation of the earth demands an ecological perspective (Southern 1970; Andrewartha 1984). These examples are therefore particularly sad because they represent attempts by ecologists to answer questions of social concern and importance. Too often, these have failed because the poor examples provided by ecological practice and training have left ecologists incapable of answering practical questions. Regrettably, the particular ecological perspective we have been able to provide has proven so ineffective that we have undermined our own credibility (di Castri & Hadley 1986; Simberloff 1987).

Summary – The costs of non-operational concepts for ecology

Although non-operational concepts play an important role in the creation of scientific theory, scientific relationships must be built upon operationally defined classes, variables and relationships, because operational definition relates the constructs of science to the phenomena of the real world. Nevertheless, ecology is dominated by complex and inadequately undefined terms which confound the development of predictive theory. As a result, ecological classifications, ecological characteristics and ecological relationships may refer to phenomena that vary with each change in focus, scale, or author, and ecologists are often not sure they are talking about the same thing. The solution to this problem is to embed concepts in a web connecting these entities to methodologically defined relations and to the confirmational tests offered by theory and prediction.

The absence of such controls in ecology allows superficial association of concept with a diverse set of concordant observation, concept, and theory such that an entire conceptual system may be erected and confirmed with observations. In ecology, such conceptual systems obscure the absence of theory and the impossibility of falsification. This same looseness makes any definition of the limits of the concept impractical, so different uses of the single term and different terms blend into another. Thus one term may come to represent a multiplicity of divergent, even opposing, meanings. The complex concept represented by this term can then be fragmented into smaller concepts, which may receive terms of their own. Terminological proliferation thus attends conceptual fragmentation and a welter of terms often signals underlying operational difficulties.

Science may always begin with a concept, and the presence of vague concepts may be essential to both operationalization and the social discourse of science. However, successful sciences go beyond this primal stage to develop variables and theories. Because ecology has been so halting in its development, it is perhaps time that we consider our current concepts more critically, removing those which have proven repeatedly futile from our teaching curricula, from our research proposals, from our future publications, and from our minds.

5 · Explanatory science: reduction, cause and mechanism

The view of science underlying the criticisms of this book (Chapter 2) is a form of 'instrumentalism' (Popper 1983), because the critique views scientific theories as 'instruments' or tools to predict and control the behaviour of our environment. More extravagant claims – for example, that theories are 'true' or 'realistic' in the sense that they capture some essence of the universe, are not made because those claims seem unnecessary, unwarranted, and confining. They are unnecessary because any finite set of observations is consistent with a potentially infinite set of theories, so predictive success cannot establish a theory to be true, even if truth should imply predictive success. The claims are unwarranted because history shows that the theories of science have been revised many times and because philosophy suggests that even the strongest theories be considered hypothetical or conjectural, and possibly ephemeral. They are confining because once a scientific construct is accepted as true, it becomes dogma, stifling theoretical improvement and scientific growth. Scientists are never entitled to conclude that successful theories are true. They can only make the modest claim that the theories which worked in the past are more likely to do so in the future than theories which failed in the past. For this reason, successful theories make more reliable tools.

Loehle (1983) recognized this dichotomy in distinguishing two kinds of ecological models: 'calculating tools' and 'theoretical models'. In a wider ecological context, calculating tools correspond to the predictive scientific theories described in Chapter 2, for they are intended only to inform us about the configuration of the world and are judged solely on predictive success. Theoretical models embody universal statements and mechanisms, and Loehle (1983) suggests they may prove useful despite predictive failure by virtue of their elegance, completeness and explanatory power. In the wider context, this class corresponds to those 'explanatory' concepts and theories that satisfy a widely felt need for plausible, causal descriptions of nature, even if these explanations are ineffective in prediction. Such explanatory constructs may be useful devices in heuristics and popularization, but they can also be dangerous to science, because they hide the shortcomings of our theories under prose that explains away rather than explains.

This chapter begins with a brief consideration of the differences between explanatory and predictive theories, and then considers several related classes of explanation at length. The first such class is based on reductionism, the view that all scientific theories are ultimately instances of very general scientific laws, likely those of physics and chemistry. The second class invokes the related idea of mechanism and explanations that analyze ecological systems as if they were complex machines. Finally, the concepts of causation and causal connection are examined to show that only a limited interpretation of these concepts is scientifically justified. In all cases, I intend to show the limitations of these explanations, so that they can better serve their proper uses without impeding the growth of predictive ecology. The discussion is continued into the next chapter which deals with the special case of historical explanation in ecology. These chapters touch on topics that have been debated so long that they have generated an extensive, and sometimes inconsistent, terminology. To avoid confusion, Table 5.1 lists definitions of these terms as they are used here, although I acknowledge the existence of other definitions that are at least as valid as those I use.

Prediction and explanation: alternate goals for science

Instrumentalism is not a popular view. Drawing lessons from physics, Popper (1983) rejects instrumentalism in favour of the view that better theories yield better explanations and predictions because they approach some external truth more closely. However, he admits that this metaphysical view is completely untestable (Popper 1983) and therefore has no practical application in separating science from non-science. Although Popper feels that predictive power is only a byproduct of effective theory, he identifies this power as the best practical demarcation between science and non-science.

Many ecologists accept the importance of predictive power, but feel that this alone is not the defining characteristic of a theory. Instead or in addition, they invoke vaguely defined concepts like 'explanatory power' or 'biological reality', ignoring questions about how these additional claims can be demonstrated. For example, Wroblewski (1983) argues that predictive power is of secondary importance to the development of 'dynamical explanation' and therefore that the falsification of a theory by comparing observation and prediction is no impediment to insightful

Table 5.1 *Thumbnail definitions of some philosophical terms used in this chapter and the next.*

Analysis: Treatment of a problem by decomposing it into components; in a larger sense, analysis is any treatment.

Causality: The view that science should address natural phenomena as a collection of discrete cause-and-effect relations.

Determinism: The view that natural phenomena are precisely decided by antecedant conditions.

Empiricism: The approach to scientific questions that gives primacy to experience and observations.

Essentialism: The view that science addresses true and palpable realities or essences behind the phenomena of nature.

Explanation: One of the aims of knowledge, explanation rationalizes observation in terms of the view or approach favoured by the explicator.

Historicism: The view that a grasp of the unique set of the many past events experienced by an entity is necessary to explain its present complexion.

Holism: The view that science can be effective if it addresses phenomena as whole entities, without a decomposing analysis.

Instrumentalism: The view that scientific theories and hypotheses are merely tools to deal with the world around us.

Materialism: The view that all phenomena reflect scientific principles and theories, without reference to the supernatural.

Mechanism: The view that natural phenomena, and especially biological phenomena, may be studied as machines, as assemblages of casually interactive parts.

Metaphysics: Non-scientific knowledge, in the non-perjorative sense that this knowledge does not predict.

Nominalism: The view that the general categories of nature to which we give names are only convenient labels for hypothetical entities, not essential classes of reality.

Organicism: The view that nature is more effectively treated as a hierarchical system in which different theories and properties are relevant to particular hierarchical levels.

Phenomenology: The scientific investigation of observables, empiricism.

Realism: The view that science captures the reality of nature in its abstractions and categories, essentialism.

Reductionism: The approach to scientific questions that holds all phenomena to be governed by, and expressions of, a small number of very basic (usually physical) laws.

Vitalism: the contention that living phenomena can only be traced by invoking special, unobservable properties, like a soul, and therefore cannot be reduced to physico-chemical processes.

Note:
Each of these terms can be the focus of philosophical debate and argument (Sattler 1986), so the definitions are at best only indicative.

theory. Caswell (1976) makes a similar claim when he suggests that a theory may be falsified as a predictor but still corroborated as a theory, and Loehle (1983) also feels that models, as explanatory theories, need not be criticized because they predict dismally. All distinguish between the model as precisely defined in its constituent equations and the theory reflected in the general, unarticulated relations these equations represent (Partidge *et al.* 1984; Taylor 1988). Thus theories can not only 'be right for the wrong reason' (Dayton 1973) because successful prediction does not guarantee truth content, but if predictive success is not paramount, they can also be wrong for the right reason. This view debilitates the criterion of predictive power and allows other criteria to confound the selection process by which science winnows theory and grows.

Rigler (1982a) tried to resolve some of these issues by distinguishing between 'empirical theories' and 'explanatory theories'. Empirical theories are tools that describe regularities in the world around us and predict likely patterns. Explanatory theories, unlike Loehle's (1983) theoretical models, also predict; in addition, they tell us why the system behaves as it does and therefore why the predictions are valid. Thus the difference between Popper's realism and my instrumentalism is that Popper is seeking explanatory theories whereas I would be content with empirical theories.

This discrepancy may be more apparent than real. Rigler (1982a; personal communication) felt that well confirmed regularities representing successful predictive tools would eventually become acceptable explanations – as one explains a stone's fall by gravity or the great appetite of elephants by body size. He presumed that eventually we would discern or construct explanatory superstructures around and between simple empirical tools to provide the missing explanatory unity, and he felt the immediate need was to construct the empirical base for this development.

Rigler's research program would leave some distance between prediction and explanation. Explanation by reference to established regularities like gravity or allometry has been called pseudo–explanation (Loehle 1987a), and realistic or representational models in ecology may support explanations involving biological reality, but they are consistently out-performed in prediction by purely predictive, empirical, statistical models. Statistical models are faster, cheaper, simpler and better predictors than their representational competitors, because their sole purpose is to make the best possible estimate (Poole 1978). A similar situation may occur in physics, where explanations are provided by the general theories, but applied decisions are made with empirically justified calculating tools (Cartwright 1983).

Acceptance of this duality reflects the maturity of physics. Because everyday physical phenomena are adequately described by well established physical laws (Nelson 1985), physicists are no longer preoccupied by the problems of falling spheres or colliding billiard balls. Nor are they concerned with airplane design or the path of a falling feather, because neither offers an opportunity to distinguish between competing theories: Aircraft design is largely in the hands of engineers who can adapt physical laws to that application, and the flight of a feather is 'understood' in principle, even if it is too complex to be predicted in practice. Theoretical physicists therefore focus on what they see as tractable problems involving the syntheses of seemingly unrelated theories or on selection among competing theories (Slobodkin 1988). Since most theories in physics agree in the sense that they predict equally well, distinctions must be based on minutiae. The major tasks are for theoretical physicists to discover what these distinguishing details might be and for experimental physicists to make the appropriate measurements. This leaves practical physics to engineers.

Ecology is far from this state. Our ability to predict the course of everyday events is small; theoretical and experimental ecologists seem to have little in common (Caswell 1988), there are few theories for an ecological engineer to apply, and ecologists are more likely to search for a single functional theory than to compare two or more highly successful alternatives. Ecology is not yet ready for its Copernicus or its Kepler, much less its Newton or Einstein or its Watt and Oppenheimer because ecology has yet to develop even the consensus about what observations are interesting, much less develop the data base to make scientists of these heroic proportions possible. We are closer, perhaps, to a lonely priest of Ur, scanning the night skies for patterns and crudely calculating the future course of the heavens, despite gross misconceptions and uncertainties.

It is therefore my contention that Popper's metaphysical realism is inappropriate for ecology because it demands too much, too soon. Instead, we should develop simple predictive tools that allow us to propose and confirm observable patterns that are relevant to the biological world. In future, such patterns may lead to more ambitious theory, but at present, they can serve as tools for environmental management, arguably a more pressing and important problem than creating a general field theory for ecology.

Many ecologists are not prepared to wait. They have instead elected to build explanatory 'theories' which stress unity, reality, cause, and mechanism, but jettison predictive power. However, because a construct without predictive power tells us nothing, this choice represents a pathology whereby the trappings of theory beguile us away from predictive power.

Reductionism: an unattainable goal

Reductionism reflects the erstwhile hope that behind the manifest complexity of the world there lurks a beautiful simplicity, such that a few very basic laws could account for all phenomena. This principle of reduction is nearly universally held by scientists, even if almost all ignore it in practice. In a sense, this principle is the hope of every scientist engaged in the search for order in nature, interpreted on the grandest scale of all.

The appeal of the principle of reduction is as obvious as the unattractiveness of the alternatives. Historically, non-reductionism rested on theological or mystical vitalism (Sattler 1986) and many contemporary appeals to emergent properties, holism, or organicism are too vague and unstructured to merit serious scientific attention (Salt 1979). In contrast, all scientists are aware of successful partial reductions whereby phenomena at one level are explained by reference to a lower level. For example, Lampert (1988) explains the early summer increase in the transparency of Lake Constance by referring to the behaviour of one component of the system, feeding by zooplankton; Belovsky (1986a,b) explains feeding preferences of terrestrial herbivores by reference to their digestive capacities; Yagil (1988) explains a camel's resistance to dehydration by reference to low blood viscosity which allows the animal to cool its core despite substantial water loss; and Gates (1980) has shown that animal behaviour and plant morphology are partly determined by the physics of heat exchange. It is easy to imagine that this process repeated over and over by generations of scientists might eventually order the apparent complexity of nature into the austere simplicity of a single scientific theory.

Neat cases of reduction are particularly appealing to biologists, but we remain a long way from reducing biology to physics, chemistry, or some other primal science. The hopelessness of literal reductionism is evident if one considers the phenomena that must be predicted from, for example, subatomic physics: the wording of the King James bible, the corpus of

English Common law, mate choices of 10 million red-winged black-birds, and the exact steps in the dance language of the bees are only a few. Despite the common success of partial reduction, no one seriously expects full reduction of every phenomenon (McIntosh 1987). Even Popper (1974b), while embracing reductionism as a methodological rule, suggests that the loss of information at each step in reduction renders complete reduction impossible.

Wimsatt (1980) resolves the debate by stating that complete reductionism is a dead issue. It is scientifically uninteresting that ecological phenomena are reducible in principle, if in practice they are not reduced. The reductionism which is possible is the demonstration of consistency of interpretation among the different phenomenological levels. This is appealing because it suggests a potential unity and consistency in our scientific constructs, but it is incidental to their theoretical status and should not be allowed to confound that status (O'Neil *et al.* 1986).

In Wimsatt's (1980) view, the traditional division between ecological holist and reductionist is false, for both inevitably accept reductionism in principle but act as holists. The contentious issue is instead the complexity of the models and the scope of the analyses required to treat ecological phenomena. Consequently, the division between empirical and representational theories in ecology is not a philosophical debate about holism and reductionsim, but a methodological dispute about the most fruitful research strategy to adopt. That controversy splits ecologists into holistically inclined, statistical empiricists on one hand and mechanistically inclined analysts on the other (Cherrett 1989). The strategy of the latter is under scrutiny here.

Mechanistic analysis in ecology

Although full reduction cannot be an active goal in ecology, most ecologists appear to be reductionists because they work within mechanistic research programs. These programs sanction the analysis or decomposition of ecological systems into component processes and structures; the subsystems are then intensely studied by appropriate specialists with a view to the eventual reassembly of the results in a coherent overview, often as a simulation model.

It is a popular strategy. Both Lewontin (1968) and Smith (1970) urge the establishment of multidisciplinary research teams to analyze ecological systems. Governmental funding agencies also encourage the development of such teams, not simply to achieve an economy of scale, but because they believe in the necessity of a mechanistic analysis of complex systems.

Two false metaphors

Mechanism

Mechanistic research interprets natural systems as 'nothing more' than complex machines, and studies the cogs and gears of this machine in an effort to come to terms with the whole. Mechanism does not entail reductionism, because it need not assume that the nuts and bolts of the machine are the most basic, universal laws. Mechanism does normally entail materialism, the view that vital forces, deities, souls, and other non-material emergent properties are not needed to explain the workings of nature, but materialism is so widely accepted in science, that it will not be examined here. Materialism seems essential to science, but mechanistic analysis is only one of several approaches that ecologists may use or ignore as required to build effective theories.

The familiarity of modern scientists with machinery predisposes them to view unknown phenomena as complex mechanisms. However, successes in analyzing man-made machines does not warrant mechanistic interpretations of other phenomena. Machines are built by men with purposes, constraints and theories that the human analyst recognizes. Machines therefore yield to mechanistic analysis because they have been constructed from a reductionist, mechanist, causal perspective and are readily explicable within that tradition (Y. Prairie, personal communication). The parts of typical machines are each distinct, they vary little in time and they bear only a few, relatively fixed relations to one another. Thus each part can be identified and studied separately or several parts can be assembled into modules. Modular construction encourages teamwork in construction or analysis, and large teams, like those of NASA or the Manhattan project, have proven effective in dealing with very complex machinery.

In ecology, mechanism depends on an analogy between a large, complicated machine and a part of nature. This analogy may be strained (Sattler 1986) because most natural systems have no obvious purposes, their parts may do many things, and some roles may be obliterated by decomposition of the system into its parts. In practice, the system is assigned a purpose – the fixation or degradation of solar energy in trophic dynamic models, successful reproduction in evolutionary population models, thermal balance or optimal foraging in some autecological models of individual behaviour – and the roles of different components are studied relative to this goal. This procedure overlooks the possibility of conflicts or compromises in design, of multiple purposes, and of the absence of a designer (Gray 1987). It requires focus on one facet of the system and is necessarily partial because the full system is never adequately known (Oster & Wilson 1978; Pierce & Ollason 1987). In addition, the whole, the parts and their interactions are poorly defined and variable in time and space. Every part has many relations to other parts, and some parts and processes may not have been recognized. In short, although natural ecological systems sometimes yield to mechanistic analysis, they are not machines. Ecologists are free to use the machine metaphor as an heuristic device, but they must keep an open mind about its utility.

The cathedral view

A common metaphor likens the scientific creation to the construction of a glorious building, like a cathedral (Cole & Cole 1972; Rigler 1982b): theorists play the role of architects who provide the design, laborers and craftsmen (field and laboratory biologists) provide the bricks, and all, working together over many generations, build the cathedral. Although this can justify the biological ignorance of a theoretician and the tunnel-vision of a specialist, the metaphor holds no appeal for me. The plans designed by our theorists are uncertain, single bricks are only paper-weights, and no edifice of scientific theory has been built to combine the two. Cole & Cole (1972) suggest that the metaphor is inept because scientific growth is less a communal activity than the product of a few builder-architects who do both jobs. Their research is not done in isolated pieces, but with a clear view to what information is needed, how this information will be used, and what this use will tell us. Mechanistic analysis clouds that view.

The metaphor of the bricks arises because scientists do not study nature in its entirety. There are too many aspects of too many things to make such a project feasible, so every scientist must select some part of nature as the focus of study. Therefore, any distinction between mechanistic analysis and empirical holism does not depend on the necessity of selecting a problem. Instead, it depends on the sort of selection made and on how the results are used.

The cathedral view implies the need for analytical decomposition whereby the system of interest is separated into subprocesses, which are then studied separately, leaving the problem of reassembly to an unspecified future. The subprocesses thus become problems for their own sake, even if the research is justified in terms of the whole program, and the program is relatively safe from evaluation until the slowest phase is completed. For example, feeding rate estimates of an animal may be required for either an analytical research program describing energy flow through the ecosystem or by an holistic program into determinants of feeding rate. In the holistic case, the treatment of the data is straightforward, the relevance of the program is determined by the interest in feeding rate, and the success of the program is judged in terms of the explained variance relative to other models (Peters & Downing 1983). In the analytic case, relevance is again determined by the interest of the audience, but the treatment of the data is unspecified and no judgement of success can be made until all other subprocesses are analyzed, the model is specified, constructed, and independently tested. In practice, this has proven so time-consuming and ineffective that ecologists have lost interest before the original problem is resolved (Rigler 1976).

An alternative view (Popper 1968; Kuhn 1977; Brown 1981; Rigler 1982b) is that theories are not assembled step-by-step by committees or by the accretion of published data, but by the insights of creative individuals. A more appropriate metaphor would depict ecologists as artisans, and each of their products as self-contained, useful artefacts with limited, and so far as possible, specified dependence on the rest of an unreliable science.

This is an holistic approach in the same sense that each scientist would develop a whole theory to solve a whole problem, rather than study bits and pieces in the hope that some other scholar will be able to use them. It would allow the development of a sizeable body of useful, predictive ecological tools that are sorely needed. It is not necessarily empirical, but empiricism has proven effective for the purpose (Poole 1978; Rigler 1982b). A similar view was advanced by Pielou (1981a) who argued that ecologists should investigate 'unit questions' to look for 'empirical answers to single clear cut questions' rather than 'modelling' which entails the construction of a plausible symbolic representation of the functioning of an ecosystem. This is also a component in a 'hierarchical' approach to ecological theory (Allen & Starr 1982; O'Neill et al. 1986) entailing independent theories for different purposes at different scales, rather than a single all-purpose model. Kalff (1989–MS) describes in some detail the difficulty of addressing problems at the ecological scale of the whole lake or macrophyte bed with studies on the physiological or autecological mechanisms used by aquatic plants.

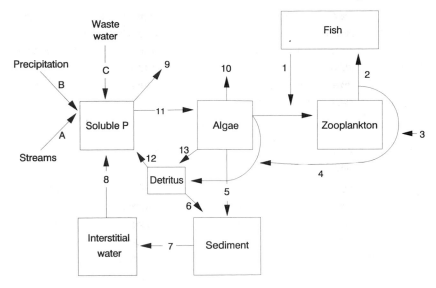

Fig. 5.1. The phosphorus cycle in a eutrophic lake, a moderately complex model of the behaviour of an ecological system. (From Joergensen 1984a).

Systems analysis

Although ecological systems are not machines and although the decomposability of ecological systems is in doubt, mechanistic analysis is probably the most common approach in the science. The most familiar and thorough expression of this is seen in systems analysis and simulation. Simulation models, like that of Joergensen (1984a) describing phosphorus cycling in lakes (Fig. 5.1), identify a number of key components (like external phosphorus supply, zooplankton, sediments) and a number of properties of these components (like ingestion and excretion by the zooplankton) and then provide a system of equations to calculate the dynamics of the system. Such simulation models usually require an initial specification of the state variables, experimental determination of the parameters governing transfers between compartments, and often the monitoring of one or more forcing functions, like external phosphorus supply. Many empirical studies in ecology are designed to provide these estimates and many others can only be applied in the context of such models.

A vast number of systems models describe parts or all of various biogeochemical cycles (Detwiller & Hall 1988), ecosystem productivity and dynamics (Patten 1968; Carpenter 1989), and contaminant flows (Griesbach, Peters & Youakim 1982; Joergensen 1984b). Simulations have been used in the study of the biosphere (Ehrlich *et al.* 1983), ecosystems (Canale 1976), humanity (Meadows *et al.* 1972), isolated populations (Moen 1973) and individual physiology (Ware 1978). Such models assumed a dominant place in ecology during the International Biological programme (1964–74), where they were often seen as the solution to the daunting problem of ecological complexity.

The limitations of simulation and the necessity of a more balanced approach to ecological systems models are now widely recognized, especially by systems modelers (Walters 1971, 1986; Cole *et al.* 1973; Biswas 1975; Watt 1975; Berlinski 1976; Clark, Jones & Holling 1979; Beck 1981; Walters *et al.* 1980; Joergensen 1986; Loehle 1987a). The calibration of the models may result in relations that violate observed patterns in component processes to preserve the whole, and the correspondence between models and observations that results from calibration is often only curve fitting (van Keulen 1974); models cannot address every aspect of the system, but must be tuned to specific problems (DeAngelis 1988), so that a single system must be represented by a suite of different models (Botkin *et al.* 1979); models representing the same reality can give diametrically opposed predictions as a result of the modelers' unsupported assumptions; and the results of alternative deterministic models may be indistinguishable from one another or from random processes (Cohen 1976; May 1976b; Oster 1981). Complex simulations are no longer touted as predictive models but as heuristic devices to explore the logical implications of certain assumptions (Wroblewski 1983; Walters 1986). Contemporary ecologists consequently treat such models with circumspection (Rigler 1976; McIntosh 1985).

Just as the failure of complex models shows that complexity does not ensure success, so the success of very simple models shows that complexity is not always necessary. For example, Vollenweider (1968) successfully treated the problems of eutrophication in lakes with a one-compartment model (Fig. 5.2) which simply ignored the complexities depicted by Fig. 5.1, and this approach has become fundamental to eutrophication control and research. Although only a massive systems model can hope to integrate the multitude of processes in ecological systems, we should recall that our first role as scientists is to predict not integrate.

If there are reasons to suspect that large-scale models involving many interacting subcomponents cannot predict in principle, then there are even more reasons to doubt that mechanistic research into some part of

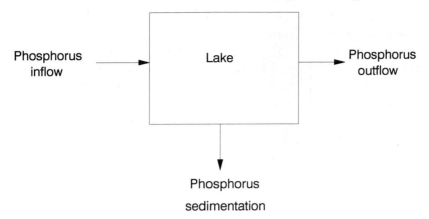

Fig. 5.2. The phosphorus cycle in lakes depicted as a completely mixed reactor. This model has proven effective in eutrophication control programs (Vollenweider 1968, 1987).

the system will ever permit predictions when the parts are reassembled. Nevertheless, the implications of the failure of systems analysis for all mechanistic research are not widely acknowledged. For example, the study of phosphorus excretion by zooplankton is normally justified as an important element in the phosphorus cycle (Peters & Rigler 1973; Joergensen 1984a), but estimates of a single flux will be useful only in the context of a comprehensive model of the P cycle. If such models are not trustworthy, there is no longer any reason to study the flux. If we study partial phenomena which will be given a context only by 'future research', we are open to all the criticisms usually reserved for systems analysis plus that of not having thought through the implications of our approach. Since this criticism holds for any process we choose to study, we must study processes which are interesting in themselves.

This section focusses on systems analysis, not because its chorus of critics needs support or because its practitioners are unaware of its shortcomings, but because those models represent a mature form of mechanism (Levins 1968; Levins & Lewontin 1980). By examining simulation models, one can show that mechanistic approaches do not realistically depict nature, but do offer a complexity which confounds specification of the model, propagates error, and renders the goal of the simulation too vast to be effective and too vague to be stated. The significance of these criticisms for the rest of ecology should be much better recognized.

The mistaken goal of realism

Some ecologists believe that the complexity of nature necessitates complex models. Model systems, and especially systems models, can be highly complex and therefore seem to offer a way to build realistic models. This is a misapprehension. Scientific theories are never as complex as nature, for if they were, they would be as hard to deal with as nature itself. Instead, theories simplify and abstract. Thus a systems model invariably lumps some organisms together, ignores others, depicts only a few of the relationships that are known to exist, assumes or 'improves' essential empirical relations and in many other ways departs from 'reality'. Systems models are not therefore depictions of reality, but abstractions made for some purpose. They can be used and evaluated relative to this purpose, but study of these models cannot substitute for the study of the original system. Models must not be muddled with reality (Hedgpeth 1977).

The problem is to decide how much we should simplify and how much we should abstract. The solution depends on the complexity of the model required to make the necessary predictions. However, if we do not know what predictions we wish to make, this criterion is useless. Ecological modeling misled us into thinking that we could avoid that hard question by building a model of reality, which would handle all questions we might have. This now seems hopelessly naive (Wildavsky 1973; Watt 1975; Biswas 1975; Clark *et al.* 1979; Dennis, Downton & Middleton 1984; Joergensen 1986).

Whether a system is seen as complex or simple cannot derive only from the nature of the system, because all systems are complex if we examine them in enough detail (Ruse 1977; Kitts 1983). For example, Wimsatt (1980) shows that even a computer as big as the universe would be orders of magnitude too small to specify the moves in the game of chess. Instead, complexity is a question about the nature of theory needed to treat the system. If a scientist believes that a very complex theory is required, then the system will be treated as complex. If a simple theory will suffice, then the system will seem simple. *A priori*, neither approach can claim to be more realistic, but simple theories can make a pragmatic claim to being easier to test and use. Once constructed, the better theory can be identified by predictive power. Thus Vollenweider's model predicts phosphorus concentration better than Joergensen's and to that extent is a better description. This example shows that we need not presume that the problems before us require complex solutions until we have considered the likelihood of simple ones.

Incidentally, the thought that complexity is as much a reflection of our theoretical view of the world as it is a characteristic of that world puts into doubt the view that the weakness of ecology reflects the complexity of its material. One could as justifiably hold that the complexity of ecology reflects the inability of ecologists to find or to consider simple theories and solutions. In any case, whether ecological complexity is intrinsic or extrinsic to the subject, the challenge for ecologists is to overcome that complexity by producing theories to treat phenomena of interest.

The problems of complexity and specification

Simulation models require the specification of a greater or lesser number of variables and parameters. For example, the model in Fig. 5.1 has seven compartments, 13 transfer functions, and three forcing functions. Specification of this complex set of phenomena in any one set of conditions requires a lot of work, and since the various unknowns change with different conditions, some of this work will need to be repeated at each application.

The magnitude of this problem is best shown by a few calculations. The number of determinations required to specify any component, parameter or variable depends on its sensitivity to other factors. In the best case, the component is invariant under all conditions and, if measurement error and the problem of identifying an invariant be ignored, one determination could suffice. In a slightly more complex case, the component responds linearly to all factors and none of the factors interact. In such a case (Fig. 5.3), one can determine the response from measurements at two levels of each factor. The first response will therefore require two measurements (one at each level) and each new variable considered will require one additional measurement at some new level of the additional variable. Thus the number of measurements required to describe linear, non–interactive responses increases with the number of factors (F) as $F + 1$. Unfortunately, experience shows that we are not justified in assuming invariance, independence, or linearity.

This modest experimental effort is greatly increased if we consider the possibilities that the responses are interactive or nonlinear. In the still improbably simple case that there is interaction but only linear responses (Fig. 5.4), the experimental effort grows geometrically with the number of factors. When only one factor is considered, the parameter must be measured again at two levels of the factor, so two measurements are needed. When two factors are required, the parameter must be measured at least four times, and if three factors are considered at least eight

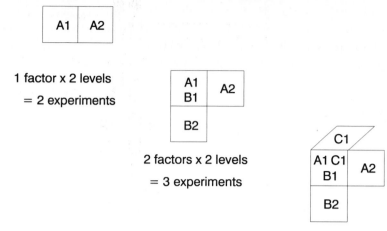

1 factor x 2 levels

= 2 experiments

2 factors x 2 levels

= 3 experiments

3 factors x 2 levels

= 4 experiments

Fig. 5.3. A diagram to show that the number of measurements needed to specify linear, non-interactive systems increases arithmetically as the number of factors considered increases. Because the response is linear, two levels of each factor suffice to specify the response of any single variable. Because the factors do not interact, the response need only be considered in one pair of conditions for each factor (from Peters 1989b).

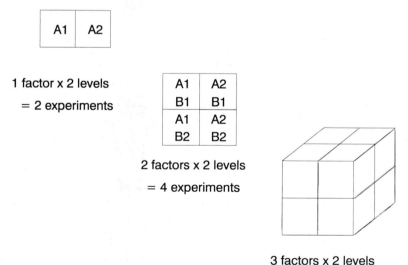

1 factor x 2 levels

= 2 experiments

2 factors x 2 levels

= 4 experiments

3 factors x 2 levels

= 8 experiments

Fig. 5.4. A diagram to demonstrate that the number of measurements needed to specify linear interactive systems increases geometrically as more factors are considered. If the responses are non-linear, more than two levels of each factor must be considered (from Peters 1989b).

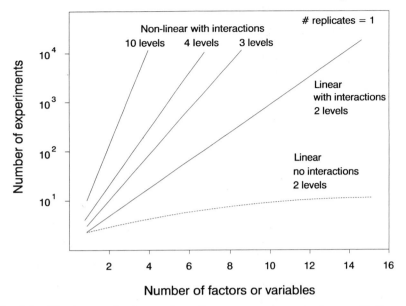

Fig. 5.5. The increase in the number of measurements required to specify linear, non-interactive systems, linear interactive systems, and non-linear, non-interactive systems as the number of factors and the number of levels of each factor required increases (from Peters 1989b).

measurements are needed. In brief, the number of measurements required in interactive linear response rises as 2^F. If the responses are nonlinear, more than two levels (L) of each factor must be measured at each combination of factors and the number of measurements can be calculated as L^F. Since we invariably encounter errors in measurement, this minimal number must be increased by the number of replicates required for each measurement (r), so the number of measurements will rise as rL^F.

These considerations show that the experimental effort needed to specify a system becomes astronomical as soon as even moderately complex systems are required (Fig. 5.5). For example, specification of 10 unknowns with three replicates at five levels in systems where these interact, will require 300 000 measurements.

It is often forgotten how demanding such determinations can be. To provide a personal example, my doctoral research was directed at measuring a few of the parameters in the zooplankton compartment of Fig. 5.1 (Peters 1972). After four years, I was able to provide an empirical model of zooplankton excretion which explained just over half of the variation in excretion rate as a function of animal size, food level, food phosphorus concentration, and ambient temperature (Peters & Rigler 1973). This remains the most complete model of the process available, but does not function well where food phosphorus levels are low (Olsen & Østgaard 1985). This likely reflects a bias in my experimental design, for the food, a yeast (*Rhodotorula*), proved phosphorus-rich (Peters 1987). Thus an extended program of what seemed dedicated and enthusiastic research only produced an imprecise model which is inaccurate under many field situations.

I believe that this level of success is typical and therefore hold scant hope for models which require several score, or even several hundred, such parameter estimates. Many of these estimates must be made at the level of an ecosystem over a restricted time period, and researchers who have tried to measure even a few ecosystem level variables simultaneously will know how impossible this chore could be.

Given that the effort in specifying complex models is unrealistically high, one would expect a movement to simpler models (Beck 1981). While this is so in models involving lake phosphorus loading (Peters 1986; Vollenweider 1987), it is not a universal trend (Carpenter 1988). Fig 5.6 shows that models for system productivity seem to have become increasingly complex, reflecting the growth of our computational skills and power. Even models with 100 unknowns among variables, transfer functions and forcing functions are relatively small and may be combined in still larger supermodels.

Error propagation

Given that we cannot specify all the components of even a simple ecosystem, we must be content with only a partial specification which does not consider the full extent of interaction among the variables, or all significant factors. Thus, we will have to make do with imprecise and erroneous estimates and we will have to adjust for this uncertainty by including error estimates in the output from the models. As a result, each variable and parameter must have an associated uncertainty and any output from the model will have associated confidence limits. This raises the problem of propagation of error.

Calculations involving uncertain values lead to results which are also uncertain. Moreover, each calculation compounds the uncertainty. The amount of this increase depends on the magnitude of the variances of the components, on the covariance among these components, and on the degree to which the variances are additive or multiplicative. These weighty statistical problems place severe limitations on simulation

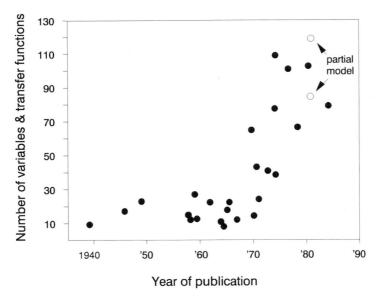

Fig. 5.6. Temporal trends in the number of parameters and variables that must be specified in order to use different models of ecosystem production. In recent years, huge models consisting of two or more partial models have been created (from Peters 1989b).

models involving high levels of interaction and many imperfectly known terms.

To illustrate the problem, Fig. 5.7 shows the increase in uncertainty of a single variable in an iterative calculation. This might represent, for example, the extent of plant damage due to trampling, the net growth of a population over a series of time intervals, or the rate of phosphorus addition to a lake. It depicts the increase of uncertainty in a mildly stochastic variable over a series of iterative calculations. When the uncertainty around a mean of 100 units is such that the variance may be added and the 95% fiducial limits at each step (a computer 'day') are $\pm$ 10 units, it is likely that after one step, the variable will lie between 90 and 110 (Fig. 5.7A). This uncertainty increases after the second step, but since it is improbable that two extreme values would occur on two successive days, the fiducial limits grow by less than another 10 units. This process continues until the spread in the fiducial limits reaches two-fold (by day 10), four-fold (day 40), 10-fold (day 80), and more. After 100 days, the compartment may not be distinguishable from zero and negative values can be expected before a year is over. Similarly if the variances are

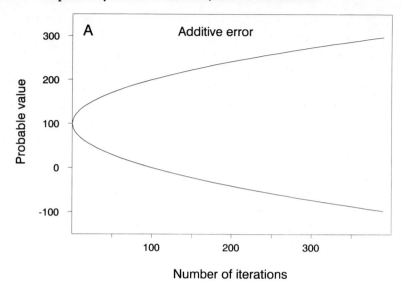

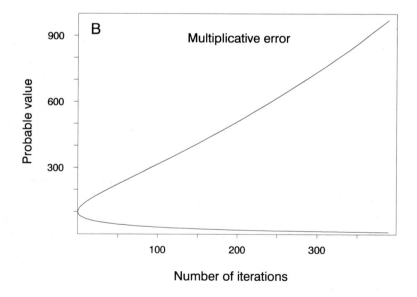

Fig. 5.7. The increase in uncertainty about the magnitude of a variable subjected to iterated calculations that introduce an error of ± 10% of the original size (100) at each iteration. Variance is additive in panel A and multiplicative in panel B.

multiplicative (Fig. 5.7B), the log-variances will be additive and the fiducial limits will diverge even more rapidly until the range covers almost two orders of magnitude by year's end.

The error propagation in Fig. 5.7 is small compared to that involved in most systems models. Few biological parameters can be determined to within 10% and all models are more complex than these calculations. Uncertainty in the predictions of systems models will depend on the error of the different parameter estimates, on the number of calculations involved in each phase of the model, on the complexity of the equations, on the length of 'time' the model runs, and on the length of each step. Since most ecological parameters are known relatively imprecisely, and since most ecological models are exceedingly complex, involve many calculations, have relatively short time steps and relatively long runs, the potential for error propagation is immense. As a result, complex models yield results which cannot be distinguished from random processes (Oster 1981).

There are a number of possible responses to this problem. One is to design the models so that internal constraints and feedbacks prevent extinction or explosion of the model, regardless of what parameter values are used. This results in tautology for it constrains the model to act as we believe it should. The construction of the model becomes an exercise in logic which may be very interesting, but which is unlikely to yield surprising information about nature.

A more interesting approach makes use of the statistical property that the uncertainty associated with a mean is less than that associated with the individual estimates (Shugart & West 1980; Detwiller & Hall 1988). The system is treated as a series of subsystems. Assuming that covariance among subsystems is negligible, the variance of the sum will be the sum of the variances of the subsystems. Since the error associated with a sum varies as the square root of the variance, uncertainty grows more slowly than the sum itself. As a result, the coefficient of variation associated with the sum or the mean is smaller as more subsystems are added. This solution to error propagation is neat, but it is subject to the nature of the covariance term which must be determined empirically or the approach incurs the risk of serious bias. It also requires that the error propagated by the complexity of the subsystem be balanced by increasing the number of subsystems considered. This limits the complexity of the models which can be profitably considered and excludes extensive use of iterative calculations.

The most common solution to the problem of error propagation has been simply to ignore it. As a result, models are governed by deterministic equations and error estimates are not required. Such models are intended not to predict but only to provide 'explanations', 'ideal models', and aids to thought (Pielou 1981a). As a result, Caswell (1976) can claim that a model may be predictively false because prediction and observation differ quantitatively, but theoretically corroborated because of qualitative similarities in trend. This is so much less than the early promise of systems analysis had led us to expect (Rigler 1976) that systems analyses in ecology are often met with a sense of disillusionment.

The problems of error propagation are well known to systems analysts. It is time ecologists working on the isolated parts of a decomposed ecological system begin to ask how the uncertainties associated with their determinations will affect the application of their results. They must recognize that the uncertainty associated with many ecological phenomena ranges over an order of magnitude and that even highly standardized physiological measurements with very tractable subjects are variable. For example, estimates of human basal metabolic rates vary about two-fold even after correction for the effects of weight (Peters 1988a). This high uncertainty limits the number of terms, equations and iterations that can be considered, but most work with parts of biological systems presupposes a far greater level of sophistication in the final model of the recomposed system than propagating error is likely to permit. If ecologists want to predict more than qualitative indices of trend, they must adopt a simpler approach.

Identification of the purpose of simulation models

A further problem presented by systems models is that they predict so much. Systems models predict the sizes of dozens of compartments and the magnitudes of scores of fluxes and processes over long periods of time. This increases the chances of getting something right, but it is absurd to pretend that the models should predict everything. Indeed, modelers recognize that if publication depended on agreement between the model predictions and existing observations, most models would never reach the press; those few that did, would do so because no relevant data exist (Caswell 1976). In practice, systems models are not expected to predict all aspects of the systems precisely and quantitatively, but this license threatens to immunize the model from scientific criticism. Unless we know what the models are intended to predict and with what success, we have have no guide to select models appropriate to our own purposes, no criteria whereby we may judge the models, and no basis to reject or replace the model.

Table 5.2 *The problem of testing the predictive powers of complex models can be illustrated with six qualitative predictions generated by a lake model of Walters* et al. *(1980).*

(1) Primary production is independent of the concentration of free nutrients.
(2) The biomass of nannoplankton is independent of nutrient concentration.
(3) Primary production per unit of lake surface is little affected by lake trophic status.
(4) Zooplankton biomass varies with nutrient concentration.
(5) Planktonic bacterial populations are low, except in eutrophy.
(6) Benthos biomass is independent of nutrient concentrations.

Note:
Since observation confirms the odd numbered predictions but falsifies the rest, we are in a quandary as to whether the model has been confirmed or falsified.

The problem is alleviated if we accept that most simulation models are expected only to provide qualitative indications of trend, but this limits their utility sharply. Moreover, although the problem of making too many predictions is reduced by this ploy, the problem is not eliminated. The prediction of qualitatively different phenomena creates the likelihood that science will be left on the horns of a dilemma when some predictions succeed and others do not.

A simulation model for lakes developed as a summary of the freshwater productivity section of the International Biological Program provides an example (Walters *et al.* 1980). In this study, the authors developed a series of six qualitative patterns which the model suggested would hold across a variety of lakes (Table 5.2). Empirical evidence supports half of these (Numbers 1, 3, and 5 in Table 5.2) and shows the remainder (numbers 2, 4, and 6) to be false (Peters 1986, 1989 b). This places us in a quandary, because different observations have falsified and corroborated essentially independent parts of the theory. One may decide that the model is an adequate predictor for some, specified types of phenomena (even though it is inadequate for others), or simply that it is false. The latter decision would require us to abandon all complex models because they are all false in some particulars (Caswell 1976). However, the former decision requires that we specify where the model applies and this specification is only made possible by making the observations required to build an empirical relation. In the case, we would not need the model.

The only tenable solution is to return to simpler problems. We must select 'unit phenomena' (Pielou 1981a) and attack these holistically. These may be the particularly important, tractable, everyday phenomena which in aggregate constitute the complexity of nature – this has been a frequent choice in allometry (Peters 1983; Calder 1984) and limnology (Peters 1986) and is likely imposed whenever practical advice is sought (Poole 1978). A more demanding choice, to select unit phenomena representing a fundamental similarity beneath the observed complexity, seems too ambitious for contemporary ecology, but it has been appropriate for dealing with the simpler, grosser forces of gravity and subatomic physics. In either case, we must then seek to build theories that predict the phenomenon of interest and that phenomenon alone.

Causality

Causality is a vast topic in both philosophy and science. It is relevant to the present discussion of realism, systems models, and analytic research, because these approaches are sometimes said to isolate the fundamental causal relations that should permit better prediction and better explanation. Causal relations explain phenomena by showing them to be the effects of prior causes and predict phenomena by describing them as effects of observed causes. If the world behaved deterministically, like a huge clockwork toy, and if we were armed with a sufficient number of causal connections, we should be able to pursue the causal chain indefinitely into the future or into the past, thereby achieving perfect predictions and explanations, respectively. Causal connections offer another advantage because they are conceived as fundamental, invariant links; if this is so and if we can identify these links, our causal models would apply despite perturbation of the system and may be transferred to other systems (Poole 1978; Loehle 1983, 1987b; DeAngelis 1988). The concept of causality also offers a possibility of limiting reduction, for further analysis is not required once the fundamental causal relations are found (Simberloff 1980b).

The list of virtues draws the questions 'What is a cause?' and 'How do we identify it?' The present section answers these questions by showing that 'cause' is so many things to so many different people, or to the same people at different times, that it has become a non-concept. There is no way to identify the cause.

The section begins with an overview of the different interpretations given 'cause' and an argument that, although regularity is a scientifically acceptable criterion for causal attribution, ecology would be less confused and more effective if it ceased trying to identify causes and instead sought to make predictions. Both instrumentalist and causalist research use the same sorts of information and neither can claim a greater truth content, yet the two approaches differ in the sort of phenomenon which they see as worthy of study. A series of examples illustrating this difference ends the discussion.

Definitions of causal connection

Commonsense notions of cause

The principle of causation states that every phenomenon has a cause. This is so widely accepted in biology (Mayr 1960) that few biologists ever consider the principle at length. They are content that the success of causation in everyday life is sufficient to sanction the extension of the principle into science. Surprisingly, when the commonsense notion of causality is examined, people prove to be very ineffective at attributing cause (Nisbett & Ross 1980). Our faith in causal attributions is not based on their everyday success, but on our everyday failure to test them critically. This is not the only time that commonsense has proven a poor guide in science (Simberloff 1983; Strong 1983).

An extensive analysis of the failure of commonsense causal notions (Nisbett & Ross 1980) identifies six factors that confound causal attribution. People rely heavily on the criterion of resemblance whereby the cause must resemble its effect: big effects must have big causes, dramatic effects dramatic ones, and so forth. People build new causal models by analogy with existing ones and are extremely uncritical in choosing analogues: popular songs, stories, parables, political biases, television dramas, proverbs, and beliefs will all serve. Once a plausible causal scheme is identified, it is maintained in the face of evidence that shows it to be wrong, and those who change or even reverse causal allegiance may be unaware of their inconsistency. People assume that whatever information at hand is sufficient and will use it to determine a cause even if the information at hand is inadequate. Usually the most salient or apparent of prospective causes is selected as the cause of an event. People are unable to use statistical information adequately and are too willing to look for single causes. Thus once a cause is found, they see no need to look for others. In short, Nisbett & Ross (1980) show that our everyday notions of cause are so shaky and heterogeneous that they cannot be a model for causal attribution in science.

Nisbett & Ross (1980) base their conclusions on experimental evidence and it could be maintained (barely) that causal inference is better suited to the real world where it evolved. In ecology, this is epitomized in the thought that the predictions of our sophisticated apparatus of science could (or even should) be replaced by the advice of a 'wily old' fisheries biologist, or wildlife manager, or farmer or whatever.

There are indeed some startling examples of causal attribution by non-scientists: The painstaking but essential process of washing and grinding to remove the poisons from manioc root which provides the starch staple for indigenous peoples of South America is a remarkable example. Nevertheless, this faith in folk wisdom is misplaced. Even in basic questions of economy and health, folk wisdom is often wrong. In northern Italy, wily fisherman have planted thousands of trout in the acid, metal-contaminated, defaunated waters of Lago d'Orta (Bonacina 1986) where the fish were soon poisoned or starved to death (G. -L. Giussani personal communication). Wily old peasant dairy farmers in the Alps refuse to adopt gentle milking techniques even though the traditional technique so damages the teats that potentially productive cows cease to yield (L. Zacchi, personal communication). On the North Shore of the Saint Lawrence River, commercial fishermen have a host of causal notions about where cod lie relative to winds, shores, icebergs and weather, but this folk wisdom changes from person to person and is not effective because the fish positions are multiply determined (G. Rose, personal communication). Wily old forest rangers in Germany reject the notion of acid rain, because their causal world is contained by the borders of their jurisdictions (R. Wittig, personal communication) and farmers all over the world predict the weather from colour patterns of caterpillars, the sleep patterns of rodents, and the hairiness of corn cobs. Finally, it was the wily old game manager and naturalist, Aldo Leopold (1943), who gave us the now discredited story of the population explosion of deer that followed predator control on the Kaibab plateau (Caughley 1970). The causal attributions of these important events are not wrong because the people are stupid, but because the events are complex, the knowledge of the circumstances affecting the phenomena is scanty, the distances between cause and effect are great, and the training and tools which would allow an effective test are unavailable. Causal attribution is a demanding process which has so little to do with the everyday process that untrained people rarely do it well.

Scientific notions of cause

The strength of causality in science cannot be based on the success of commonsense causal notions, but on the ability of scientists to extract more consistent and more rigourous causal concepts from the welter of commonsense notions. In so doing, they have identified a series of attributes which may indicate causal connection. Thus the position of cause is that of a concept cluster and its effective use requires that the particular concept be specified at each use. In his overview of causality, Beauchamp (1974) recognizes five sorts of cause.

'Singularist theories' invoke causal connection to explain the course of unique events after they have occurred. This view is particularly historical in outlook and will be examined at greater length in the discussion of historical explanations in ecology in Chapter 6.

Humean cause is identified by 'constant conjuction' so the cause and its effect always accompany one another, temporal 'priority' of cause relative to its effect, and temporal 'contiguity' of the cause and effect. Hume believed that conjunction, priority and contiguity were sufficient to produce a psychological (but logically unfounded) belief in an unobservable causal connection whereby whenever a cause was observed, the effect would be expected. Thus, we would presume that given enough causal models, we would be able to produce causal chains which would allow extensive explanation and prediction. Humean causation is largely antiquated (Hilborn & Stearns 1983): constant conjunction is rarely achieved in science, so scientists rely on significant correlation and multiple causes; modern causal concepts no longer require temporal contiguity, as the biological distinction between ultimate and proximal causes (Mayr 1960) demonstrates; and the cause need no longer be prior to its effect, as when the pollution of a system coincides with its effects. Hume's lasting contribution was to signal the importance of the causal connection and his contention that its validity, like that of other of inductions, is undemonstrable (Popper & Miller 1983).

Further development of the concept of cause divides around the issue of the importance of the unobservable connection to science. 'Necessity theories' propose that causal connection is essential to connect the parts of our experience. They see causation as 'the cement of the universe' and consider it 'real'. Unfortunately, causal connection cannot be observed and the only criterion which can distinguish between a 'cause' and some other regularities is predictive success following perturbation. Thus, even if we believe that a pattern represents a causal necessity, we can never be sure that the connection will hold in new contexts. This denies any benefit from the concept of causal necessity and makes the concept appear superfluous. Causal necessity is however not a neutral term because the concept induces us to have greater faith in our theories than the evidence warrants. It leads to dogmatic assertions about what observation should be made and what theories should be permitted; it substitutes preconceptions for predictive power as the deciding criterion for scientific acceptability.

Views of causation which do not invoke necessary connection are termed 'regularity theories'. These are derived from Humean conjunction and hold that the causal connection between two phenomena implies only that whenever one is observed (the cause), there is a significant probability that the other (the effect) will also be observed. Regularity implies simple correlation between cause and effect, but not that this correlation is causation. A causal regularity also requires that changes in the 'cause' produce appropriate changes in the effect. In other words, regularity theories require that the correlation apply to both comparative (or observational) and experimental (or manipulated) phenomena. When no experimentation is possible, as in astronomy or meteorology, causal regularity can imply no more than comparative correlation. Regularity theories thus assimilate the empirical content of necessity theories without residuum. They differ from necessity theories because they deny the necessity of the connection. Thus, they hold that there can be no assurance that the observed regularities will hold in new contexts.

Manipulability theories are similar to, but more restricted than, regularity theories. Like all models of causation, they require regularity, but they reserve the attribution of cause to phenomena we control. In this sense, a cause is a handle by which we can change and manage our environment. Consequently, causal attribution is restricted to relationships which are also amenable to experiment. Manipulability is a relative concept, for what can be changed or controlled depends on the means at our disposal. The concept's originator, R. G Collingwood (1940) illustrated this with the example of an automobile accident: to the driver, the cause may have been excessive speed, to the highway engineer a fault in the roadbed, to the mechanic, wear of the brakes, and so on.

Of these alternatives, the Humean cause is so rare as to be useless; necessity theories introduce an unobservable and therefore unnecessary term into science, 'causal connection'. Invoking parsimony leaves regularity and manipulability theories of scientific causation. For the scientist – who presumably has access to any available controls or handles and, if possible, requires both experimental and observational corroboration before making causal attribution – these are almost identical. They differ because the manipulability criterion reserves causal connection to those things within our control and this may be too restrictive. The most justifiable notion of cause is therefore that based on regularity.

No such simple a solution is possible. Cause has so many meanings and nuances in so many areas of human endeavour that it cannot be restricted by declaration. Moreover, even if we try to restrict ourselves to one definition, other associations would likely creep in. Instead, the claim of causal connection, like that for realism, obscures the simple evidence for regularity on which the claim must be based and gives the claim a greater validity than it merits. A safer route is that already followed in physics, simply to avoid the term (Russell 1917; Kuhn 1977). It adds nothing to ecology but confusion.

Infinite regress and causalist research

In principle, predictive power depends on regularity or manipulability. Thus, the success of a research program aimed at the causes of a phenomenon could be judged under the same criteria as those aimed at prediction (Chapter 2). In practice, research into causal connection is not the same as the search for predictive power and causal research programs assume some very different characteristics. The crux of this difference is that causalist research seeks first to place the effect in an explanatory model which will also provide predictions. As a result, causalist research seeks first to build a causal web, relegating an evaluation of predictive power to such time as the web is described. Instrumentalist science strives for predictive power from the first.

This search for the cause of a phenomenon involves an infinite regress. The causal connection is usually described naively as a Humean relation between a single cause and a single effect (Fig. 5.8,A). This might be represented by the example that phosphorus loading to lakes causes eutrophication, a premise well supported by both comparative (Dillon & Rigler 1973; OECD 1982) and experimental (Edmondson & Lehman 1981; Schindler 1971, 1978) studies. However, the causal connections underlying these empirical observations are actually so much more complex, that they defy scientific analysis (Quinn & Dunham 1983).

Unlike a single cause–effect relation, causal loading response models invoke a series of other contributing causes which are usually described as preconditions – the lake must be phosphorus limited, the light levels and temperature must be sufficiently high, flushing rates must be sufficiently low, the phosphorus must be in a biologically available form, the lake must not be poisoned by other pollutants, and so forth. In

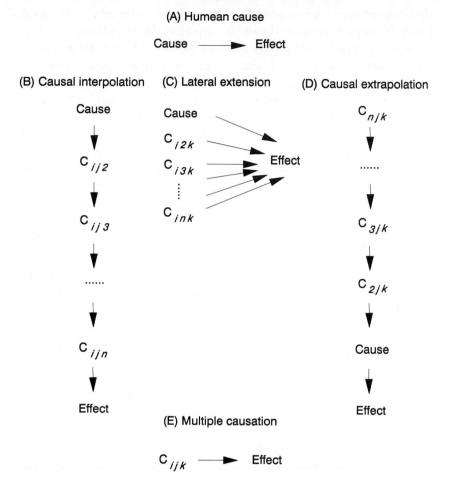

Fig. 5.8. Close examinations of (A) a simple Humean cause–effect relation leads to an infinite proliferation of causal pathways because one can always (B) interpolate more causal steps, (C) discover more contributing factors, or (D) extrapolate to more distal causes. As a result, every causal study attempts to unravel some portion of a vast web (E) of interacting causes and effects.

general terms, causal statements are subject to an infinite lateral extension (Fig. 5.8,C) as more and more factors are considered and at no time are we entitled to say that all the causes have now been determined.

In addition, causal research is not restricted to a single scale, so many causalist programs seek causal connection by identifying more proximal or intermediate causes. In the example of the phosphorus cycle, most of the processes depicted in Joergensen's diagram (Fig. 5.1) are proximal with respect to the increased phosphorus load initially posited as causal. However, this does not exhaust the amount of interpolation possible. Joergensen's model does not begin to address the intricate problems involved in transferring phosphorus among compartments, but other, equally complex models do (Kirby 1978). The still more proximal stage where phosphorus is utilized inside the cell involves much of modern biochemistry and therefore is at least as complex as any less–proximate processes (Williams 1978), and these cycles and pathways are themselves but coarse models representing the molecular activities of enzymes and reactants. In short, causal models allow and encourage an infinite interpolation of causal processes (Fig. 5.8,B).

Finally any causal model allows an infinite extrapolation to less-proximal causes (Fig. 5.8,D). For example, phosphorus loads, a forcing function in Joergenson's model, may be the effect of a further suite of models involving sewage treatment (Collingwood 1978), land-use (Harrison 1978), geology, geography, economics, and politics (McClellan & Hignett 1978) interpreted at ever-increasing spatial and temporal scales. This is not an isolated example: every phenomenon sits in a web of interacting, multiple factors. Attempts to describe this web lead one back to a mechanistic approach to ecology and to an infinite research program.

This peculiar extendibility of causal analysis has several advantages. Because one can be sure that there are always more proximate, more distant, and more collateral causes, the principle of causation serves heuristically by directing research towards other determinants of the effect (Roughgarden 1983). Moreover, because causal attribution is infinitely varied, research into a problem is never finished. Generations of graduate students can take up part of the question, dividing that problem among their graduate students and so on, yet none need fear the dislocation that would result if the task were finished, nor need they ever confront a basically different set of problems. Causal analysis also allows simultaneous approaches to a problem on many levels and scales; this organizes science into separate schools that work on different components of a causal interaction without the ill-feelings that result when personal fiefdoms are violated. The same process encourages the intense specialization of modern science.

These advantages, if such they be, are balanced by a series of abuses to which causal research is subject. Specialization leads to tunnel vision. The development of schools of thought may diminish healthy competition and criticism by allowing each school a different niche in academia, and may encourage the pursuit of trivial problems simply because a large school considers them interesting. Infinite extendibility has allowed the detergent industry, the tobacco industry, the American National Rifle Association and many other powerful lobbies to resist remedial action and government control, because causal connection has not been shown. Instead, causal analysis offers government a relatively inexpensive, but face-saving, alternative to action: research into various causal pathways. Researchers use the same ploy to explain the failure of their studies to yield answers by pointing to the lacunae in their flow diagrams and demonstrating the need for 'more research' and especially 'more funding'.

Because causal analysis is never over, it is heuristic in the sense that it always leads to new problems and new publications. If ecology were a government make-work project to occupy as many ecologists as possible for the longest possible time, causal analysis would be an ideal vehicle. If ecologists instead wish to provide answers to the pressing problems of contemporary man, causal analysis is a singularly ineffective tool.

Instrumentalist research

Instrumentalist research contrasts with causalist research because the former must be directed to better prediction. This may involve any or all of the factors used in a causal explanation, but the criterion for consideration is not plausibility of the causal explanation, but reduction in the residual error. The interest of the prediction may be defined by an academic school, but it could also be identified by societal needs. Since the variable to be predicted is not subject to analysis, the possibility of division of the research area is smaller and the scope for competition and criticism is much larger.

Instrumentalist research has a recognizable endpoint defined by the pure error associated with measuring the response variable. Since predictions that are more precise than measurements cannot be evaluated or tested, a new problem should be sought when the error on the prediction is equivalent to that involved in replicate measurements. This new problem could be a methodological solution to the variability of the measurements or it may involve a change of focus to equally pressing problems that cannot yet be predicted so effectively. In practice, new problems will often be sought before this degree of prediction has been achieved.

Instrumentalism and experiment

The rejection of causal connection as a useful scientific concept in favour of instrumentalism first seems an unacceptably extreme proposal, but few drastic changes would actually result. An instrumentalist perspective would affect the choice of phenomena to study and this would represent a serious challenge for ecology. However, since both approaches ultimately depend on the same sorts of evidence, comparative observation and experiment, this shift would not alter scientific practice.

Instrumentalism may also require a change in our attitudes to the results and theories. Instrumentalists should always accept results and hypotheses tentatively and be ready to accept a better alternative when it is available. In contrast, causal links are often so convincing that causal connection seems better founded than the results alone would warrant. This is inappropriate since a theory must be judged on the evidence, not on plausibility of the prose in which it is couched.

Whether the suspicion of the instrumentalist is more justified than the enthusiasm of the causalist depends on the quality of scientific evidence. This evidence is assembled by one of two basic methods, comparative observation and experiment, collected in the laboratory, in field experiments and in correlational studies of natural patterns. The quality of this evidence is therefore a critical issue.

Laboratory experiments

Laboratory experiments effectively standardize experimental conditions so that the effect of a single factor is thrown into relief against a constant background. This facilitates the identification of even small effects and reduces the possibility of confounding variables through strong controls. Therefore laboratory study is especially useful in dissecting various causal paths.

The problem with the laboratory is that its artificiality may simply swamp processes of ecological relevance. The laboratory setting may evince effects which are not experienced in the field. The restrictions of the laboratory may not allow the natural behaviour of free-living organisms (Hilborn & Stearns 1983) or the laboratory setting may magnify incidental and trivial factors. Because most of the variation which the specialist confronts in the laboratory is that induced by the experiment, laboratory studies tempt researchers to assume that they study a process of particular importance. Since researchers working on different processes in the same system make identical claims, this cannot always be the case (Rigler 1975a).

These limitations led A. C. Redfield (1960) to suggest that an overemphasis on experiment stalled the development of marine biology at Woods Hole for fifty years. The deleterious effects of experiments can easily be found in other fields as well. Laboratory studies (Goldman 1965) have identified many limiting nutrients for algal growth, thus suggesting that multiple limitation is likely in nature. However, in the field, phytoplankton biomass almost always reflects phosphorus, nitrogen or light. Laboratory experiments may be irrelevant to nature because they present an unnatural constellation of environmental conditions which constrains the organism to unnatural, irrelevant responses. These sorts of experiments slowed or even stopped the implementation of phosphate abatement programs which are essential to eutrophication control (Schindler 1971).

In summary, because the conditions of controlled experiment are so refined, laboratory experiments are insufficient to show that a factor will have a similar effect in nature or that the effect observed in nature represents a response to that factor. Indeed, laboratory experiments can likely show some effort of almost any factor simply by using sufficiently extreme conditions. Laboratory studies are effective in isolating a response to a factor but the response may not be ecologically relevant and the number of potential factors that could be investigated is so large that study of any isolated factor may be futile.

Field experiments

Although there is no qualitative difference between experiments in the laboratory or field (Mertz & McCauley 1980), many ecologists are attracted to field experiments and particularly to controlled field experiments (Connell 1974; Paine 1977) because of differences in degree. Field experiments achieve less standardization, but include a greater proportion of potential influences. Field experiments show the manipulated factor may have its presumed effect in a more natural setting, despite the uncontrolled variations of other factors, but these results are suspect because uncontrolled variation may induce chance colinearities and erroneous attribution of the significance. Since there is an infinite number of factors and variables in field experiments, such experiments always risk confounding the manipulation with some correlate. This danger is reduced by replication and controls, but there are simply too many potential factors and too few degrees of freedom to eliminate this possibility in field studies (Quinn & Dunham 1983; Bender *et al*. 1984).

Dayton & Oliver (1980) discuss some of the limitations of field experiments in studies with benthic invertebrates and most of these reservations apply equally well to any field problem and to laboratory experiments (Mertz & McCauley 1980). Field experiments consider only part of the system for part of its existence, they are therefore subject to a complex of interrelated problems involving representativity, scale effects, 'bottle effects' and 'edge effects'. All experiments introduce edge effects because they must choose to include or exclude portions of the natural environment. Thus large predators may appear unimportant because the experimental conditions exclude them from a site or particularly important because the experiment confines them to an unnaturally small resource base. Temporal edges exclude some relevant part of the life cycle, like reproduction or larval settling, or some seasonal factor, like winter storms and cold. Bottle effects refer to artefacts that result from the experimental manipulation in addition to the factor(s) under examination. For example, marine enclosures may allow periphyton growth which may provide extra food, reduce water movement, increase shading, or decrease desiccation rates, all of which may have both dramatic and subtle effects on the process under study. Scale effects result from the small size of all experiments and lead to the under-representation of rare events (like the damage caused by logs to mussel beds) and events involving larger geographic areas (like oyster spawning) or larger scale spatio-temporal heterogeneity, like the availability of rare refuges.

Field experiments are also cumbersome and costly so that logistic constraints may influence the results. Replication is frequently minimized to reduce costs and sampling effort, whereas the number of factors studied is expanded to consider more possibilities. This reduces the probability of identifying the significant effect of any single factor, but increases the number of 'significant' effects simply because, if 20 factors are considered, one is likely to prove significant by chance alone. Moreover, logistic considerations place the researcher in a dilemma since the better the replication, the less effectively will the experiment represent the range of natural conditions. Often the threat of vandalism or questions of access require additional compromise in the choice of sites, so that field experiments are often unrepresentative of the larger scale.

These problems make the interpretation of any field observation ambiguous. Underwood & Denley (1984) describe a series of alternate causal hypotheses which could fit published field observations as well as those proposed by the original authors. This is the case even in laboratory experiments, since alternate hypotheses are always possible; but in field studies, the alternatives seem more apparent and more plausible than in closely controlled laboratory studies. This ease with which field data can be fitted to alternate causes of hypotheses and the impossibility of testing causal connection lie at the heart of the bitter debate over the significance of competition in ecology (Diamond 1978; Connell 1983; Lewin 1983a; Roughgarden 1983; Strong *et al.* 1984; Simberloff 1982, 1983; Schoener 1983).

Field correlations

Natural experiments and gradient analyses do not suffer the artefactual ambiguities that manipulation imposes on laboratory and field experiments, but the increased likelihood of artefacts in sampling and design have substantially similar effects. Moreover the chance of spurious correlations is the greatest of all these techniques. As a result, a correlation between two static properties over a range of natural situations provides only circumstantial evidence that new observations will fit this pattern, and can never support the assertion that a change in one variable will be accompanied by a change in the other. These contentions are simply preliminary hypotheses that require further test.

Correlations among factors in nature may therefore serve to suggest variables for subsequent experimental analysis. Ideally these experiments should be conducted at different levels: in the laboratory because laboratory experiments are fast, inexpensive and easily interpretted, in small-scale field experiments because they increase the relevance of the test while reducing the logistic problems of larger-scale experiments, and finally in large-scale field tests which are usually restricted by cost but more relevant to the problem. However if a choice must be made, large-scale manipulations will probably be the most convincing and relevant, since the value of an ecological hypothesis about a system is best judged by the behaviour of the system. Thus if smaller-scale and full-scale experiments prove inconsistent, the full-scale tests are likely to be most ecologically meaningful.

To recapitulate, all of our observational tools are flawed to a greater or lesser degree and interpretations of the results are likely to be biased by these flaws (Diamond 1986). The appropriate tactic, given this unreliability, is to combine techniques and to suspect hypotheses which are not consistently confirmed by different techniques. Nevertheless, even this confirmation would not guarantee that the manipulation will have the desired effect in novel conditions, so any important manipulation will require an 'adaptive management policy' (Walters & Hilborn 1978; Walters 1986; Hilborn 1987) wherein planned environmental interventions are reassessed at intervals to determine the degree to which the intervention was successful, to evaluate the need for further intervention, and to collect more data at the scale relevant to application. This is the nature of hypothetical knowledge.

Instrumentalist responses to causal problems in ecology

For many years, ecologists have claimed to seek the interacting causal mechanisms which produce the patterns we observe. This strategy would lead to research programs which are doomed to fail because they try to observe an unobservable ('causal connection'). Such programs result, not in solution, but in infinite regress of research problems. In fact, ecologists are not so naive. The grand phrases about reality, truth and cause presumably represent a short-hand for a more tractable goal, the identification of regularities in nature and the development of models to predict and control these patterns. These could then be called 'causal relationships' without harm, and the goals of instrumentalist and causalist ecology would be the same.

This synthesis is not possible because causal analysis has been so emphasized that causal research has become an end in itself and the instrumentalist goals (prediction of the effect) have been forgotten. Thus a feasible research program into prediction of a phenomenon has become confused by the varied commonsense definitions of cause and with rhetorical devices to render ecological work exciting and satisfying. If ecology is to emerge from its academic slumber to play an important role in the management of our natural resources, it must shed these metaphysics and assume an instrumentalist position. A few examples show how easy and advantageous this could be.

Animal diversity

Many questions in ecology begin with an explicit or implicit 'Why?' Such questions demand an explanation in terms of cause and mechanism, rather than a concrete statement about what will be observed. Thus Hutchinson (1959b) could ask and seem to answer the question 'Why are there so many kinds of animals?', even though we did not and do not know how many kinds of animals there are (May 1988). This appoach is attractive in the early stages of a scientific field because it allows one to experiment with explanatory models, even though the hard work of making the observations that require explanation has not been done. The popularity of this approach in ecology is demonstrated by the recurrence of titles like 'Why are biennials so few?' (Hart 1977), 'Why are biennials sometimes not so few?' (Silvertown 1983), and even 'Why biennials are not so few as they ought to be.' (Thompson 1984). This whimsical series suggests that ecologists should learn that 'Why?' questions lead to an infinite regress or to the empiricist's conclusion 'Because that's the way things are'. Parents of persistent children make the same discovery.

Animal abundance

Instrumentalism avoids infinite regress because it does not pretend to answer 'Why?' questions. It answers questions about 'How many?' 'How much?', 'When?' and 'Where?'.

For example, one of the most widely known themes in ecology seeks to explain the commonness and rarity of different species (Andrewartha & Birch 1954; Rigler 1982b; Price 1986). After sixty years of research, the successes of this program are represented by a species-by-species accumulation of sound investigations. No general rule to explain population abundance emerged from this causal research.

In contrast, a simple regression of animal density on body weight accounts for three-quarters of the variation in reported abundance of all animals (Peters & Wassenberg 1983) or selected groups (Damuth 1981; Peters & Raelson 1984, cf. Fig. 2.1). This general rule only quantifies a well known pattern (Mohr 1940; Hutchinson & MacArthur 1959). It is not causal. It explains only in the statistical sense, and the residual error is so large that improvement is still greatly needed. Moreover, much of variation that has perplexed generations of population biologists remains unexplained. Still the relation provides a better, more general answer to questions about 'How many?' animals there are than all the 'Why?' questions that have ever been asked. It is a start.

This is not to say that all processes will be as simply determined as abundance, nor that this model has resolved all questions about animal abundance, nor that improvement of the regression models will be easy. My position is only that, if improved estimates are the goal, experience has shown that we are more likely to achieve that goal by addressing it directly, empirically and statistically, than by beating about the bush looking for contributing causes.

Experiment, correlation and trophic cascades

The appeal of mechanisms and causes is so great that some researchers place an unwarranted faith in experimental analyses. A vocal group stresses the importance of field experiments (Connell 1974; Paine 1977). This stress seems valid in context where it is directed against claims that comparative observational studies alone can show that competition caused observed patterns (Roughgarden 1983; Quinn & Dunham 1983). Although only perturbation studies can distinguish between correlation and manipulability, a poor experiment is less relevant to manipulation than a well executed correlation. In particular, perturbation experiments that are not conducted on the same scale as the intended manipulation may be trivial in terms of the whole system. In such cases, the likelihood of experimental artefact is large enough that differences between microcosmic and macrocosmic patterns only suggest the irrelevance of the microcosm experiments (Heath 1980).

For example, experiments in field mesocosms support the view that 'cascading trophic interactions' (Carpenter et al. 1985) could help control chlorophyll development in lakes and therefore that 'biomanipulation' (Shapiro & Wright 1984) of the food web should allow 'top-down' control (McQueen et al. 1986) of eutrophication. The mechanistic argument supporting this process is a familiar one that recalls elements from Hairston et al. (1960); Brocksen et al. (1970), Fretwell (1977, 1987) and Wiegert & Owen (1971). In brief, it holds that, although the upper limits of algal development may be set by nutrient supplies, further limitation may be imposed by the herbivores. Thus abundant piscivores will lower populations of planktivorous fish, releasing populations of herbivorous zooplankton from predatory control and depleting the phytoplankton. The converse holds if piscivore populations are depressed.

This is an attractive theory. It offers a far less-expensive way to control algal development than reducing phosphorus inputs; it restores the traditional foci of lake biology (fish, zooplankton, algal types) to a central position in eutrophication research; it offers a mechanistic explanation of the residual error in nutrient response models; it can be simulated (Carpenter & Kitchell 1984; Carpenter 1989), and it is consistent with many laboratory (Elliot et al. 1983) and field (Andersson et al. 1978) experiments, and with a variety of observations, experiments and anecdotes from whole lakes (Benndorf 1988; Benndorf et al. 1988). In short, the hypothesis is a good candidate for systematic testing.

Given that fish populations are extremely hard to census, a reasonable test of this hypothesis would be to determine if zooplankton abundance explains a portion of the residual scatter around regressions of chlorophyll concentration on nutrient level in a set of data collected systematically across a number of lakes. When this is done, the additional variance explained by zooplankton abundance is negligible (Pace 1984; Yan 1986). On this basis, the theory that top-down control influences chlorophyll development in lakes must be rejected.

Carpenter *et al.* (1985) reject the rejection. They are so convinced by simulation modeling, microcosm experiments and the plausibility of the mechanism that they wish to pursue this cause-and-effect relationship in 'experimental manipulation of food webs' despite the negative evidence in the preceding paragraph and their own arguments that the mechanisms are likely undetectable in natural patterns of phytoplankton and zooplankton abundance. Many such decisions could swamp ecology with a multitude of inconsequential, process-oriented investigations which are studied for themselves because they cannot be related to any observable phenomena outside of the experimental system. Pramer (1985) calls this terminal science since it leads only to more science.

Nature–nurture

The relative importance of genetics and environment in determining biological processes has long been a bone of contention in evolutionary ecology, in animal behaviour, and especially in sociology. The 'nature–nurture controversy' arises because all biological phenomena result from the interaction of genetic potential with environmental constraint, such that it would be impossible to identify a purely genetic or a purely environmental process in a living organism. The multiplicity of causal pathways, and therefore of scientific interpretations and emphases, is further confounded because 'genetics' and 'environment' are such broad concepts that they can be infinitely subdivided. In short, searching for the cause of a biological process is an infinite process and there is no reason to suppose that this will be simplified by division of all potential factors into genetics or environment.

Instrumentalism would solve (or rather dissolve) the problem by rephrasing it to ask what regularities involve the process of interest: what are its correlates and which of these allow us to manipulate it. Since science cannot go further, this is where we should stop.

To take a particular and highly charged example, Kempthorne (1978) discusses the balance between nature and nurture in determining human intelligence quotients. Because the environment cannot be defined or controlled, because genetics always exist in an environment, because genetic and environmental factors are always colinear, and because we lack relevant manipulation studies, Kempthorne concludes that the question cannot be resolved. This is an instrumentalist position. IQ-scores have many correlates including wealth, social status, race, parental score, degree and quality of schooling and so forth. We need therefore ask what are the results of manipulations of those factors which can be manipulated and then use the most effective one to alter levels, if this is our political will. Since we cannot decide if these responses reflect innate or environmental causes, the question of causal attribution and the sense of shame or stigma that arises when cause is attributed to racial or social stereotype should not arise.

There is an even more crucial question in the IQ controversy which is also subject to an instrumentalist solution: 'What do the scores mean?'. For the instrumentalist, this question can only be reinterpreted to ask if IQ is a good index of other characteristics of greater concern. That scientific questions must be answered before the thorny moral and political questions involved in manipulation need be confronted. In Kempthorne's opinion, such regularities do not exist so the questions of morality and policy can be relegated to science fiction.

Density dependence and independence

The roles of density-dependent and density-independent control in population ecology have been the focus of bitter and protracted debates in ecology. Like many ecological debates, the arguments are characterized by fuzziness of the basic concepts (for example, the terms are sometimes associated with a division between biotic and abiotic, respectively). Like many causal debates, observed regularities are reinterpreted as 'really' density-dependent or 'really' density-independent (despite appearances to the contrary) by analyzing some other part of the causal nexus in which the phenomenon is embedded. Finally, like many pointless arguments in ecology, the issue can never be decided because it requires observation of an unobservable: the real cause of population regulation.

For the instrumentalist, the problem is simple. Assuming that one wants to predict population abundance or rate of population growth, questions about density-dependence only ask if present population size is a useful predictor (Pielou 1974). Density-independent relations do not use population abundance to make the prediction (e.g. Kajak 1980), density-dependent relations (e.g. Roff 1983) do. The cause of such predictive success is irrelevant, but the predictive success is real. Similarly, biotic relations invoke biological properties as preconditions and predictor variables (e.g. Oglesby 1977) whereas abiotic relations do not (e.g. Dillon & Rigler 1973). However, since our interest should be in making a prediction rather than classifying the kinds of regularities, our more successful predictions are likely to involve both biotic and abiotic factors (Morin & Peters 1988; Godbout & Peters 1988). Debate arises only when we ask not 'What works?' but 'What really happens?'.

Summary – The twin perils of mechanistic and causal explanations

Sir Peter Medawar (1967) called science the art of the soluble, because one of the secrets of successful science is to ask important questions that can be answered. Ecologists have ignored this advice for fifty years and asked seemingly important questions that can never be answered. Perhaps, it is time to try another tack.

Non-predictive constructs in ecology are often defended as causal or mechanistic explanations because such explanations, despite weak predictive powers, successfully reduce the observations to the processes that really underlie appearances. This chapter challenges that view by arguing that theories are only tools to describe, predict and manipulate nature. The chapter argues against the view that effective research must break observed phenomena into separate mechanisms or cause–effect relations, because it is unlikely that these can ever be reassembled into a functioning whole.

To demonstrate the unlikelihood of reassembly, system models are examined in detail because they represent the only device that could integrate the many different processes now under study. Integration into systems models is implied by many narrow ecological studies, but modelers have already realized that limitations of the approach, like unwieldy complexity, error propagation, and goal confusion, make whole ecosystem synthesis impossible. Many specialists have yet to assess the importance of this discovery for their own work.

Although the interpretation of scientific theories as tools seems to invoke the notion of cause, causality has so many, easily confused interpretations that it is easier to avoid causal terminology. This would involve some changes in research problems, strategies and interpretations, but not in methods of study. In compensation, some of the longest-standing and most contentious issues in ecology evaporate when cause, mechanism, and explanation are ignored.

6 · Historical explanation and understanding

There is a fundamental difference between the intellectual endeavors of history and science. Science develops theories for prediction and scientific explanation. History develops historical explanations not to predict but to understand selected phenomena as events in the ontogeny of the universe.

The distinction between history and science should not be confounded with the smaller difference between historical and non-historical sciences. All phenomena have a history, and all sciences deal with that temporal dimension when it allows more effective prediction and explanation. Nevertheless, some sciences are more likely to use this dimension in theories than others. Thus ecology, like paleontology, geology and evolutionary biology, is often referred to as an historical science, in contrast to molecular biology, physics and chemistry which make fewer references to the history of the system under study. The difference between historical and non-historical sciences represents a distinction in what has been found to constitute a sufficient basis for prediction and scientific explanation in different disciplines. The gulf between history and science instead represents a difference in the kind of explanation acceptable to the two disciplines (O'Hara 1988).

This chapter maintains that contemporary ecology invokes both scientific and historical explanations, and that the latter confound scientific theories. The purpose of the chapter is therefore to describe both categories, to distinguish different types of historical explanation, to demonstrate these types with examples from history and ecology, and to provide a rationale for the coexistence of historical and scientific explanations by distinguishing between a science of ecology and an art of natural history, explaining the domain of each and describing the dangers of their confusion. Historical explanation represents a precious part of the human experience, but it debilitates science if it replaces predictive knowledge and confuses scientific explanation.

Scientific explanation and understanding

It is a commonplace that when we explain observed phenomena, we come to understand the world around us. Nevertheless, the meanings of

147

these phrases are obscure. We rarely ask what we mean by explanation or how we know when we have achieved understanding.

In science, these questions can be resolved by references to predictive ability. An explanation explains an observation by showing that the observation could have been predicted by an existing theory or law. Thus an explanation consists of showing that a phenomenon is a particular case of a known regularity. Explanation differs from prediction only by the order in which the theory and the observation are invoked. In prediction, the theory is used to identify the probable observation and subsequently this prediction may be compared with the actual observation. In explanation, the observation is already in hand and we seek to show that this observation could have been expected by referring to the appropriate scientific theory. This reference simultaneously explains why some other logically possible but theoretically improbable observation did not occur. Thus scientific explanation explains both the observation of the probable event and the non-observation of possible but improbable events as instances of known theories (Hempel 1962, 1966; Popper 1960). In this context, understanding is the command of theory that allows one to make relevant predictions and explanations.

This view of scientific explanation is consistent with Popper's demarcation between science and non-science, for it insures that any theory which explains a phenomenon can also predict it. Such explanations are referred to as 'covering law explanations' because they explain by reference to widely accepted theories or laws (Dray 1964).

The terms 'explanation' and 'understanding' are used far more broadly than this in both everyday speech and the ecological literature, but their use as scientific terms should be restricted to that outlined above. Any wider definition would introduce a double standard in which both constructs that predict and those which cannot are called scientific theories. Constructs that cannot predict but explain in some other sense cannot eliminate logically possible observations as improbable. As a result, such constructs are not subject to the objective tests of science: they are not asked to resolve uncertainties about the external world and thus cannot be culled from the literature as unpredictive (Chapter 2). As a result, they persist, confusing our standards for acceptable theories, and diluting our tiny body of predictive tools.

Scientific explanation in history

Covering law explanations are well known in the historical literature, because Hempel (Hempel 1942; Hempel & Oppenheim 1948) recommended them as the model for explanation in all areas of human experience and particularly in history. Historians replied that this model could not apply to history because there are almost no general historical laws on which to base such explanations. They argued that historians offer different sorts of 'explanations' which have been almost universally accepted as satisfying bases for understanding, but not prediction (Scriven 1959b, 1962; Donagan 1964; Dray 1964, 1966).

Critics of covering law explanations also showed that many explanations in science did not seem to fit Hempel's model (Scriven 1959b; Hull 1981). Hempel (1966) maintained that such explanations in history were only 'explanation sketches' which made reference to half-formed, qualitative generalizations in the absence of general, historical laws. In science, he held that such explanations contain implicit references to theory and boundary conditions, even if these elements were not specified. Otherwise, Hempel felt that historical and scientific explanations should fit the same model.

Hempel was overbearing in his insistence of a single form of explanation, and the debate often took on semantic overtones as both sides argued about the nature of 'true' explanation. A less-confrontational approach accepts that very different constructs are termed 'explanations' and determines what the differences are, what advantages the different constructs offer, and where these constructs are best applied. Those aspects of the debate are examined here.

Hempel's covering law explanations preserve predictive power which is the hallmark of science. This primacy must be protected if the science is to remain informative (Chapter 2). Therefore the first step in this chapter is to examine the nature of explanation in the science of ecology. The historians' alternative forms of non-predictive explanation are then examined to show that homologues are common in the ecological literature where they may confound scientific explanation if used carelessly. However, non-predictive explanations also form a basis for the care of unique systems and can promote that more personal and aesthetic form of understanding that is 'natural history', a phrase which merits more literal interpretation than is usually the case.

Scientific and historical laws in ecology

The covering law explanation invokes 'general laws' which are defined as universal statements, unbounded by space or time. Such laws would also have explanatory force in the historical sciences, but these sciences also rely on several types of 'historical laws' (Thompson 1983). One type is represented by 'empirical rules': Patterns observed in the past that allow both prediction, on the assumption that the patterns persist, and explanation, by showing that a particular case is consistent with known regularities. A second type consists of 'developmental laws' which hold that certain temporal sequences recur, as early successional stages lead to later ones and that both explanation and prediction can be made by reference to this recurrent sequence. A last type employs entities that invoke temporal concepts, like ancestor or relict; but since variables or theories built from such concepts seem no more dubious than others, this class of theory holds little of special interest for this discussion and is not considered further.

Universal laws

Ideally, the laws that Hempel would see invoked in historical explanation are universals (Hempel 1942) that apply at all places and times and are therefore outside any historical context. Although the operational meaning of 'universality' is never quite clear (the term seems metaphysical), the great laws of physics and chemistry are presumably such laws. Since historical material is always constrained in time and space, a universal law is rarely the focus of history or historical sciences; these disciplines are more likely to consume general laws, than to produce them (Joynt & Rescher 1961). For example, the defenestrations of Prague, whereby the regime disposed of undesirables by pushing them out a window, depended upon the law of gravity, but historians rightly ignore that law as historically uninteresting. Similarly, the fundamental role of phosphorus in the productivity of freshwaters undoubtedly reflects the chemical properties of the atom but these properties are never explicitly considered in studies of eutrophication. Physical and chemical laws are invoked in ecology, for example, in physiological autecology (Alexander 1982; Gates 1980) or in discussions of the fitness of the environment (Henderson 1913), but distinctly ecological laws are often more frequent. In the context of the role of historical explanation in ecology, universal laws serve as a point of reference in the examination of other explanatory modes, rather than as examples of ecological practice.

Empirical laws

Perhaps the most familiar class of historical laws simply restate empirical patterns which are presumed likely to recur because they have recurred in the past. In history, such a rule might claim that repression encourages civil unrest. An ecological analogue might state that increases in plant biomass lead to increases in herbivore biomass (White 1978). An evolutionary example is Dollo's law; phylogeny is irreversible (Rensch 1960).

If tested, these brief statements would all prove false. Some acts of repression, like the destruction of Carthage, helped impose order, in that case the Pax Romana. The reestablishment of the lighter phase of the peppermoth (*Biston*) following reductions in air pollution is an exception to Dollo's law, and the rise in herbivores with primary producers is a statistical trend not a hard, deterministic rule (Lynch & Shapiro 1981) and likely does not apply in comparisons involving grossly different plant life forms, like forests and grasslands (Peters 1980a). Thus these should perhaps be seen as incomplete 'explanation sketches' which would necessarily be replaced by more explicit models before sound explanations or predictions could be made.

This process of redefinition and precision of empirically derived hypotheses is a case of the hypothetico-deductive process in action (Chapter 2). In the case of the putative plant–herbivore relation, this process led to a series of statistical relations that describe different relations for different community types (Coe *et al.* 1976); Andrzejewska & Gyllenberg 1980; Kajak 1980; McCauley & Kalff 1981; East 1984), but such redefinition is considered less necessary in history or evolution, because the patterns are presumed to have many exceptions.

Empirical rules seem unacceptable as universal laws because they are based on restricted samples in space and time, so there is always a possibility that the regularity will cease to apply. For example, we cannot assume that dinosaurs were poikilothermic on the basis of evidence from contemporary reptiles (Farlow 1987). In future, the present, positive relations between plant and herbivore biomasses may change or cease to exist as plants improve their defenses. Empirical rules in both biology and history can be called 'laws of limited generality' (Dray 1964). However, the laws of thermodynamics are accepted as universal laws for scientific explanations, even though they are 'only' warranted by a very restricted period of observation. Regardless of the name one applies, empirical ecological rules can serve in scientific explanations, because such explanations need only relate particular observations to known regularities.

The question of universality is not important in practice, for both empirical rules and universal laws must be treated similarly. Laws, whether universal or empirical, are always tentative and subject to reformulation. In fact, none has remained unchanged for more than a few hundred years so that the threat of obsolescence by evolutionary processes (e.g. pesticide resistance in insects) is only rarely important at this stage of ecological law-making. Moreover, even explanations involving the laws of physics require some additional information, like measurement of the initial conditions, ascertainment of boundary conditions, and frequently specification of ad hoc rules and premises. Consequently, the explanatory capacity of physics also depends on particular spatio-temporal conditions (Hull 1981). In short, limited generality does not bar the generalizations of history or ecology from use in a modified form of covering law explanations.

Developmental laws

Developmental laws seek to explain by reference to a regular pattern of development (Drury & Nisbet 1973). Such a regular sequence allows both the prediction of future observations and the explanation of past ones, because each stage is presumed to be replaced by the next member of the series. The consistency of the developmental pattern is therefore critical.

When disruptions of these patterns are rare and the time scales of the development short, the complete pattern may have been observed many times. The developmental theory need simply describe this empirical regularity. As a result, many common phenological and ontogenetic processes, like the leafing of trees in spring (Lechowicz 1984), the succession of phytoplankton (Sommer *et al.* 1986; Marshall & Peters 1989) or the development of an organism (Taylor 1968; Case 1978), serve in ecological explanations. Such simple cases are not considered further in this section.

Developmental theories for incomplete or frequently disrupted phenomena bear closer examination. When a developmental pattern is incomplete, parts of the patterns are often inferred rather than observed. When disruptions or novel phenomena occur at the same temporal scale as the main pattern, frequent exceptions to the rule must be explained away. Despite these potential problems, such laws have been invoked for many long, or disrupted, temporal sequences. The increasing size of evolving horses might be explained by Cope's rule that individual size tends to increase with phylogenetic descent; the great antlers of the extinct Irish elk (*Megaloceros*) can be explained by an allometric relation between the size of cervids and their antlers (Gould 1974); the climatic theory of succession, that all plant communities in an area are moving towards the local climax, explains the diversity of plant communities in a region; and teleological and teleonomical arguments relating current animal behaviour to future goals (Thompson 1987) could be construed as developmental laws. Such rules predict only to the extent that they specify the probabilities and time limits for observing subsequent members of the sequence. If this is not the case, no prediction can be made and no covering law explanation offered. Unfortunately, this specification is rare. Each phylogeny is considered independent and unique, so that exceptions to Cope's rule present no difficulty in explanation (Rensch 1960; Stanley 1973), however much they vitiate prediction; the concept of a single, climatically controlled climax succession has proven ineffective because exceptions are common (Whittaker 1953; Drury & Nisbet 1973), and behavioral goals are difficult to define without anthropomorphism and tautology (Thompson 1987), whereas the patterns themselves depend on an unobservable state, motivation.

Drury & Nisbet (1973) suggest some adverse affects that reliance on developmental laws may have for ecology. Such laws invoke the developmental pattern as an endogenous process or one imposed by higher elements in an environmental hierarchy. This decouples the system from part or all of its environment and may discourage study of the interactions involving the developing entity. For example, succession theory had little place for the impact of animals on the plant community (McNaughton 1984; Owen-Smith 1987), nor of plants on climate (Geiger 1965; Charney, Stone & Quirck 1975). The presumed existence of an orderly sequence of stages enhances a tendency to typological thinking, thereby encouraging analysis of continuous processes as a succession of discrete types. The models which emerge from such analyses are frequently highly qualitative, infinitely detailed, typologically specific, and weakly extensible. They narrow the vision of researchers and therefore hinder the development of alternative theories. The potential for study of individual components of the types becomes irrelevant and the study of seemingly anomalous sites is discouraged. In addition, developmental laws discourage human intervention, because opposition to the natural tendency of events is either unnatural or futile, or because no action, short of preserving the whole, is likely to prove effective.

In practice, many non-empirical developmental theories have been so ineffective that they are no longer thought to predict, even though they are still used to explain. Animal ecologists still use successional language, even if botanical ecologists have been circumspect in this area for over thirty years (Drury & Nisbet 1971, 1973). As insidiously, succession still forms a part of the teaching of ecology at all levels of the educational system.

Historians and philosophers (Gardiner 1959; Popper 1960; Dray 1964; Kochanski 1973) have generally rejected developmental laws in history, like those of Karl Marx, Oswald Spengler and Arnold Toynbee, and for similar reasons. The predictions have either proven to be wrong or the theories have been distorted to make them tautologically true: the patterns hold, except where they do not. Moreover, at a larger scale, exceptions to evolutionary and successional patterns are particularly interesting because of the novel elements they reveal over the course of development. These novelties are not predictable, but they are still targets for historical explanation. Therefore, to the extent that developmental laws must treat these novelties, developmental laws for historical processes cannot exist (Popper 1960; Kochanski 1973; Thompson 1983).

Drury & Nisbet (1971) recommend that developmental laws be replaced by a form of mechanism they term 'kinetic relationships'. These relations would deal with interactions and changes of the parts of a system on circumscribed time scales without reference to a proposed endpoint, so that the descriptions remain testable. Horn's (1981a) successional models based on tree-by-tree replacement might be one example. The appropriate distinction is not, I think, between mechanistic (or kinetic) and organismic models, but between patterns that are empirically established and those that are assumed. Small-scale developmental laws can be said to succeed because they have passed the empirical testing that large-scale developmental laws either avoid or fail.

Historical explanations and ecology

The first section in this chapter maintains that scientific explanations are possible in historical sciences if these explanations depend on universal covering laws or limited generalizations. However, such scientific explanations do not represent the full range of explanatory modes in ecology. Ecologists must learn to identify these non-scientific explanations so that their science will not be confused. This section characterizes six facets of non-predictive, historical explanation in ecology and then looks at the dangers they present.

MacArthur (1972a) distinguished historical from non-historical explanation as the distinction between researchers who look for similarities and those who look for differences. MacArthur (1972a) included ecologists who seek patterns, and thereby do science, among the former and many paleontologists and biogeographers, among the latter. This characterization seems unnecessarily provocative for MacArthur would recognize a number of biogeographical patterns, including his own analysis of island biogeography, as science. On the other hand, so much of ecology involves the study of exceptions and differences that the status of all ecologists must be open to critical scrutiny. In any case, MacArthur's dichotomy removed historical patterns from scientific consideration in ecology (Kingsland 1985), because particulars and unique characteristics cannot be reduced to the patterns of similarity that constitute science and can only be understood through historical explanations. This is a distinction I would support.

Six historical modes of explanation

Historical explanation is a vast topic (Gardiner 1959; Dray 1964, 1966), so eight classes of historical explanation identified by Dray (1964) are used to structure the present discussion. Dray's categories include the covering law explanations and limited generalizations discussed in the previous section, plus normic, genetic, colligative, narrative, individual and rational explanations. The first two modes have scientific counter-parts, but the last six are non-scientific because they promote an understanding based on historical explanation not prediction. This subsection characterizes these six modes, relating them to examples from history and ecology.

Normic explanations

Normic explanations have the form of a covering law explanation but use a 'norm' or truism in place of the law (Scriven 1959a). In history, such truisms use our knowledge of human nature to explain historical behaviour so that it becomes plausible that humans behaved as the historical record shows them to have done in the circumstances they encountered. For example, we might sometimes explain past examples of parental sacrifice as instances of familial affection, but in other circumstances, we might invoke the dehumanizing effect of extreme poverty to explain parental infanticides. Truisms may also be invoked to explain the history of human events. The aphorism 'haste makes waste' may partly explain the premature and tragically costly raid on German installations at Dieppe on the Normandy coast in 1942, but failure to recognize that 'a stitch in time saves nine' may as well explain the unwillingness of the world community to contain Hitler as Germany rearmed and expanded prior to 1939.

Such explanations differ from covering law explanations because they do not suppose that the observed behaviour could have been predicted: Parental sacrifice is not a law, and many children survive despite excruciating poverty. Inconsistency is no impediment to the use of aphorisms and commonsense notions in explanation, because the events themselves are known and this knowledge helps us decide which truisms are appropriate. However, because we do not know *a priori* which truism to use, truisms cannot predict. Normic explanations succeed because they show that the facts of history fall within the normal range of human behaviour, even if that behaviour is unpredictable.

Weaver (1964) argued that much of non-scientific explanation depends on the use of appropriate metaphors and similes to make us feel 'intellectually comfortable' with the unfamiliar. He termed this feeling 'understanding'. As an example, he described how we might explain interference in the transmission of radio waves by reference to the patterns of overlap in water waves meeting in the lee of an island. Ecologically, we might try to explain competition by reference to market practices or compare population growth to compound interest. Although analogies can explain in this sense and may be useful heuristically, analogy is no basis for prediction (Weaver 1964; Ollason 1987), because we do not know which analogy is appropriate. Instead, explanations based on analogy are a form of normic explanation in which an unfamiliar phenomenon is explained by reference to the norm provided by an unrelated, but familiar, phenomenon.

Because commonsense notions can be applied to the living world beyond man and because ecologists have in addition a vast literature describing potentially paradigmatic analogues from other ecological systems, ecology can avail itself of this same explanatory mode. For example, Southwood (1981) explains the gliding flight of albatrosses, condors, and *Morpho* butterflies as instances of the energetic efficiency of K-selected species, although he does not presume that all K-selected fliers are gliders or that all gliders are K-selected. Thus Southwood explains unpredictable phenomena by reference to a putative common property of K-selected animals, energetic efficiency. Lowered fitness in inbred lines or of some inbred offspring can be easily attributed to 'inbreeding depression' but the accomplishments of selective breeding show that some inbred offspring and their lines are spectacularly successful (Horn 1981b). Inbreeding is better used to explain depression when it is observed, than to predict it beforehand.

Adaptationists have proven particularly zealous in offering normic explanations based on the truism that everything is adaptive. For example, *The American Naturalist* once featured a debate which sought to explain the protuberance of human breasts (Leblanc & Barnes 1974; Kuttner 1975), although none of the participants pretended to predict this phenomenon. Similarly, Moynihan (1985) offered an engaging adaptationist explanation of deafness in cephalopods, but it is improbable that this could be predicted, for cephalopods may not be deaf after all (Hanlon & Bedelman 1987).

Normic explanations are also invoked as the 'ghost of competition past' (Connell 1980) which explains present differences in terms of past competitions. For example, the differences between sympatric species of rock nuthatch (*Sitta*) has been interpreted as competitively driven, character divergence (Brown & Wilson 1956), and non-overlapping local distributions of chipmunks (*Tamias*) as competitive exclusion (Brown & Bowers 1984). These explanations can be advanced, even though the components of the explanation are irretrievably lost in the past, because they are part of normal ecological discourse.

Sometimes the implications of these hypotheses can be evaluated: the difference between co-occurring species of *Sitta* is more parsimoniously interpreted as a coincidental overlap of geographical trends than as character displacement (Grant 1972a, 1975), but competitive exclusion in chipmunks is consistent with direct interference observed among different species (Brown & Bowers 1984). Because the processes which actually led to the present patterns are no longer observable, both 'tests' invoke modern norms to explain the patterns. Such tests are only as strong as our belief in Ockham's razor and uniformitarianism. This is not a major difficulty in normic explanation because these assumptions can be set aside if need be. Otherwise they add to the normality and plausibility of the explanation. This flexibility allows the development of norms for any set of observations, but makes any predictive use of the norms impossible.

Genetic explanations

Gallie (1959) proposed a related form of explanation which he called 'genetic' explanation, because he believed that a similar form was used in the genetic or historical sciences. Genetic explanations explain an observation by relating it to a particularly crucial aspect of the prior state. Thus genetic explanations focus on one or a few necessary preconditions, even though these preconditions are insufficient to allow prediction of the observation. Such explanations bind series of events into units comprising both the observation and the proposed precondition(s). These units form temporal wholes and are one source of norms for subsequent normic explanations.

The most common type of causal attribution in mature sciences, like physics (Kuhn 1977), is a form of genetic explanation. When an experiment fails due to some ineffectively controlled, extraneous factor, as humidity or static electricity affected the rate of fall in Millikan's oil drop experiments (Franklin 1986), the physicist attributes cause. Thus, in physics, causal attribution occurs when expectations are not fulfilled, and genetic explanation is used to give any associated blame or merit to one of many contributing causes. Ecologists often use this sort of argument in explaining the results of their field programs by reference to 'unusual' events (Weatherhead 1986).

Gallie (1959) provides an evolutionary example of genetic explanation in the familiar story of the phylogeny of the giraffe which is presumed to have developed its long neck because long-necked individuals held a selective advantage in collecting food from trees. Obviously this is far from a complete account of the long neck of the giraffe. It makes no mention of the other necessary adjustments, for example in the circulatory system or the legs; it does not explain why only giraffes and no other African herbivores evolved this adaptation; and it does not explain why the giraffe developed this adaptation rather than another, like the trunk of an elephant. In short, this is a genetic explanation because it focusses on only one salient part of the process and offers this partial story as an explanation. It claims no predictive power.

Scriven (1959b) takes a very similar position in his own analysis of evolutionary explanation; he argues that, although evolutionary predictions are impossible because we can predict neither the genetic potential of the organisms nor the relevant changes in their environment, this should not impede our interpretation and explanation of phylogenetic patterns after the fact.

Genetic explanations include all those which apply only if 'all else is equal'. Since this is never the case, such explanations serve not to predict new instances of the observation but only to view past observation in the light of certain processes. They are often used to explain observations from the relatively complex situations in the field in terms of carefully controlled laboratory experiments with single factors or in terms of abstract models of isolated mechanisms. To cite one example from 20 'Darwinian predictions' about human social behaviour, Alexander (1977) claims that 'all else being equal, close relatives are favoured'. In

Table 6.1 *Summary of a model of prey choice (Stephens & Krebs 1986).*

Assumptions

Decision:
   The predator invokes this model to determine if it should attack each prey type upon encounter.
Currency:
   The predator seeks to maximize its long-term average rate of energy intake.
Constraints:
   (1) Searching and handling are mutually exclusive.
   (2) Prey are randomly distributed and sequentially encountered.
   (3) Net energy gain, handling time and encounter time for each prey type are fixed.
   (4) Encounter without attack takes no time and has no energetic costs.
   (5) The predator knows the model's parameters, recognizes the prey types and does not learn. It has complete information.

Deductions

   (1) The zero-one rule: either prey are taken upon every encounter or never taken.
   (2) Prey types are ranked by 'profitability', the ratio of energy gain per prey to handling time, and are added to the diet in order of these ranks.
   (3) The inclusion of each prey type depends only on its ranking.

Note:
The deductions are logical consequences of the assumptions, and should not be described as predictions (Chapter 3).

practice, observations that fit this pattern are taken as evidence that all else was equal enough and observations that do not fit the pattern are explained away as instances where all else was too different.

   Simple optimality models provide a further example of genetic explanation. These models abstract the system by assuming that no constraints prohibit optimal solutions, that the process under consideration is independent of other exigencies of existence, that the optimal solution can be measured in a stated currency, and so forth (Table 6.1). When the model fits observation, this is evidence that the assumptions have held and the model is taken to explain the observations. If the fit is poor, other assumptions are introduced until the fit is improved; then the modified model is taken as a new explanation (Gray 1987; Ollason 1987).

Because we do not know what model to apply without determining which describes the observations, the models cannot predict (Beatty 1980). Instead they provide a simple, logical framework within which the observations can be interpreted and, in that sense, explained. Because the failure of some optimal foraging models is acceptable and expected in any particular situation, the models are not falsified by failure. Thus optimal foraging theory, like theories invoking the principle of evolution by natural selection and theories based on the Lotka-Volterra equations (Chapter 3), is a tautology because it admits all possible observations. This weakness is harder to see in more complicated models but it is not removed.

The difference between scientific and genetic explanations does not lie in the need for abstraction or simplification, for all explanations ignore many situational details. The essence of the difference is that genetic explanations explain only when prior events are shown to fit the explanation. They accept the possibility that other instances will not fit. Genetic explanations therefore neither explain why the inevitable contextual differences of similar events should not always yield different results nor indicate when contexts are so similar that similar results can be expected. Scientific explanations instead presume sufficiency and therefore the explanation holds not just if all else is equal but also if all else is subject to normal variations (i.e. 19 times out of 20).

Colligative explanations

Historians sometimes explain an historical period or event by showing 'what it really amounts to' (Dray 1959). The explanation achieved thereby casts seemingly unrelated events as parts of a single process. In history, this might be achieved by showing that contemporaneous changes in art, religion, technology, political institutions, and trade could all be related to a general movement, like the renaissance of the fifteenth century or the enlightenment of the eighteenth. In such colligations, the historian explains by an interpretive narrative that shows 'what' constitutes the event, rather than 'why' the event occurred. Hull (1981) has discussed this sort of explanation as 'integrating explanations' – the term is from Goudge (1961) – and argues that, although integrative or colligative integrations may include reference to general laws, the explanatory force results at least as much from other elements the author provides in the historical account.

Colligative explanations are common practice in popular, nonfiction. Books like Charles Reich's (1970) '*The Greening of America*',

Alvin Toffler's (1970) '*Future Shock*', and even Charles Darwin's (1859) '*The Origin of the Species*' succeed in large part because they are colligative. They show that a vast number of seemingly disparate and independent events can be interpreted as elements of a single process. Darwin's contribution, which is by far the greatest in this purposely heterogeneous listing, was recognized not simply because it described the principle of natural selection. This principle had been adumbrated many times, as Darwin acknowledged in later editions of this book. Darwin succeeded where others, including Herbert Spencer and Alfred Wallace, failed because, in the *Origin* and in his later work, Darwin showed over and over again how the principle could be used to explain the varied facts of nature (Ghiselin 1969). Darwin himself made the point in the last chapter of the *Origin*, 'It can hardly be supposed that a false theory would explain, in so satisfactory a manner as does the theory of natural selection, the several large classes of facts above specified.'

Antonovics (1987) provides an in-depth analysis of the ill effects of colligative explanations in his call for an evolutionary 'dys-synthesis'. He maintains that the very success of 'the new synthesis' in evolutionary ecology during the last four decades has, in some respects, proven a disservice to ecology and evolutionary biology by entrenching a monistic and complacent orthodoxy which discourages the separation of science from history, suppresses alternate views, approaches, mechanisms and techniques, and stifles scientific research by offering facile explanations and non-quantitative, non-rigorous model solutions.

In contemporary ecology, a number of academic schools reinterpret sets of diverse phenomena around a given theme and thereby use colligative explanation. For example, hierarchy theory has been advanced to explain inconsistencies among behaviours at different scales and applied to problems as different as stability–diversity relations and scale in algology (Allen & Starr 1982). Sociobiologists have treated phenomena as diverse as altruism, parental behaviour, mate selection, and, perhaps less happily, human homosexuality and warfare (Wilson 1975). Optimality theory has been applied to problems in foraging (Stephens & Krebs 1986), digestion (Penry & Jumars 1987), mating competition (Parker 1978), and life history phenomena (Stearns & Koella 1986), competition theory to problems of resource use (Tilman 1982), biogeography (Diamond 1975), morphology (Grant & Schluter 1984), and habit selection (Brown & Bowers 1984). Frequent recourse to the same explanatory model gives these constructs greater credibility than they could merit individually, yet the more frequently such explanations are advanced the less critical we are likely to be about accepting them (Stephens & Krebs 1986). This leads to the development of 'just-so' stories which appear to confirm and establish a generality for phenomena like competition and natural selection, out of all proportion to the strength of the scientific evidence (Gould & Lewontin 1979; Connell 1980; Simberloff 1982; Wiens 1983).

One characteristic of colligative explanations in ecology is the presentation of tables of essentially qualitative or bibliographic information describing the range of phenomena that have been explained by the approach (e.g. Schoener 1983; Peters 1983; Endler 1986; Stephens & Krebs 1986; Stearns & Schmidt-Hempel 1987). This is an effective means of summarizing a large literature, but since the tables are too condensed to allow much evaluation, these tables also serve as elements in colligative explanation.

Colligative explanations may also suffer from the overenthusiasm of their proponents which leads to explanations that are either forced or precious. For example, when a scientist is confronted by falsification of an hypothesis, advocates of hierarchy theory suggest that 'the scientist either changes the hypothesized observed holon filters (he erects a new model) or changes his own observation filter with a new experimental design, having presumed that the signal was there but he missed it' (Allen & Starr 1982). For myself, neither 'erecting a new model' nor 'trying another method' gains by these reformulations.

Colligative explanations draw their strength from the range and diversity of the phenomena they can explain. This power also makes them difficult to criticize because they can involve the critic in such different areas. For example, one reason that Velikofsky's (1950) heretical cosmology lasted so long was that he invoked evidence from so many fields (astronomy, geology, history, ethnography, biology) that only a co-ordinated team could expertly refute more than a fraction of his claims simultaneously. As colligative explanations become more diverse, an assemblage of weak individual cases may provide acceptable evidence for the argument as a whole, because precision in individual explanations may be sacrificed for the simplicity of a single general explanation. This renders the colligative argument difficult to reject because a miscellany of alternative, case-by-case interpretations of many components can rarely refute strong colligative explanations convincingly. Instead, criticism of colligative explanations that treat each part to independent analysis usually seem cavilling and picayune.

Despite these drawbacks, colligation plays an important role in organizing knowledge. For the critic, the task is to see beyond this framework to judge each theoretical element in terms of its predictive power and to use the colligative explanation as an organizer of good theories rather than a defense for bad ones.

Narrative explanations

History, unlike science, finds the unique aspects of different events at least as interesting as what they have in common. Thus we may be as attracted to read histories of Alexander the Great, Julius Caesar, or Napoleon by what distinguishes these men from one another and from the common man as by their similarity as warriors and leaders. Oakeshott (1966) focussed on this uniqueness in developing his alternative to covering law explanations in history. In his view, the historian should try to explain a period of history by describing it in greater and greater detail until no lacunae remain. The goal of history is therefore an unseamed account of the past. Goudge (1961) used the term 'narrative explanation' for a similar concept.

Narrative explanations differ from genetic explanations because they deny the distinction between crucial and other necessary conditions. Instead, narrative explanations presume that it is the historians's task to draw the whole together. Similarly, they differ from colligative explanations by resisting the temptation to distinguish among parts of the whole. Thus the freedom of the renaissance and its artistic glories should not be allowed to obscure the barbarism of its internecine wars, its primitive sanitation and medicine, or the dismal state of women and the peasantry. Narrative explanations differ from normic explanations and from appeals to either limited generalizations or covering laws, because narrative explanations treat history as a unique sequence rather than the composite result of general laws, special conditions and chance events. For Oakeshott, any attempt to apply covering laws in history was an abandonment of history.

Ecologist use generalities more often than do historians, nevertheless some modes of ecological explanation approach narrative explanations in their attention to detail. Ecological narratives attempt to account for an observed ecological sequence or system, and thereby to make sense of all our observations, by finding them a place. This approach is particularly likely when ecologists deal with a single site or system, because the amount of information available about the system is large, but the number of degrees of freedom associated with any observation is small so that there is no way to isolate significant from insignificant casual factors.

As a result, everything that was observed is reported under titles like '*The Ecology of X National Park*' or '*The Limnology of Lake Y*'. Some simulation models similarly strive to find a place for everything, not because everything needs to be considered for some end, but simply because each thing is there and to leave it out would make the model incomplete (e.g. White 1984; Goda & Matsuoka 1986).

The desire for completeness of narrative explanations holds special problems for predictive ecological theories in general and for environmental science in particular. Predictive models necessarily ignore many aspects of the ecology of a system to make a specific prediction and therefore offend those who feel ecology can consider everything, all at once. This belief in the possibility of complete knowledge is reflected in criticism of ill-fated environmental decisions, as the early decisions to apply DDT globally were criticized because they did not consider the then unknown dangers of bioaccumulation, or the faulting of eutrophication models for their inability to predict other phenomena, like shoreline erosion, mercury mobilization, and fisheries collapse in a newly constructed subarctic reservoir (Lehman 1986). The presumption that sciences can offer a predictive analogue of narrative explanations is implicit in requirement for 'comprehensive assessments' in environmental impact statements. In practice, any study can assess only some aspects of the system and data only gain meaning when predictive theories are available to relate those data to selected aspects of the system and to alternate management policies.

Individual explanations

Individual explanations preserve some of the uniqueness of narrative explanations, but cast this in the form of a covering law explanation. However, in the place of general theories, individual explanations use statements about the propensities of historical entities. Thus, the Danish conquest of England at the beginning of the second millennium may be related to the ineffectiveness of the English King, Ethelred the Unready, the rise of Britain as a naval power to the resolution of Elizabeth I, and the last crusades to the piety and bigotry of Louis IX of France. Entities other than individual men can also be considered, if their characteristics can be individuated. For example, the Jewish holocaust may be explained by the anti-semitism of Hitler, the National Socialist hierarchy, the party members, the Germans, or the international community, depending on the level of historical entity one wishes to invoke.

Ecology also has suites of entities with individual characters which can be used to provide explanations. For example, colonizing ability is often related to taxonomic differences and each hierarchical level has some individual characteristics. At a gross scale, flight explains the greater abilities of birds, bats and insects to invade oceanic islands; drought resistance explains the advantage of reptiles and mammals over amphibians in rafting, and low food demands that of reptiles over mammals. Within these groups, certain species are considered more capable colonizers and others less; the supertramps, tramps, and high-S birds of Diamond (1975). Indeed, a large part of ecology is founded on the presumption that different species have different potentialities that determine their fates in colonization, succession, competition, evolution and so forth. If we were to possess a complete listing of these capacities, we would presumably be able to predict or explain any interspecies encounter. In the meantime, we can cite those characteristics we have identified or presumed in explanation sketches.

Ecologists also offer explanations in terms of individual characteristics. When one works with a small number of animals for a period of time, their differences are apparent even to the human observer and it is easy to assume that these are more profoundly felt by the animals themselves. Indeed, much of the behaviour and ecology of higher vertebrates would be chaotic without reference to alpha males, dominance hierarchies, territoriality, mate choice, and other terms which reflect individual characteristics. This can lead to overinterpretation, and anthropomorphic explanations of animal behaviour are one of the pitfalls involved in teaching large-animal ecology and behaviour. This is less a problem at the professional level, although those who work with faceless composites like biomass or productivity often express skepticism about the objectivity of studies in which the individuality of each subject is integral to the study. Nevertheless, even these skeptics sometimes offer individual explanations in terms of the distinguishing characteristics of other entities, like this river, this valley, or this wood. Such individual explanations often tempt tautology, because the observation that is explained (Elizabeth's resolute action, high levels of production in this lake) often underlie the individual generality on which the explanation is based (Elizabeth's resolution, a eutrophic system).

There is no philosophical objection to the use of individual characteristics as the basis for limited generalizations, perhaps about a geographical location, or even as a universal statement, for example about a given species. However, when the explanations are based on loosely held or poorly enunciated beliefs about possible individual propensities, it may be fairly claimed that characteristically historical, individual explanations have replaced scientific theories.

This substitution carries three major costs. When available, historical, 'explanation sketches' hide the need for scientific explanation and the shortcomings of the science are less appreciated. When crude beliefs are accepted as theories, the effort of restating our beliefs as testable theories appears unnecessary, so theories remain half built, ready to explain what is already known, but useless in predicting what is not. Finally, the beliefs that sustain individual explanations can be the germs from which theories grow, but if these starting points seem sufficient, they are less likely to develop their greater potential. In short, reliance on historical explanation is likely to thwart the development of prediction and science.

Rational reconstruction

The last explanatory mode described by Dray (1964) is the method championed by R. G. Collingwood (1946) whereby historians use the fact that they deal with other human beings with whom they can share ideas through the medium of history and so come to understand why historical figures behaved as they did. Historical understanding is achieved when we can think the same thoughts as the historical actor and realize that, in the actor's place, the rational action was that which the actor took. For Collingwood, the aim of history is the vicarious reliving of past thoughts, plus the critical judgement of this past in terms of our own values. The role of the historian is therefore to facilitate this empathy by reinterpreting and re-explaining history for each new generation.

Collingwood's conception of explanation may seem so strictly historical that it offers little opportunity for application outside of history. Collingwood (1946) himself denied that it could have any place in treating material which was incapable of sustained rational thought. However, Collingwood's intellectual predecessor, the Italian essayist Benedetto Croce (1960), encouraged application of the empathetic approach to fields as different as anthropology and biology:

Do you wish to understand the true history of a Ligurian or Sicilian neolithic man? First of all try if it be possible to make yourself mentally into a Sicilian or Ligurian neolithic man; and if that not be possible or if you do not care to do this, content yourself with describing and classifying and arranging in a series the skulls, the utensils, and the inscriptions belonging to those neolithic peoples. Do you wish to understand the history of a blade of grass? First and foremost, try to make yourself into a blade of grass . . . (Croce 1916)

Few historians, including Collingwood (1946) and Gardiner (1959), who provided this translation, were prepared to accept that people could think like blades of grass.

Ecologists are more intrepid. Thinking like other organisms, or in more conventional terms, discerning strategies for finding nutrient, avoiding predators, growth, mating, reproduction, and defense of the offspring, is the stock-in-trade of contemporary ecology. The approach is used whenever we attempt to discern a design or a mechanism in nature (Gray 1987; Ollason 1987). The process is not dependent on whether we think that animals 'really' plan strategies or whether natural selection 'really' designs animals; it depends on how we act. When we try to see the rationality of our subject, we act as Collingwood suggests. The empathetic approach, which begins with the assumption that all biology is purposive and rational and ends with the demonstration of the rationale, necessarily uses the method of rational reconstruction.

Historical explanation and ecology

Historical explanations are never restricted to one mode, but blend all eight forms of explanation according to the taste of the historian and the nature of the subject matter. Thus vicarious re-enactments of rational reconstruction (Collingwood 1940) are made possible by the fullest possible narrative account or, since our knowledge is never complete, by the sum of colligative work; from these, the crucial genetic elements that weigh upon the historical actor are isolated and used in a variety of explanations certainly including universal laws, limited generalizations, and truisms about human behaviour. It is the historian's task to blend these diverse elements into a convincing amalgam.

The end-product of all this effort is not intended to be a testable theory about the past, although the explanation may be subject to test. Instead, the goal of history is a state of mind made possible by the historian's ability to command the evidence, to see connections between the historical actors and the world as they perceived it, to know the rules which govern human action then and now, and to wield the critical faculties of modern man to evaluate the action that occurred. In this state of mind, the historian and the historian alone can claim historical understanding.

The rest of us need not be left behind. While few may attain the perfect understanding Collingwood sought, we can gain something of the same sensation. With imperfect information, makeshift laws, and a blend of explanatory modes, we can claim to understand ecological phenomena just as the historian understands the past. However, our ability to do something does not imply that we should. Among the questions to be asked are 'Where are historical explanations used in ecology?', 'What dangers does the use of historical explanation hold for the science of ecology?', and, in the next Section, 'What roles can historical explanation legitimately play?'

The place of historical explanation in ecology

Historical explanation can be used whenever ecologists interpret a set of past observations. This use is particularly likely if the data seem or are taken to be exceptional, because this implies that the observations do not fit with some preconceived norm. Historical explanations are therefore likely whenever we deal with unique systems, because these cannot be placed in a scientific context, or well known systems, because our knowledge of such systems makes them individuals. Historical explanations are encountered in histories of the ecological changes of particular natural systems (Edmondson & Lehman 1981; Scavia & Fahnenstiel 1987; Whitney 1987), and in discussions of phylogenies (Wilson 1987), biogeographic patterns (Ball 1975; Endler 1982) or other unique patterns (Grant 1975; Winterhalder 1980). Historical explanations may also be invoked in overviews of data and opinion in the literature, but they are most frequently encountered in the 'discussion' sections of the primary literature. There, they explain away outliers (data that do not fit the expected pattern), relate proposed mechanisms to selected past observations, and fit present observations into previously accepted patterns, processes, and causal models. These *ad hoc* explanations thereby give the set of observations a unity and a consistency both with themselves and with the existing literature.

Most discussions are retrospective. They relate the present work to what has been done. Much more rarely, the discussion builds on anomalous observations to create a new synthesis that reduces old and new data to instances of a better theory. Such prospective discussions point to new questions, new predictions, new applications, and new research directions that result from the study. One challenge for ecology is to use historical explanations so as to encourage this development, rather than defend the status quo with retrospective justification.

The dangers of historical explanation for ecology

Historical explanations present problems for any science which employs them. The ecological problems associated with developmental laws and different historical explanations have been described earlier in this chapter. Nevertheless historical explanations often colligate different modes of explanation, so several explanatory problems can be woven together. This introduces a range of single and compounded dangers for ecological science.

As an art form, historical explanations depend more on the effectiveness of exposition than do the explanations of science. This gives an advantage to those with a particular command of language. Given that English has become the language of international science, Anglo-Saxon writers are advantaged, so general ecology, where the particularities of predictive power are less stressed, is ever more dominated by Britons, Americans, and those northern Europeans (Simberloff, personal communication) whose schooling and personal history permit a good command of English.

Historical explanations can be seditious. They are so facile that they are formulated far faster than they can be tested. A scientific community which uses them extensively must habituate to untested explanations and therefore must accept explanations uncritically. If this is frequently allowed, the effort required to formulate rigorous explanation and prediction will seem unnecessary, so the standards of the science will fall. Moreover, because historical explanations can be more precise than predictions, they give an unjustified sense of understanding about the phenomena which have been explained, and simultaneously eclipse the cruder predictions of which we are capable.

Historical explanations in ecology and elsewhere reinterpret selected observations in terms of particular constructs and jargon. This selection process is particularly open to criticism. When a range of models could be invoked to explain a phenomenon, theory must specify which particular model will apply or neither prediction nor scientific explanation is possible. For example, Stephens & Krebs (1986) present models to describe the behaviour of 'rate-maximizers', but they do not provide criteria whereby rate-maximizers can be identified; thus optimal solutions, like other models in theoretical ecology (Chapter 3), are invoked only to explain observations that we have already determined to fit the model (Beatty 1980). Stephens & Krebs (1986) interpret this procedure as evidence for the autonomous standards of ecological science, but the procedure might also indicate that optimality models do not represent science as it is usually intended.

Even if historical explanations are correct, in the sense that they describe what really happened, this need not entail predictive power over what happened or the testability of the explanations. For example,

Stephens & Krebs (1986) begin their monograph on foraging theory with a simple model to describe prey choice under certain circumstances (Table 6.1); this model suggests, among other deductions, that predation should follow a 'zero–one rule' whereby there is a threshold for prey quality such that profitable prey should be eaten in the proportions in which they are encountered and less-profitable prey omitted from the diet. However, Stephens & Krebs state that this theory is not intended to predict the proportions of prey in the diet nor food preferences in dichotomous choices, therefore predictive failure is incidental, but the theorem is still used to explain observed shifts (Krebs & Averey 1985) and dichotomies in diet choice (Krebs *et al.* 1978). Historical explanation drives a wedge between prediction and explanation that should not exist in science.

Application of a scientific explanation to a particular phenomenon invokes a series of properties that can then form the basis of future tests to demonstrate that the proffered explanation applies. In the case of the zero–one rule, one could test that any of the assumptions in Table 6.1 apply. If not all assumptions are accessible to testing, the explanations are only subject to partial tests. Since much of contemporary ecology consists of testing these parts, ecologists must distinguish successful test of some part from confirmation of the whole. A homologous difference exists between predicting the observation to be explained and predicting some element in the explanatory model. Both confusions may give an unearned authority to incompletely and ineffectively evaluated theories.

The fitting of explanation to observation may be an early stage in developing predictive theories, for the theorist may use these explanatory modes to determine which measurements are required for predictive power or where particular assumptions apply and where they do not. MacArthur (cited by May 1986) held that this might be the future form of ecological theories. However, until such models are advanced, the explanations neither fit the covering law model nor count as scientific knowledge.

Legitimate roles for historical understanding in ecology

One current of thought in contemporary ecology holds that the first task of science is not to predict but to explain our observations and thereby make sense of our universe. Prediction is considered important, but it follows from understanding (Begon *et al.* 1986) or it is valued more because it permits independent testing of explanations, than because it provides knowledge of future events or a basis for practice (Popper 1983; Stephens & Krebs 1986). For those scientists, explanation in ecology is analogous to explanation in history. Its reward is understanding and insight.

Understanding and insight are nebulous concept clusters. In one of their best senses, they refer to the intuition and wisdom of the individual scientist in addressing the subject material appropriately, so that the research and thought leads to effective approaches, meaningful variables and scientific solutions. Thus, valid understanding and insight are judged retrospectively, once they have proven themselves. The researcher can claim those virtues when they have developed a theory, and their claim can be evaluated by testing the predictions of the theory. In the absence of predictive power, claims to understanding and insight cannot be evaluated independently and depend instead on the personal feelings of the scientist. These feelings may be important in developing the confidence of the individual, but they must be discounted in judging the product of that confidence.

I do not share the view that scientific explanation can exist without prediction. Without predictive power, there can be no check on the validity of explanation and without prediction, ecologists cannot address the concerns of environmental practice. Scientists who strive for explanation without prediction leave questions of practice in the hands of shadowy, sometimes mythical, technologists, whereas ecologists themselves are the best equipped group to deal with these pressing problems (Southern 1970; Bliss 1984; Schlesinger 1989; Slobodkin 1988). Whatever support ecologists enjoy partly reflects the public's hope that ecologists are the scientists who can address the problems of the age. If we are unwilling to do so, perhaps we only merit the support the public purse provides historians.

Many applications of historical explanations in ecology are gratuitous. Some are only strings of commonplaces, others are transparent attempts to explain away rather than explain, and a few are no better than filler for an essential section entitled 'discussion'. No doubt an heuristic role will be claimed for historical explanations and perhaps there is justice in such a claim. On the whole, however, historical modes of explanation are overused in ecological science.

The confusion of scientific and historical modes of explanation obscures what science there is in ecology. Nevertheless, there are at least two areas of human endeavour which depend on historical explanations of ecological phenomena for insight and understanding, and which are too critically important to be ignored or dismissed as 'unscientific'. These two topics end this chapter.

Singularist causation and unique events

There is a disparity between the precision with which we can know past events and the crudity with which we can predict the future. This difference is a reproof to science, for it shows clearly how much remains for science to do. Until science addresses these legitimate questions, non-scientific explanations will continue to be sought.

The attraction of historical explanation results in part from the failure of science to meet the needs of its audience. The complexity of historical and ecological processes has resisted reduction to common, simple patterns, so if we wish to explain this complexity we cannot look to science. Scientific laws explain an observation by showing it to be a member of a set of probable values, but are silent on the further question of why this particular value was found rather than at least equally probable alternatives; many times we wish to know this level of detail. In addition, we are fascinated by what is different in our experience, so exceptions make a greater call on our attention than expected patterns.

Since scientific relations always simplify, one obvious advantage of non-scientific explanation is to consider as many factors as are needed to specify the exact observation. For example, one could consider each effect as the sum of many contributing causes, or each event as the sum of a number of different cause–effect relations, or one could consider that the exact form of the effect depends upon both the cause–effect relationship and the circumstances in which this relationship occurs (Ducasse 1974). Because each of these approaches requires that we see cause in unique events, they are called singularist theories.

Singularist theories assume that causal relationships are revealed through intuitions about individual events, not as regularities revealed by repetition (Beauchamp 1974). Because the number of potential causal factors and circumstances is infinite, such intuitions are not hard to form. If a phenomenon is examined in isolation, there are always enough potential causes to provide many possible explanations and because phenomena are never exactly repeated, these intuitions are not subject to test or capable of prediction. Therefore, singularist theories of causation are doomed to fail as scientific explanations (Chapter 5), and doomed to succeed as historical explanations.

Because singularist explanations cannot be assessed scientifically, they would not concern us if we were not asked to confront unique events in the course of our personal and professional lives. When we do experiments, the data themselves are invariably unique and scientists must distill pattern and generality from this uniqueness. When we manage a lake or a forest, we must manage it as a unique entity. And most grimly, when we consider the problems of global carbon balance (Detwiller & Hall 1988), global albedo (Ehrlich *et al.* 1983; Turco *et al.* 1983; Marshall 1987), or the stratospheric ozone loss (Kerr 1987), we invariably consider unique changes in the unique system on which all life depends. We cannot afford to ignore unique events, or the singularist causes and historical explanations they entail.

Escaping uniqueness

Most scientists do not address the problem of uniqueness, they try to escape it by collecting more data. However, complete escape is not possible because every data set was inevitably collected in unique circumstances and may not represent the phenomenon as supposed. In the case of normal scientific experiments, each new application or experiment involves some extrapolation beyond the limits set by the original data and any subsequent confirming instances. This extrapolation may fail, but the more times it succeeds the more evidence we have that different patterns do not hold and the more conviction we have that the expected pattern will. That is the nature of hypothetical knowledge.

Sometimes, as in the case of large-scale and long-term studies like Hubbard Brook (Berkowitz *et al.* 1989), we do not have the resources or the time required for replication. Sometimes, as in the nuclear disaster at Chernobyl, we wish to learn from a unique event, but not to repeat it. In these cases, we try to set special circumstances aside and place the unique event in the context of some larger set of observations which we judge to be similar, despite obvious differences. Thus, the theories about mineral budgets for whole watersheds arising from experimental clearcutting in the Hubbard Brook Forest might be compared to less precisely measured and controlled data from forestry operations and forest fires (Likens 1985). This would sacrifice the very high resolution which the experimental studies provide, for we would only be able to generalize about characteristics the scientific study shares with the 'found' experiments. However, this approach allows a normal evaluation of scientific theories derived from exceptional studies.

Alternatively, we could decompose the theory dealing with the large-scale into a composite of smaller-scale processes and test these partial models with experimental work in the field or laboratory. Such tests are unlikely to be very convincing. Differences in scale would justify the dismissal of contrary results, for we are unlikely to discard reliable information relevant to large-scale systems on the basis of small-scale analyses. On the other hand, confirmations of small parts of a large model might encourage more faith than the overall model merits, but they could never establish the predictive ability of the whole model.

Accepting uniqueness: a medical analogy

When we cannot place the system in a context of similar systems, or where we suspect that the uncertainties and risks associated with predictions from current theories are too great, we must deal with the system as a unique entity. This is how we hope to be treated by physicians, this is how we should approach questions of management of individual systems, and this is how we must approach the problems of the biosphere.

No doctor can assume that the next patient's ailment will be influenza simply because the 'flu' is going around. Instead, the doctor prescribes treatment based on an informed comparison of general medical knowledge, particular symptoms and special knowledge about the patient. This assessment may be modified in subsequent visits, but it can never be tested because the patient is treated as a unique case, without controls and without replication. If we recover, we can never be sure if the doctor helped or not. This is a weak point in medicine: it allows the proliferation of quacks, it forestalls the effective evaluation of common practices, it leads to the over-medication of the population of developed countries, and sustains an unsubstantiated belief in the efficacy of contemporary medicine. Nevertheless, the unique treatment of each case seems inevitable for individualized medical care.

A similar approach must be used in applied ecology. When we manage a forest or a lake, our first intervention will represent the applications of the best available model(s), tempered by our insight into the special circumstances of the situation. For, example, when phosphorus abatement was begun for Lakes Erie and Ontario, initial target concentrations for phosphorus were purposely set higher than desirable because of the immense costs involved in full abatement and the uncertainty of the prediction. When the observed reductions were close to expectation, further abatement was authorized (Chapra & Reckhow 1983). Taking this as a model, ecological management will base initial policy decisions on theory and specific information, but responses to this initial program must be monitored so that further adjustment can be made on the basis of existing theory and additional information. Ecological management cannot depend on a blanket policy or a single 'fix'. It will involve extended monitoring and intervention into the indefinite future. This is already accepted in human health and agriculture; it is policy in Europe where the long-standing impact of man is accepted as part of the system (Jacobs 1975) and will become essential in North America and elsewhere as we push against the limits of our biological resources.

This 'adaptive management' approach (Walters 1986; Hilborn 1987) must also become biospheric policy. We have only the one world and its characteristics at any point in time are the entrained components of a unique historical sequence. As a result, global models can be compared only on the basis of their reconstruction of the past and this performance may be irrelevant to their performance when extrapolated into the future (Caswell 1976; Deevey 1987). Such models are especially likely to fail if applied to novel conditions and it is about just such conditions that we need information. Because the probabilities of future success cannot be assessed, the risks involved in any management strategy can only be subjectively established. Because global models cannot be tested effectively, the status of models predicting nuclear winter, global warming, or lethal increase in ultraviolet radiation can have no stronger status than models predicting less serious disruptions. They are all plausible historical scenarios. However because the worst scenarios are so unspeakable, we must manage our world with these in mind. Since there is no room for error in this unique system, any decision about the future of the entire planet must be extremely conservative.

Natural history

Although Elton (1927) once described ecology as 'scientific natural history', the two fields differ. Ecology is a science intent on the development and assessment of objective scientific theory. Natural history is an art, the goal of which is the personal and subjective development of the individual practitioner (Hutchinson 1963). Ecology and natural history may be confused because of their common roots and common material, but this confusion detracts from both disciplines (Peters 1980b).

Natural history achieves its goal by an activity like the rational reconstruction of a Collingwoodian historian. The naturalist, like the historian, lives a vicarious existence which is made possible by a deep and extensive knowledge of the laws and norms involved in the circumstances and activities of other beings. The difference is that, for one, these other beings are human and, for the other, they are not. Like the historian, the naturalist's achievement is first and foremost a personal one, a state of mind representing a profound understanding of and with the natural world. Again, like the historian, the naturalist's empathetic state does not involve the loss or submission of the human personality, but its enrichment and evaluation in light of the greater breadth of experience, albeit vicarious experience, that natural history provides.

This understanding is for individual consumption, like the thrill of great music or the repose of a cloister. It cannot be shared like the predictive power of a scientific theory. Nevertheless, this understanding is among the precious and extraordinary qualities which justify human existence. As an ethical base, natural history could sanction the deployment of scientific theories to protect both the high quality of life needed to achieve our potential humanity and the biosphere in which to enjoy it. Because historical explanations of ecological phenomena promote this understanding, they are to be cherished as part of the art of natural history (Peters 1980b).

If natural history is an art, it is a folk art because of its vast public following, because of the absence of a recognized professional cadre, and the absence of rigid standards. As a result, the quality of information required for natural history is often weaker than that for academic history or scientific ecology. Indeed, sometimes the misapprehensions of natural history are so gross that the empathy achieved is closer to that produced by an historical novel or a Hollywood epic than that produced by a real grasp of events. However, for the elite naturalists, an understanding based on a misapprehension of nature or on evidence which will not stand the tests of science is a false understanding, a cheap and tawdry imitation. For them, as for the professional historian, the canons of evidence are stringent. It is with this select group that we must class those scientific naturalists who claim an understanding of nature as the prime goal of their research. They are to be honored, but they are not to be models for scientific ecology.

Summary – Explanations in ecology

The confusion of scientific explanation with other modes of explanation, here termed 'historical explanations', has debilitated ecology by stressing the rationalization of observation to the detriment of prediction. We therefore need to affirm that only understanding based on covering laws or limited generalizations can provide explanations that invoke standards consistent with the need for predictive power in science. If we ignore this stricture, we can never develop the tools we need to manage and protect our environment. Historical explanations can serve to organize knowledge, to inspire us to theory, and to describe potential scenarios where our predictive power is weak, as it is when we wish to distinguish among equally likely predictions, or when we deal with unique systems. Historical explanations also constitute a form of natural history, which can prove ethically powerful and personally rewarding. In short, historical explanations have their legitimate purposes but we gain nothing by confusing these with science or allowing them to supplant predictive power.

science (Price 1986) make obsolescence of much of the literature increasingly likely and rapid, and perusal of any ancient scientific literature shows that parts are now dated and unusable (Edmondson 1987; Hrbàček 1987).

In short, scientists are willing to invoke the criterion of relevance, if they or their peers are arbiters of relevance. It is in this insider's sense that this section argues the irrelevance of many current ecological constructs to the science and to society. To support this charge, the section begins where the criterion of relevance is most frequently used in science, in the comparison of data and theory, and then considers the more controversial areas of larger scientific and societal relevance.

Irrelevant data and irrelevant theories

Theories are used to explain observations that are already in hand and to predict more. Therefore theories are irrelevant if they do not explain or predict the type of observation we made or wish to make. Conversely, data used to test the predictions of a theory would be irrelevant to this purpose if the theory actually predicts something else. This seems so self-evident that it scarcely needs elaboration, yet the comparison of irrelevant theory and data is not extraordinary; it is a common theme of 'letters to the editor' in scientific journals.

Plant–herbivore coevolution

Owen & Wiegert (1976) suggest that herbivores and plants have coevolved so that grazing actually increases the fitness of plants by encouraging nitrogen fixation and nutrient conservation. One test of a mechanism within this theory involves the experimental application of herbivore exudates to plants and soils. This has shown that mammalian salivary factors can increase plant growth rate (Dyer 1980) and that sugars in aphid honeydew increase nitrogen fixation rate in the soil (Petelle 1980). Although these observations seem consistent with the theory of Owen & Wiegert (1976), one can show easily that they do not test a general theory that herbivory enhances fitness. If the converse results had been obtained, if these experiments had not shown an effect on growth rate or nitrogen fixation, the test could be dismissed as irrelevant because, in nature, grazers could influence the fitness of their prey in many ways besides increasing growth or nitrogen fixation (Choudhury 1985; McNaughton 1986a). If falsifying data are irrelevant to the theory, then confirmatory data must also be irrelevant. Both are consistent with the theory that grazers can enhance fitness, but neither shows that grazing and plant fitness are positively related.

A less dubious experiment would ask directly if the presence of herbivores increases some index of fitness. Choudhury (1984) performed such an experiment with peas and aphids, and found that the numbers of flowers, pods and seeds were depressed by aphid infestations, an observation consistent with the experience of gardeners and farmers. Owen & Wiegert (1984) responded that these data were irrelevant, because the theory applies to natural conditions, not domestic plants and their pests. Choudhury (1985) replied that his experiments were relevant because they were consistent with studies of a native plant and its aphid pests (Foster 1984). In a very similar exchange, McNaughton (1986b) maintained that observations on cattle (Westoby 1985) were irrelevant to his theories (McNaughton 1984) about the mutual benefit achieved by coevolution of natural mammalian grazers and their prey.

Questions about the shared relevance of particular theories and data sets are common in ecology. They are encouraged by vague terms, like fitness, and by implicit or incomplete specification of important boundary conditions, like the requirement that the plants and herbivores be 'coevolved'. In any case, 'sufficient opportunity for coevolution' is too vague to serve effectively as a precondition for this or any other theory.

Sufficiency, necessity and immediacy

Ecology encourages the construction of irrelevant theories and the collection of irrelevant data by proposing and testing underlying mechanisms which might contribute to the hypothesized pattern, although they are neither necessary nor sufficient for that pattern. When one's interest in the mechanism derives from the significance of the overall pattern, these efforts are misplaced (Kalff 1989–MS). For example, if we are interested in the general hypothesis that native grazers benefit their plant foods more than exotic domestic animals or use their resources more effectively than domestic animals (McNaughton 1984), we must assess the status of the grass or grazers directly. We cannot test these hypotheses by examining the effect of salivary factors on growth (Dyer 1980) because the general pattern involves more than this single process.

Although such partial tests may seem to verify the hypothesis, they are actually irrelevant. Any claim to have confirmed the hypothesis will only be inferential, because, at the level of the general hypothesis, many other processes (Chapter 5) could alter or offset the implications of our experiments. Belsky (1987) describes this difficulty as a confusion of scale in which we try to test a theory about the effects of animal communities on plant communities by examining the effects of individual, parts and processes on other individuals, parts, and processes. It is actually an example of a logical fallacy called 'affirming the consequent' or conclusion, in order to affirm the antecedent proposition (Chapter 8; Flew 1975; Y. Prairie, personal communication).

When one affirms the consequent, observations that are consistent with deductions from a proposition are mistakenly thought to confirm the truth of the proposition. Thus the proposition 'cyanide ingestion induces death by asphyxiation' is enough to deduce 'death by asphyxiation' of those who ingest cyanide, but not to warrant the conclusion that someone had eaten cyanide simply because they were asphyxiated. For the same reason, the proposition 'grazing induces better plant growth through natural animal exudates' implies that the application of exudates increases plant production, but does not warrant the deduction 'grazing enhances plant production' from the observation that 'plant growth increases after experimental application of animal exudates'.

Similar confusion is the root of much of the irrelevance of theory and data in ecology. It is addressed by proposing and preferring tests that are immediately relevant to the problem at hand, whether this be the lethality of cyanide ingestion or the benefits of grazing. Such tests involve theories which treat necessary and sufficient conditions for an effect, rather than potentially contributing factors. If no immediate solution is possible, we should entertain approaches in which a finite number of specific steps suffice to achieve the goal of the research. For example, Dillon & Rigler (1975) chose not to relate lake response to land use and geology in a single equation, but instead developed a sequence of equations to relate nutrient loading to watershed characteristics, lake nutrient concentration to loading, and biological response to nutrient concentration (Chapter 10). Thornthwaite & Mather (1957) developed predictions of evapotranspiration from a tabular analysis involving sequential considerations of temperature, rainfall, runoff, soil type, and rooting depth. Such approaches are statistically less desirable than single relations, because of the accumulation of error in sequential calculations. In compensation, sequential analyses allow independent tests of various intermediary relations, and are more immediate than partial tests because they specify what is necessary and sufficient to make the prediction. Such cases are exceptional in ecology: they involve a clear view of what problems the overall sequence of relations were intended to address.

Scientific irrelevance

The scientific relevance of a particular construct is the degree to which it bears on the central theories of the science. Scientifically relevant theories resolve or promise to resolve the common doubts and questions of a substantial proportion of the research community. Scientifically relevant data are those which can test the predictions from such theories. Ecological constructs that cannot predict are not relevant to any observation nor can any data be relevant to them, at least not in the sense of relevance intended here. One of the many adverse effects of the theoretical weakness of the central constructs in ecology is that standards of scientific relevance are inapplicable.

In the absence of a widely acknowedged body of central theory, the purpose or goal of ecology could identify what is relevant to the science. Such a definition was offered in Chapter 2: Ecology seeks to predict the abundance, distribution, and characteristics of organisms in nature. Although this statement offers the advantage of more operationally definable goals, it gives little direction in selecting particular organisms or characteristics as more scientifically relevant. Thus this general definition is a weak criterion of relevance because it does not distinguish the appropriateness of many phenomena. Nevertheless, some theories do not predict the characteristics of organisms in nature at all, either directly or sequentially, and this subsection is concerned with those ecologically irrelevant theories.

Such a distinction is dangerous if drawn too closely. Many ecological theories predict biological properties from non-biological information. For example, evapotranspiration affects plant life form (Box 1981), primary productivity (Lieth & Box 1972), decomposition (Meentenmeyer 1978) and species number (Currie & Paquin 1987), so the prediction or estimation of evapotranspiration from precipitation, temperature, day-length or radiation (Thornthwaite & Mather 1957; Penman 1963) is ecologically relevant. Similarly, the octanol–water partition coefficient, the ratio of the equilibrium concentrations of a chemical in each phase of a biphasic mixture of water and an organic solvent (octanol), is increasingly important in aquatic toxicology because bioaccumulation and toxicity of organic pollutants are predictable from this variable (Bysshe 1982; Connell 1987, 1988a,b). Finally, many environmental problems require the management of essentially physical and chemical processes, like acid rain, stream hydrology, and atmospheric dust. Ecology therefore includes a variety of non-biological models and predictors, plus the non-biological properties that are an integral part of the larger questions of conservation and protection of both the biotic and abiotic environments. Nevertheless, a number of contemporary ecological constructs seem irrelevant to the science, even in this wide sense.

Indices and multivariables

One class of ecological theories is particularly prone to the charge of irrelevance. These are the theories that deal simultaneously with many different response variables, usually by some form of multivariate statistic, like principle components analysis or correspondence analysis.

The charge here is not that constructs which incorporate multivariate statistics cannot make predictions or cannot be subject to falsifiable tests. They can. For example, Miles *et al.* (1987) show that there is a consistency in the relation between passerine morphology and foraging behaviour (because correlations between multivariate descriptions of morphology and foraging behaviour are robust to translation to different communities). Principle component analysis has been used to test for 'convergence' (correlations between axes from geographically separated sites) in different Mediterranean-type ecosystems from around the world (Cody & Mooney 1978); and many other applications exist (Gauch 1982).

My reservations about multivariate analysis arise from the opacity of the results. Knowing that an organism is high or low on an axis which seems dependent on (but not the same as), for example, a distillate of fast-to-slow gradients in life-history traits (Stearns 1983) says very little about the organism. Any information that could be used, beyond the analysis itself, is irretrievably lost in the multivariate scaler. Apparently, this is recognized and accepted by those who promote and use multivariate statistics since they rarely provide sufficient detail to allow the reader to fit new observations onto the scale, and even more rarely show how these scales and statistics could predict interesting properties of the systems under study.

A further problem involves the uncertainty of multivariate statistics. Wartenburg *et al.* (1987) compare the ability of two multivariate techniques (principal components and correspondence analysis) to order successfuly a set of artificial, one-dimensional data. They conclude that the ranking, even in this simple case, is dependent on the statistical model and the assumptions of the analysis, not on the data. They recommend against any analysis which obscures the units of measurement, because the effects of such multivariate approaches are poorly understood and therefore hard to interpret. This seems sound advice. Analyses of multivariate responses may help in private data exploration and theory creation, but not in the public phase of science where theories and predictions are evaluated and used.

This problem of interpretation is not restricted to advanced statistical procedures. Many multivariate indices in ecology arbitrarily and even subjectively combine incommensurate ecological information to yield numbers that are as difficult to interpret as a principle components axis and are subject to the same imprecision in depicting natural gradients. For example, Hurlbert (1971) found that different diversity indices are neither concordant nor biologically meaningful; Guhl (1987) showed that a number of limnological and biological indices were insensitive to an obvious gradient in water quality in his comparison of three German lakes. Electivity indices, which are usually used to describe feeding preferences in situations where some choice is made (Lechowicz 1982), can also obscure the process they are intended to describe. This is especially likely if more than two food types are involved, for such complex behaviour can never be faithfully depicted by the one-dimensional scale of an electivity index (Peters 1984).

In summary, scientifically irrelevant constructs are those that do not address the concerns of the science. Multivariate indices provide an example of this precisely because they are indices. They are surrogates for an interesting phenomenon which we cannot measure and because we cannot measure this phenomenon, the indices do not represent it. This inadequacy may be disguised by statistical sophistication but it is not remedied. Indices are descriptions, but descriptions are only informative when their elements are applied to make predictions through scientific theories. Many ecological indices, especially those dependent on multi-variate statistical descriptors, fail to make this transition and remain irrelevant.

Invisible colleges

Price (1986) suggested that the scientific world is composed of many 'invisible colleges' consisting of the body of world scientists who are interested in similar problems, their lesser colleagues, and their students. These colleges promote recognition and interaction among their members and Crane (1972) considers them essential to the growth of science. Invisible colleges present a special problem because these groups can act as though relevant science is the science which has traditionally interested the college's senior members. This autodefinition encourages the development of an academic science done for its own sake ('terminal science'; Pramer 1985) and may leave scientists 'talking to themselves' (Biswas 1975). This is especially likely as one generation trains the next, reinforcing their interest in and dedication to a series of established questions (Box 1976; Hall 1988).

Invisible colleges may be recognized in the themes, the composition of individual sessions, and workshops at large meetings. For example in my own field, they include, in order of increasing scope, groups interested in particular methods to measure the feeding rates of zooplankton (Peters 1984), the size-efficiency hypothesis (Hall *et al.* 1976), the trophic cascade (Carpenter 1988), optimal foraging (Stephens & Krebs 1986), life history analyses (Stearns 1976, 1983), sociobiology (Wilson 1975) and the new synthesis (Antonovics 1987). In another dimension, different invisible colleges have focussed their attention on each of a succession of paradigms which dominated ecology over the past century (Simberloff 1980a; McIntosh 1985; Gimingham 1987). Such colleges may be essential to a healthy science, to promote sharing of information, to ensure communality of standards, to focus attention on key points of interest, and to reassure members of the college that their work interests others. However, this self-affirmation must be held in check by demanding that the goals and programs of the college be demonstrably relevant beyond the confines of group interest.

Societal irrelevance

Societal relevance refers to the applicability of the science to the problems of society. This form of relevance is that requested and expected by non-scientists, and partly as a result, it is the relevance held most suspect by scientists. It also seems the most difficult form of relevance for researchers at the edge of knowledge to achieve. There are usually many steps before scientific work can be applied to the problems of society and applications often require skills the research scientist lacks: those of the engineer, the technologist, the entrepreneur, the politician. In consequence, many scientists cannot concern themselves with the societal relevance of their work. Regrettably this has sometimes demeaned application and made good applied science seem second rate.

The division between pure and applied ecology is unnecessary and dangerous. If there are pressing environmental problems, then the world needs a science to manipulate and control the environment (Peters 1971; Schindler 1987) and part of ecology must be applied to treat those problems. To be useful, such a science must achieve the very highest standards of scientific excellence, for its theories will be tested repeatedly in an unforgiving world.

Applied ecology is a particularly challenging field, but it remains a part of ecological science. It selects its theories from the larger body of all ecological theory in light of its particular problems (Vollenweider 1987), just as ornithologists or oceanographers select only a part of ecological theory as relevant to their intersts. Contributions to applied ecology, like contributions to ornithology or oceanography, are also contributions to the whole of the science and answers to applied questions can also be fundamental advances. Indeed, the grist provided by such questions may lead to greater scientific productivity (Box 1976). There ought to be no difficulty in accommodating applied ecology within ecological science.

When the various threads of these ideas on relevance are wound together, they provide a composite of the desirable interrelations of relevance and application in science. The decision of which research area to pursue must be that of the researcher, since only the researcher has sufficient knowledge of the peculiarities of the field, personal capacities, and beliefs to make such a decision. Moreover, the sense of intellectual freedom involved in selecting a research area is important and perhaps essential to good science. Certainly, the single minded dedication, almost obsession, needed to pursue major scientific problems can only be achieved with a deep sense of personal commitment to the research on the part of the scientist. However, the freedom to chose a research topic does not imply a right to funding. Funding opportunities are the result of societal decisions and reflect the societal relevance of the research. Ideally, enough individual scientists will respond to the attraction of better funding opportunities, to the sense of moral responsibility to society, and to the enlightened self-interest of environmental science to enlist in societally relevant research and to promote these programs aggressively. Those who succeed in their research should be justly honored by their scientific peers for the quality of their science and by society for the effectiveness of their solutions.

Difficulties arise because some scientists claim that their work is relevant to societal needs when this is not so. As a result, they lay claim to a false applicability that confounds both scientist and funding agency. My interpretation of this is that 'pure' researchers desperately want to be useful but have been so badly trained in applied science and are so rarely part of applied management decisions, that most ecologists are incapable of very practical advice. They cannot distinguish socially relevant and irrelevant science and instead offer a naive academic exercise which may yield sterile 'solutions in search of a problem' (Biswas 1975; Clark *et al.* 1979). Because scientists are often educators, this ineptness is perpetuated by examples and paradigmatic cases laid before students. Mistraining of one generation in the solution of societally relevant problems has repercussions on the next (Box 1976).

Island biogeography

A case in which the claim for societal relevance has proven empty on deeper consideration is the application of 'The Theory of Island Biogeography' to the design of nature preserves. The theory of island biogeography has stimulated a great deal of other research which is not intended to be relevant to preserve design. That work is not addressed by this section and must be judged on its own terms. Some part is discussed later in this chapter.

The original, novel, and legitimate suggestion that island biogeography was relevant to the design of nature preserves (Simberloff 1974; Wilson & Willis 1975; Cole 1981; Diamond & May 1981) had great appeal. Theory suggested large, contiguous reserves to incorporate the largest number of species and minimize extinction, round in shape to reduce species losses due to 'peninsular effects', and connected with or at least close to other sources of fauna to encourage immigration and so maintain a high 'equilibrium number'.

Subsequently, flaws have appeared in some lines of evidence (Simberloff & Abele 1976, 1984). All components of the composite theory of island biogeography have not been equally tested – the relation between colonization and island or reserve size is particularly hard to address (Simberloff 1976b, 1978) – and these untested aspects should be distinguished from patterns with more empirical support, like the species–area relations (Connor & McCoy 1979). Further study has falsified some elements in the theory, like the peninsular effect (Busack & Hedges 1984; Due & Polis 1986). Further reflection has shown logical flaws in others: although species–area relations indicate that large preserves are always preferable to small ones, the relations are such that several small preserves may contain fewer, as many, or more species than does an equally large, single reserve, depending on the proportion of the total species pool shared by the smaller areas. The theory is neutral on the question of whether 'single large or several small' (SLOSS) is better (Simberloff & Abele 1976, 1984).

Two aspects of island biogeography seem relevant to preserve design despite these criticisms. The contention that extinction is more likely in smaller preserves seems to hold (Abbot 1980; Wilcox & Murphy 1985) and this suggests that larger preserves are preferable. This is also consistent with species–area relations. However, this conclusion is reached with a variety of arguments (Janzen 1983; Willis 1984) and the imprecision of the relations from island biogeography limits their utility (Murphy & Wilcox 1986). In any case, no one proposes that a small preserve is better for conservation than a large one (Simberloff & Abele 1982, 1984). This choice is imposed by economic limitations. Thus, although the prestige of island biogeographers may be useful in promoting large reserves (Willis 1984), the major theoretical contributions of island biogeographic theory to this applied problem may be redundant.

More damaging to the claim that island biogeography addresses the questions of preserve design is the charge of societal irrelevance. Society does not turn to ecologists so they can outline the desirable size and shape of a future preserve on a map. Instead, proper design is a balance among the competing non–ecological exigencies of economic forces, available funds, defensibility of the preserve, availability of land, and social disruption and demands plus a series of biological considerations concerning the ranges of available habitats in different areas and the demands of the target species for preservation. These questions are independent of Island Biogeography.

Island Biogeography is relevant to preserve design only when 'all else is equal' (Murphy & Wilcox 1986): when preservation is concerned with species numbers, not with particular species, and when all areas are homogeneous. However, since these things are never equal (Lahti & Ranta 1986), because species and regions differ in ways that are an essential part of preserve design, most of contemporary island biogeography is not relevant to practice. Until ecologists accept this failure, they are unlikely to abandon what has proven a false start or to address effectively the critical problems of reserve design which island biogeographers brought to the attention of the ecological community.

Accuracy

A theory is considered scientific if it is potentially falsifiable, and falsified when it repeatedly proves inaccurate. As a curious result, even theories which have proven wrong are still scientific (Flew 1975). Indeed, falsification is the strongest available evidence that phenomena excluded by the theory as improbable were indeed possibilities. Usually, a repeatedly falsified theory is ignored by the scientific community. If this is not so, one has reason to be concerned for the standards of the science.

The contention of this section is that ecology may retain falsified theories. Sometimes, this laxity is condoned on the argument that even an inaccurate theory is better than none. Such a defense is one interpretation of Lakatos' (1978) position that theories are only abandoned if there is a better alternative. However, at the level of the common-or-garden-variety of theory that concerns most ecologists, such alternatives are always readily at hand.

Repeated falsification should at least result in a weaker restatement of the original theory. For example, after falsification, a precisely quantified, deterministic relationship might be replaced by a probabilistic statement with broad fiducial limits or simply a statement of likely trend. Alternatively, the general theory might be restated in terms which explicitly exclude the class of the falsifying observations, and these observations would be restated as a highly specific theory in their own right. For example, birds do not seem to fit a general regression describing population density as a function of body size (Brown & Maurer 1986; Juanes 1986): their populations are too sparse and density is almost unaffected by body size. Therefore, the general relation is subject to the restriction that it applies only very approximately to birds, which are better treated with their own relation, even if this is only a simple geometric mean: bird densities average about 10 individuals per species km^{-2}, with 95% confidence limits that range between 0.4 and 250 km^{-2} (Peters 1983; Peters & Wassenberg 1983). An inappropriate response to such a falsification is to cling to the discredited theory, but the remainder of this section details several cases where ecological theories are retained despite predictive weakness and failure.

Biogeographic rules

Some inaccurate theories in ecology are quite simple. For example, Bergmann's rule, that boreal animals tend to be larger than their southern relatives (Fleming 1973; Lindsey 1966), has an exception rate of 68% (Geist 1987; McNab 1971). Allen's rule, that boreal animals also have disproportionately reduced extremities, has an exception rate of 36% (Simpson 1963). Nevertheless, these rules persist in both the primary and secondary literature.

The reasons for the durability of biogeographic rules are likely complex. In part, the rules endure because there is a strong, seemingly logical reason for them to exist: Both Allen's and Bergmann's rules are thought to reflect energy conservation strategies in extreme environments, although quantitative support for that position is unconvincing (Geist 1987). If the expected pattern is not found, we have no difficulty accommodating exceptions as 'trade-offs', reflecting other constraints and strategies.

The advantage of self-deception about the strength of biogeographic rules is less obvious. Perhaps the need for an attractive example for teaching and divulgation is a strong enough rationale to interpret the available data selectively, by stressing confirmatory examples and ignoring falsifications.

The broken stick

An equally well known example of the retention of a falsified hypothesis is the broken stick model of MacArthur (1957). The model suggested that the frequency of niche sizes, and by extension population densities, of different species in a community should decline as do the lengths of segments of a stick simultaneously and randomly broken into pieces. This would give a particular shape to frequency distributions of different species in a community. Despite the tangibility of the image, few communities are arranged in this way, a deficiency which was pointed out long ago (Hairston 1959), and fits to the model are dependent on sample size (Hairston 1969). Nevertheless, the model remains an active, if minor centre, of discussion in the literature (Pielou 1981b).

Fisheries models

Fisheries management is so accustomed to inaccuracy in its basic models that striking differences between model and observation are scarcely noticed. For example, one basic assumption is that the relationships describing the effect of parental stocks on subsequent recruitment to the fishery are dome-shaped or humped (Ricker 1954). The argument begins with logical necessities: without stocks, recruitment is zero, and when there are stocks, there will be recruitment. Since the fecundity of fish is so great that replacement could be assured by modest to moderate stocks, there will be a range of stock sizes over which recruitment is independent of stock. At still higher population densities, density-dependent processes will reduce recruitment. Taken together, these assumptions create a dome-shaped relationship between recruitment and stock size. Beverton & Holt (1957) used a similar rationale, with less-intense density dependence at high densities, to generate an asymptotic relationship.

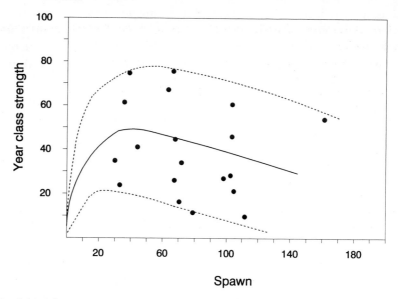

Fig. 7.1. The relationship between apparent recruitment and parental stocks in the North Sea herring, showing the stock-recruitment regression fitted to these data (from Ricker 1954). The inaccuracies of such fits have not been sufficient to discourage their use.

This entirely plausible scenario has enjoyed a long life in fisheries biology even if stock-recruitment data do not support the model (Beverton *et al.* 1984; Hall & DeAngelis 1985; Hall 1988; Fig. 7.1). Nevertheless, fisheries biologists fit data to models that are clearly inaccurate and make management decisions on that basis.

Given this standard, it is scarcely surprising that other fisheries models are also inaccurate. For example, the model for maximum sustainable yield proposes a dome-shaped relationship between catch and fishing effort. Before the extension of territorial limits to 200 miles, every one of 39 international agencies used the maximum sustainable yield concept as the basis for marine fisheries management (Clark 1981). National control has made the bases of subsequent fisheries management harder to identify but it seems likely that maximum sustainable yield continues to play a role as a 'fundamental principle' of management both in fisheries and other renewable resources (Sissenwine 1978; Clark 1981).

It is easy to see why fisheries biologists were attracted to these models. The available data for stock size, recruitment, effort and catch are weak, and a number of other factors are assumed to affect the variables and their

Table 7.1 *Desirable characteristics of successful 'enemies' or control agents in the biological control of pests.*

(1) The enemy induces a stable, low, equilibrium population density of the pest.
(2) The enemy (a) is host-specific and (b) is synchronous with the pest, (c) can rapidly increase in population when the pest erupts, (d) needs only one or a few pest individuals to complete its life cycle, (e) can search effectively, and (f) aggregates in regions of high prey density.
(3) The enemy is normally stenophagous.

Note:
Successful control programs have, collectively, breached each of these theoretical desiderata and some breachings of conditions (1), (2a), (2b) and (2d) seem critical to the program's success. (From Murdoch *et al.* 1985).

relationships (Walter 1986). As a result, considerable scatter is expected and easily explained. Moreover, fisheries biologists are dealing with an important problem for which an answer is required. Simple protestations of ignorance will not suffice. Finally, although the data are weak, the logic behind the models is strong and so, in the absence of anything better, stock-recruitment diagrams and maximum sustainable yields became the order of the day. Unfortunately, these models have proven an inaccurate, perhaps dangerous, basis for fisheries management (Welch 1986).

Biological control
The same combination of factors has had similar results in other areas of applied ecology. When Murdoch, Chesson & Chesson (1985) examined the empirical support for several principles of biological control (Table 7.1), the data were equivocal. Again the persistence of these principles can be attributed to the ease with which contrary cases can be accommodated by casual reference to other factors and the logical appeal of the models. Unfortunately, logic is an appropriate guide only if the assumptions hold and this seems not to be the case. At the very least, insect systems are not at equilibrium and so the equilibrium models of Lotka-Volterra which underpin conventional biological control theory do not apply.

Forest production
Forestry provides a third example of the preservation of an inaccurate theory. It is 'common knowledge' that plantations are more productive

than natural woodlots (Tillman 1978) and forestry practice incorporates this presumption. Yet when the data are standardized for differences in biomass, stand age, plant size, density, precipitation, temperature, latitude and plant type, the annual increases in above-ground biomass harvestable from plantations and natural stands do not differ significantly and the total annual increment in biomasss in plantations may even be reduced; no analysis showed plantations to be more productive than natural stands (Downing & Weber 1984). If plantations have appeared to do better than natural stands, this may reflect ineffective control for the effects of other variables in limited, pairwise comparisons, perhaps because plantations are often planted on better sites than those occupied by nearby natural forests.

Again the composite effects of many factors frustrate direct comparison of precept and data, and the absence of proper controls allows precept more influence. In the cases of fisheries, biological control and forestry, this tendency is surely enhanced by a desire to bolster current practices and the necessity of some management decisions. One can only wonder how much the insistence on inaccurate, but widely accepted, formalisms has impaired creative research into these and other critical issues of our time.

Ad hockery

In the jargon of philosophical dispute, the reinterpretation of apparently falsifying data to preserve a theory is said to involve ad hoc hypotheses (Popper 1985). Such a tactic will always succeed at least temporarily because the number of potential explanations for departure from expectation is infinite. Eventually however, repeated ad hoc hypotheses so degrade the initial theory that it is ignored. The ease and rapidity of this process reflects the strength of faith in the basic theory and the availability of alternatives.

The danger in ad hockery is that it hides the need for alternatives. Superficially satisfying explanations offered by ad hoc theories cloak weaknesses in, or even the absence of, predictive ability. At the same time, this approach leads us to mistrust data in favour of precept and therefore reinforces precepts in our minds. If faith has a place in science, it should reflect the degree to which the theory has been successful in previous applications, but too often it seems to depend on plausibility, familiarity and the status of the theory's defenders.

The factors that bolster precept discourage radically different approaches and theories. For example, it is difficult to imagine an alternative to stock–recruitment models because of their ubiquity and logical force. There must be no recruitment when stocks are zero, and more recruitment when stocks exist. However, this indicates neither that there is a consistent relation between stock and recruitment, nor that we can establish this relationship, if it exists. Moreover, the models may abstract the wrong aspects of the fishery, for there is no logical reason why future recruitment to fishable stocks need vary with the current stock size. Fish are so incredibly fecund that it is at least as plausible that recruitment to the fishery is dependent on survivorship of eggs and larvae, rather than simply the number of eggs laid (Hjort 1914). Nevertheless, the dominance of stock recruitment scenarios may over-shadow alternative models and the story the data might tell to open minds. The availability of other fisheries yield models (Sutcliffe 1972; Hanson & Leggett 1982; Ryder 1982; Roff 1983; Leggett *et al.* 1984; Crecco, Savoy & Whitworth 1986; Ryan 1986) shows that stock-recruitment models need not be the only choice. The failures of marine fisheries biology in protecting the pacific sardine, the Peruvian ancho-veta, and the North Sea herring (Clark 1981) show that traditional management practices cannot be our only choice.

Stephens & Krebs (1986) make the point that seeming ad hockery may instead represent reformulation and refinement of the original theory after the failure of its predictions. In this case, the new theory should explain the disparity between existing data and theory, but also suggest further observations that could test the reformulation, consisting of the original theory modified by a heretofore ad hoc adjustment. Thus the seemingly ad hoc explanations of failure in fisheries management, forestry, and biological control mentioned in the preceding section could be interpreted as novel, incipient theories.

This position is correct, but it should not become a facile excuse for ad hockery. Since the number of successive modifications is infinite, there must be some limit to our willingness to consider ad hoc adjustments. If such cases are seriously intended as new theories, then they should be tested expeditiously. Presumably, the conditions that dictated the need for the original experiment still hold and therefore the researcher who wishes to be consistent and to escape the charge of ad hockery, should feel obliged to test the modification. If this is usually not the case, the defense that apparent ad hockery is actually theory development seems disingenuous. Too often, ecologists act as if they were offering ad hoc hypotheses, not alternative theories.

Imprecise and qualitative predictions

Inaccurate theories provide no useful information about the world around us and, when inaccuracies are acknowledged, theories should be modified or discarded. One of the easiest modifications of all achieves accuracy by sacrificing some precision. For example, a highly precise theory might predict that woodpeckers change trees every 15.36 seconds. Such high precision makes a theory potentially very informative within its domain, for it identifies most possibilities as unlikely. Regrettably such theories are also likely to be wrong. If they prove false when tested, they may be replaced with a less–precise formulation, such as woodpeckers change trees after 10 to 20 seconds. This can again be tested, and again falsification might result in further relaxation: woodpeckers change trees every minute, or at least every hour, or at least once a day. With each relaxation, the theory provides us with less and less information about what woodpeckers do, but is more and more likely to be confirmed by observation. A final formulation, that woodpeckers change trees, is eminently confirmable but so uninformative that nobody cares (Krebs 1980).

Good theories are risky (Popper 1985). They have a high 'predictive information value' (Krebs 1980) because they are, among other things, very precise. Few ecological theories can be described in this way. Our most precise theories are often false, whereas our most accurate theories exclude so few possibilities that they are uninformative. This raises the further difficulty that such undiscriminating theories are confirmed by most possible observations. Imprecise theories therefore protect the status quo and impede scientific growth.

The imprecise theories of ecology include many verbal models, trend analysis, graphical analyses with unscaled or relatively scaled axes, purely algebraic manipulations, and regressions with very broad confidence limits. In short, any theory that provides only qualitative predictions or which refuses to stand by the details of its predictions may be dangerously imprecise. In ecology, such theories are so common and so diverse that they must be illustrated at some length.

Verbal models

Verbal models state a theory in prose. Often these are preliminary stages of theory construction which organize thoughts prior to further definition and quantification. The difficulties of verbal models are that few

Table 7.2 *Community assembly rules (Diamond 1975).*

(1) If one considers all the combinations that can be formed from a group of related species, only certain ones of these combinations exist in nature.
(2) Permissible combinations resist invaders that would transform them into forbidden combinations.
(3) A combination that is stable on a large or species-rich island may be unstable on a small or species-poor island.
(4) On a small or species-poor island, a combination may resist invaders that would be incorporated on a larger or more species-rich island.
(5) Some pairs of species never coexist, either by themselves or as part of a larger combination.
(6) Some pairs of species that form an unstable combination by themselves may form part of a stable larger combination.
(7) Conversely, some combinations that are composed entirely of stable subcombinations are themselves unstable.

scientists write well, that words are notoriously slippery vehicles for precise thoughts, and that preliminary stages in theory development are often confused. As a result, the thrust of verbal models is often blunted, the terms of the models are open to interpretation, and confirmation or rejection of hypotheses is very much a matter of judgement. Moreover, it is easy to become enmeshed in the verbal net we weave to explain our own ideas.

Community assembly rules
Diamond (1975) summarized his observations on the birds of the Bismarck Island in seven 'community assembly rules' (Table 7.2). These rules represent a verbal model, expressing a widely held belief that community composition reflects a natural order. Whatever the validity of this belief, the assembly rules are decidedly unsatisfying. I remember vividly the sense of expectation on first reading Diamond (1975), hoping to discover how to predict the assembly of communities, and the depth of disillusion when I realized what the paper provides.

Connor & Simberloff (1979, 1984a) examined these assembly rules in detail and claim that the rules are 'either tautological, trivial, or a pattern expected were the species distributed at random'. Tautological clauses in Table 7.2 are those that identify certain species combinations as permissible (clause 1), forbidden (clause 2), stable or unstable (clause 3) or resistant (clause 4) apparently only because these combinations were observed, unobserved, variable or persistent respectively; because the rules are said to apply within guilds (Gilpin & Diamond 1982, 1984), clause 1, that some species combinations within guilds cannot coexist, also implies clause 5, that some species pairs never coexist. Moreover it is logically necessary that more combinations of species be found on islands that have more species (clause 3) and that fewer combinations will occur on species-poor islands (clause 4). That these patterns hold true for large and small islands, as well as for species-rich and species-poor islands respectively, restates qualitative, imprecise patterns established in *The Theory of Island Biogeography* (MacArthur & Wilson 1967). Clauses 6 and 7 seem ad hoc attempts to save the argument by admitting exceptions.

Diamond's community assembly rules have become a focus for bitter debate, but most argument involves the technique of generating 'null models' to explain community assemblages as casual processes (Connor & Simberloff 1979, 1984a, b; Strong 1980; Colwell & Winkler 1984; Grant & Schluter 1984; Schoener 1984). Null models try to establish the prior probability of observing what was observed. For example, a response to the chance meeting of an old friend far away from home might be to exclaim, 'What's the chance of meeting you, here, today?' Treated in this way, the chance of doing any particular thing at any particular time becomes infinitely small, because each personal history is unique in some sense. A more appropriate, if less effusive, question would be to ask 'What are the chances that I would meet some old acquaintance, someplace else, sometime?' The prior probability of the second question, which more accurately reflects the situation to be explained, is obviously much higher (Nisbett & Ross 1980; Downing in press).

Connor & Simberloff (1979) try to make the same distinction with regard to island fauna by asking, not what are the chances that a particular pair of species will be allopatric or sympatric, but what are the chances that any pair will be allopatric or sympatric. In other words, they ask what pattern would be expected and then seek to compare this expectation with observation. Calculating the expected probability *a priori* is difficult in both the case of the old acquaintance and that of the animal communities. Connor & Simberloff (1979) used the observed number of bird species on each island, the observed number of islands on which each species appears, and the observation that some species are more likely to occur in depauperate communities than others. They then used Monte Carlo techniques to simulate the expected number of co-occurrences of all possible pairs and trios of species on the islands, under these constraints, and compared these expectations with the observed patterns. They conclude that they see no reason to reject the null hypotheses that species assemblages are casual in three of five cases they examined.

The models appear to show that a realistic distribution can be generated without abstracting much specific biological information and they circumvent the temptation to explain selected individual observations by reference to specific, but still speculative, biological processes (Diamond 1975, Diamond & Gilpin 1982; Gilpin & Diamond 1982, 1984a,b). Debate concentrates on questions about how 'neutral' such null models can be, because the data used to generate the null model could incorporate 'non-null' processes and because the rules used in the simulations are ad hoc (Colwell & Winkler 1984; Gilpin & Diamond 1984b; Harvey *et al.* 1983).

Regardless of the outcome of the argument about null models, the unanswered charges that the community assembly rules in Table 7.2 are trivial and tautological is more damning to the scientific pretensions of the rules than is the competition offered by null models. These telling criticisms have received scant attention (Diamond & Gilpin 1982; Gilpin & Diamond 1982). Debate about community assembly has ignored the inutility of the rules in predicting the composition of the bird communities on these islands or of other communities elsewhere. Insistence that scientific differences be assessed by reference to predictive ability rather than mechanistic argument might have forestalled this divisive scientific conflict or at least allowed a less ambiguous resolution.

Patterns in food webs

Pimm (1982) provides a second example of the poverty of verbal models in ecology in his summary of patterns among food webs (Table 7.3). Like Diamond's (1975) assembly rules, some of the patterns are tautological, either by definition (clause 1 – no predators without prey) or because Pimm imposed the condition (clause 1 – no loops), or because some rules restate others (clauses 7 and 8). Simplification (minor flows are suppressed, so singularities and omnivory disappear – clauses 1 and 3) and aggregation (Paine 1988) of the original observations may produce or enhance some patterns (clause 2 – webs are not complex; clause 9 – predator species outnumber prey species). Some rules are ad hoc devices to remove exceptions involving insects (clause 5) and detritivores (clause 6), but since insects and detritivores dominate many systems (Odum 1971; Cousins 1985), these clauses also make the rules exceptions in

Table 7.3 *Food web patterns (Pimm 1982).*

---

(1) There are no loops in which an animal directly or indirectly feeds on itself, there are no predators without prey, and there are no 'singular' systems in which all connections are made.
(2) Food webs are not too complex.
(3) Omnivores are scarce, usually only one per top carnivore.
(4) Omnivores usually feed on adjacent trophic levels.
(5) Food webs dominated by insects, their predators, and parasites have more complex patterns of omnivory.
(6) Donor controlled (e.g. detritivorous) webs have more complex patterns of omnivory.
(7) Compartmentalized webs are rare and where they exist they represent different habitats.
(8) Webs within habitats are not compartmentalized.
(9) There are more predatory species then prey species.
(10) Predators that have many species of prey have fewer species of predator.
(11) Prey overlaps are more common than might be expected by chance.

---

themselves. At least one of the claims seems erroneous (clause 3 – omnivores are rare; Fig. 4.3). Without considering the problems of identifying trophic levels, complexity, or habitats, the eleven rules are therefore of low predictive content. This is most apparent if we ask what they tell us about any particular organism, system or set of systems which we are likely to face in nature.

Pimm (1980) had previously reported that omnivory is rare in most food webs, but this also reflects his definition of omnivore (an organism that derives at least 20% of its food from non-adjacent food levels in the same ecosystem). The rarity of such omnivores may represent real trophic structure, but it could also reflect a tendency of ecologists to ignore omnivores in drawing food webs, because they are less tractable, because numerous trophic relations inevitably reduce the relative sizes of individual relations until most are ignored, or because they render the resultant food web unaesthetic (Paine 1988). Pimm & Lawton (1978) suggest that food chains rarely exceed five or six steps, but this may simply reflect a convention among ecologists; it is certainly a logical consequence of the convention of removing 'infinite loops' in which organisms feed cannibalistically, or on their own predators. Such loops are a common feature of aquatic systems (Hardy 1924; Larkin 1978), but their extent is generally unknown because they are unevaluated in trophic analyses (Rigler 1975b).

The rules for community assembly and food webs (Tables 7.2 and 7.3) represent qualitative, verbal models. They fail as useful ecological tools because they indicate only weakly what we may expect from nature. Because the predictions are not risky and because exceptions are explicitly allowed, the rules are protected from falsification. Because some rules are tautological restatements, one confirmation can be counted several times, thus inflating the apparent corroboration and hiding predictive weakness. Finally, both formulations suffer from operational problems that make most tests ambiguous and apparent refutations inconclusive.

Trend analysis

Monotonic trends constitute another type of qualitative theory. They can yield more precise predictions than verbal models, but are still only weakly informative. Trend analyses are frequent in ecology, and one may hope that they will eventually lead to more rigorous formulations. However, as long as the theory remains a simple claim that other values will usually be higher (or lower) than a given case, we cannot claim very much information. This may not always be apparent, for many trends seem to promise much more than they deliver.

Succession

E. P. Odum's (1969) analysis of trends in succession provides a case in point (Table 7.4). Stated as trends, these patterns have little impact. The difficulties of identifying different seral stages impedes application of the trends across ecosystems and the uncertain, but long, time needed to follow succession in a single system limits the trends' utility there. Moreover, to apply the theories represented in Table 7.4, we must be sure that the system will indeed undergo succession, but this is no longer considered a certainty (Drury & Nisbet 1971, 1973). Several trends depend tautologically on each other, e.g. the complex of diversity, evenness, information and entropy (Horn 1974; Peters 1976) and others are consequences of body size (e.g. changes in life span, yield, P/B, energy flow/biomass, stability, population growth curves, nutrient exchange rates), size class diversity (spatial heterogeneity, food web connectivity, diversity; Peters 1983) and biomass (e.g. intrabiotic nutrient, spatial heterogeneity, organic matter, detritivory: Peters 1976). Increases in biomass and size might in turn be part of the definition of succession, trivializing much of the argument. The critical tests of our

Table 7.4 *Two sets of trends in ecology.*

	Increasing	Decreasing
Succession to climax (Odum 1969)	Biomass	Production/biomass
	Organic material	Nutrient exchange rates
	Intrabiotic nutrient	Mineral export
	Body size	Energy flow/biomass
	Life Span	
	Stability	J-shaped growth
	Spatial hererogeneity	
	Information	Entropy
	Diversity and evenness	
	Connectivity	
	Symbiosis	
	Life cycle complexity	
	Detritivory	
	Quality of production	
	Specialization	
r-selection to *K*-selection (Pianka 1970)	Successional stage	Population variation
	Size	r_{max}
	Competitive ability	Survivorship
	Social behaviour	Rate of development
	Density dependence	Age of reproduction
	Efficiency	Productivity
	Life span	Colonizing ability

faith in these trends is to ask what characteristics of a particular site or set of sites are unlikely in a finite period of time. Any prediction on this basis is either foolhardy or vague. Analysis of trends associated with 'stress' in ecosystems (Odum 1985) has many of the same drawbacks.

The *r–K* continuum
Pianka's (1970) classic paper on *r-* and *K*-selection provides a second example of trends in ecology (Table 7.4) and similar dichotomies or trichotomies have been suggested for other selection gradients by Grime (1974, 1977), Southwood (1977, 1988) and Greenslade (1983). These continua can rarely be used directly because the continua themselves are operationally undefined: species cannot be unambiguously classed independently from the suite of characteristics that the classes are said to have. Thus the only trends that are identified by the *r–K* continuum are

weaker relations among characteristics. For example, long-lived species are more likely to occur in predictable sites. Moreover, the trends are thought to be loose so that no characteristic increases in step with every or any other. Thus exceptions to any trend are expected, but the frequency of exception is unknown. Once again, the list is lengthened by logical interdependencies − J-shaped growth curves are implied by catastrophic, density-independent mortality and high rates of population growth; rapid development may be synonymous with early reproduction; colonizing ability surely reflects occurrence in early succession; and if 'environment' is defined relative to the community, then life-span and environmental predictability are also logically interdependent, since organisms cannot survive in environments that are 'unpredictable' over their lifespan. Of the trends that remain, it is clear that many are simply correlates of size (Southwood 1981; Calder 1984). Indeed, because size is a much better predictor than the continuum, the only place that the $r-K$ concept should be used is to explain residual variation around allometric trends (Stearns 1983; Saether 1987). In short, Pianka's scheme has the same weaknesses as Odum's, for it too associates size with seral stage, and therefore increasing biomass and size, and most of the consequences of the $r-K$ continuum follow from this association.

This criticism of the content of two influential works does not deny their heuristic roles in the growth of empirical studies of the quantitative effects of body size on life history phenomena. That is not in dispute. The point is instead that these crude and speculative summaries of trends have also had a major influence on the status of trend analysis in ecology. They have been referenced and reproduced in most ecology texts. They have led to a cycle of modified, additional, improved and imitative analyses of trend. They have become a norm for ecological achievement that advocates qualitative statements, even when more precise alternatives are available or easily produced (e.g. Peters 1983; Calder 1984). They encourage us to accept weak theory, when stronger theories are possible.

Graphical analysis

Graphical analyses involving unscaled or only relatively scaled variables are another common form of qualitative argument in ecology (Schoener 1972). They have advantages over verbal models in that they have a high visual impact, their implications are very clear and they can offer more detail than a simple statement of monotonic trend. However, like verbal models, predictions from qualitative graphical models are easy to

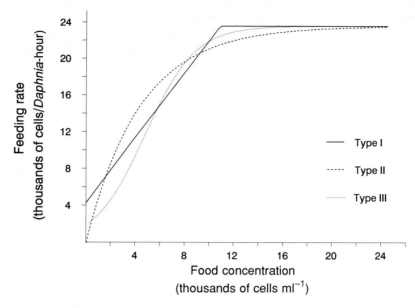

Fig. 7.2. Holling's (1959) three functional response curves representing invertebrate predation (Type I), satiation (Type II) and a vertebrate response (Type III). These curves represent a form of qualitative graphical model which is widely used in ecology, although the general model provides little information and specified models are of limited application because the three types are often indistinguishable. These curves were generated from the equations of Porter *et al.* (1982).

confirm and hard to falsify. For example, the hump-backed curves relating fish recruitment to stock, maximum sustainable yield to effort, or species richness and diversity to disturbance regime are consistent with data showing positive, negative, or flat responses to change in the predictor, for all these responses correspond to some part of the hypothesized 'hump'. The only logical possibilities that these models exclude are therefore that a rising leg of the response curve will not follow a declining leg as fish stock, effort or disturbance increases. In other words, qualitative graphical models are very accommodating because they are confirmed by most observations and falsified by very few.

Functional response

Three types of functional response curve (Fig. 7.2) describe the increase in ingestion rate with food supply. The curves are presumed to represent

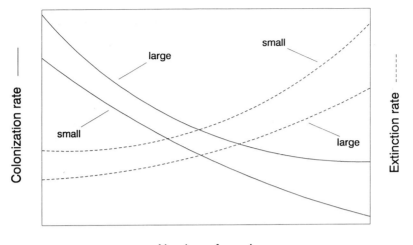

Fig. 7.3. The effect of island size on immigration rate, extinction rate and the equilibrium species number. Despite the appearance of precision and generality, this qualitative model actually says very little because the curves are arbitrary and the axes unscaled (from MacArthur & Wilson 1967).

a primitive satiation response, said to be typical of invertebrates (Type I), a satiation curve like that produced by the Holling (1959) disc equation (Type II), and a learning curve (Type III) thought typical of vertebrate predators. Although this interpretation of these curves has found wide acceptance, they differ so little that they can rarely be distinguished in practice, given the quality of even very good data (Rigler 1971; Peters 1984; Porter, Gerritsen & Orcutt 1983; Lampert 1987). Moreover, these qualitative models do not exhaust the repertoire of functional responses in even primitive invertebrate predators (Downing 1981). The clarity and appeal of the graphical analyses of these virtually undetectable among these curves have encouraged many theoretical and experimental studies, but simultaneously their appeal has discouraged the use of alternative models, the growth of different approaches, and the appreciation of contrary data.

The equilibrium theory of island biogeography

Figure 7.3 provides another qualitative, graphical analysis. This familiar diagram shows that large islands have higher immigration rates, lower extinction rates and more species at equilibrium than small islands (MacArthur & Wilson 1967). In principle, the equilibrium number is predictable from the rates of immigration and extinction, but this is rarely used in practice because the model's parameters are unavailable. These qualitative patterns are not particularly surprising. The graphical model has therefore gained acceptance, not because it makes risky predictions, but because it offers an engagingly simple, mechanistic explanation for a number of phenomena and potential application to a range of theoretical and practical problems (Rey 1984). Moreover, a number of observations and experiments have qualitatively confirmed various aspects of the model (Simberloff 1974; Connor & McCoy 1979; Diamond & May 1981; Rey 1984).

The success of quantitative derivatives from this model must be distinguished from the predictions of the model itself. The shape and magnitudes of the relations in Fig. 7.3 are unspecified and depend on a number of factors including the size and vagility of the source fauna or flora, the range of habitats and environments offered by the islands, and the distance of the islands from the source. Without controlling or quantifying these factors, the basic model holds only very few, imprecise and banal implications: larger islands tend to have more species and larger populations, and these larger populations are less prone to extinction than small populations on small islands. Quantification of the relations implied by Fig. 7.3 would replace the general graphical theory with empirical, statistical relations that predict by reference to previous observations on a particular fauna and island group. This development depends on the heuristic or inspirational power of the original model, but we must be careful neither to credit that model with the predictive success of its empirical replacement nor to allow the original model to outlive its usefulness. It would imply no disrespect for the seminal contribution of MacArthur & Wilson (1967), if a modern assessment of the status of island biogeography showed that the scientific community could move beyond their initial vague and qualitative model.

Resource competition

A more complex graphical analysis has been developed by Tilman (1982) in his analysis of 'resource competition' (Fig. 7.4). So long as the models are based on unspecified parameters they can only yield qualitative patterns: the failing competitor is the species which cannot survive at the equilibrium resource level established by initial resource levels, resource supply, resource consumption, and minimal resource requirements of the consumers. A large simulation (Tilman 1988) shows that, under the assumptions of his model, a number of predictions could be made if the parameters can be specified for all species in the plant community. Unfortunately, specifying the parameters results in a drastic loss of generality, because it limits the model to a single system, and is so large an undertaking that it likely confines predictions to highly controlled laboratory systems. Until these difficulties are resolved, the original qualitative forms are only mildly informative.

Algebraic models

Almost paradoxically, purely algebraic models represent a further instance of qualitative analysis in ecology (Wiens 1983). Algebraic manipulations of conjectured parameters have the appearance of quantitative rigour but in fact, yield only qualitative patterns. Like graphical models, algebraic analyses identify a small number of processes as important and are particularly appropriate for investigating relations among these processes. However, algebraic models have an advantage over graphical analyses because they specify the parameters to quantify and the form of the model; this should encourage the transition to quantitive treatments and theories. On the other hand, graphical and especially algebraic models may be more specific than is justified by our current knowledge (Schoener 1972) and thereby unduly restrict our imaginations.

Algebraic analyses are frequently more challenging for biologists than graphical treatments, so the two are often combined. Both analyses are less ambiguous than verbal models, but their terms are open to misinterpretation and qualitative patterns inherent in the models can be established without operational definition of the parameters and variables. As a result, many algebraic models are used to develop the necessary logical relationships among variables and the qualitative patterns these would induce.

Plant–herbivore relations

Riley (1965) used a simple algebraic equation to describe the change in algal biomass (B) as the sum of gains through production (p) and losses through zooplankton grazing (g), algal sinking (s) and water mixing (m):

$$dB/dt = B(p - g - s - m) \qquad (7.1)$$

This equation states only that the rate of change of phytoplankton biomass is increased by growth and reduced by three sources of mortality. Caughley & Lawton (1981) provide a similar equation for

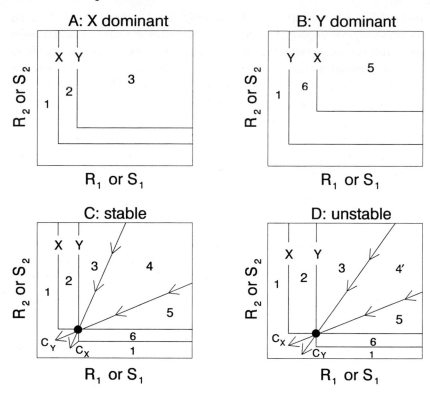

Fig. 7.4. Resource competition between species X and Y. The solid lines parallel to the axes represent the 'zero net growth isoclines' of each species as functions of the levels of two resources R_1 and R_2. At its isocline, a population has just enough resources to persist. Below the isocline, resources are insufficient and the population declines to extinction. Above the isoclines, each species grows, consuming resources, and therefore moving resource levels closer to the isoclines.

In panel A, species Y reaches its zero growth isocline while species X continues to grow, ultimately reducing resource levels to the X isocline. Since this is everywhere below the Y isocline, species Y will be driven to extinction. Panel B shows the reverse situation, where species Y survives and species X does not. Where the zero growth isoclines intersect (panels C and D), neither species dominates at all resource levels. Instead, species X dominates at resource levels where its zero isocline lies below that of species Y (region 2) and where the trajectories of resource depletion by both species (C_x, C_y) drive the amount of resource 1 below the Y isocline (region 3). Species Y dominates in the corresponding regions (5 and 6). The outcome in the intermediate region (4) depends on the rates of resource consumption and the trajectories of resource depletion they produce. In panel C, consumption patterns are such that each species limits itself more than its competitor; as a result, consumption drives the resource levels towards a stable equilibrium point (solid dot) where both species terrestrial herbivores, without the terms for sinking and mixing. In both cases, the algebra is so elementary that the models assume interest only when the terms are replaced with quantities so that predictions may be made. Caughley & Lawton (1981) see this as an unrealistic starting point for further analysis in terrestrial systems, but Riley (1965) was able to make astonishingly good predictions when he replaced the algebraic terms in his simple model with empirical estimates. The distinction between utility and failure of the algebraic approach depends on empirical definition, because only this gives the equation sufficient precision for interesting tests.

Holling's disc equation

An equally familiar example is the disc equation (Holling 1959) that models the rate of energy gain (R) as a function of the rate of encounter with individual prey (E), the average energy content of the prey (e), the cost of searching between encounters (c), and the average handling time per prey (h):

$$R = (Ee - c)/(1 + Eh) \qquad (7.2)$$

This states that, if a predator is concerned only with hunting and if food supply is strictly a problem of encounter and capture, then R will rise at a decreasing rate with E, but decline as searching costs and handling time rise. Its conceptual appeal lies in describing a concave 'functional response curve' relating R and E; its empirical attraction lies in its ability to describe some feeding responses (Ware 1978; Stephens & Krebs 1986).

Like the $r - K$ continuum of Pianka (1970) and the successional trends of Odum (1969), these qualitative graphical and algebraic models have had a major influence on ecology. Part of this influence has been positive, for the models have clarified thinking, served as sources of inspiration, and prepared the ground for more precise approaches to these particular problems. The point of this critique is, however, that these benefits are not unmixed with negative effects reflecting the qualitative nature of the

Caption for Fig. 7.4 (*cont.*)

can coexist. In panel D, patterns of consumption are such that growth of one species inhibits the other more than itself, and depending on initial population size and resource levels, consumption in region 4′ will result in resource levels moving toward one region of dominance or the other, not to the equilibrium point. Thus one species will dominate except when the population is precisely at the equilibrium point (from Tilman 1982).

models. These may have slowed the broad development of quantitative theory by offering qualitative role models which are only weakly predictive and have been disproportionately credited with the successful predictions of their quantitative successors.

Simulations and regressions

Quantification is no panacea, because numerical relationships can still be extremely imprecise. Despite the mathematical rigour of deterministic simulation models, their quantitative output is usually interpreted as a trend (Walters *et al.* 1980). Regressions can be so imprecise that their estimates of the mean are useless, and their confidence limits so wide that the prediction is likely to be consistent with almost any informed guess. For example, the fiducial limits around the allometric relationship for mammalian carnivore density in Fig. 2.1 cover more than three orders of magnitude. Even widely accepted empirical relationships, like the allometric relations for life history phenomena (Blueweiss *et al.* 1978; Calder 1984) and predicted chlorophyll concentrations (Dillon & Rigler 1973) typically range over an order of magnitude. Although these rank among the best available ecological relationships, they are frankly only a step beyond the qualitative models that preceeded them (Rigler 1982a) and many steps away from precise prediction.

Costs of qualitative theories

An emphasis on qualitative trends has a number of disadvantages. Qualitative science feeds on itself. It trains us not to look at numbers, not to use numbers, and not to appreciate the added strength of quantification. Birks (1987) points out that in biogeography, qualitative science has allowed us subconsciously to bias types according to our precepts and led us to see groupings where none exist. Moreover he warns that once a typology has been erected, new observations are easily fitted to that model. Thus the contrast between *r*- and *K*-selection accommodates almost any contrast in life history phenomena (Parry 1981). It is difficult to believe that the 'permitted combinations', 'related groups' and 'guilds' of Diamond (1975), or the food webs of Pimm (1982), have escaped the same trap (Paine 1988).

The difficulties of qualitative science reappear whenever ecologists attempt to influence governmental policy in favour of a conservationist option. When confronted by the demands for cost-effective measures, for political compromise, and for realistic advice, ecologists are in a weak position to make practical recommendations for environmental action (Schindler 1987). Because we cannot say how much action is enough, we often offer all-or-none solutions. Because we cannot predict what will happen if there is too little action, we cannot effectively assess our options in terms of dollars, jobs, human health, or even environmental damage. Since the opponents of environmental action are rarely so vulnerable, policy makers must choose between tangible benefits and vague ecological threats. Ecologists can scarcely be surprised if their hand-waving is ignored.

This failure of the qualitative approach has had several unfortunate results. We have undermined our own credibility in our areas of special expertise and our opinions about global problems carry less weight than they once did. Failure to influence policy effectively has eroded the optimism with which scientists approach governments. Many ecologists, and particularly student ecologists, now presume that government has no ear for their views. This cynical presumption can rationalize withdrawal from public affairs and immersion in purely academic problems. It also removes the best available sources of sound advice from the councils that decide environmental policy. Consequently, environmental leadership has fallen to well intentioned, but too often ignorant, political activists who seek their ends by appeals to emotion and on occasion even by terrorism.

Governments are not always the puppets of vested interest. Given sufficient information, they have proven willing to invest great sums in environmental projects. An example I know well is the institution of eutrophication control by phosphorus abatement (Forsberg 1987; Schindler 1987; Chapra & Reckhow 1983; Reckhow & Chapra 1983). Another involves the liming of lakes and rivers to offset some of the effects of acid rain in Sweden and Norway (Sverdrup, Warfvinge & Bjerle 1986; Forsberg 1987). Such programs are expensive, but the quantitative predictions concerning the costs and benefits of abatement left little doubt as to the advisable policy. Perhaps if we could give similarly detailed and credible advice in a range of other areas of concern, our opinions would again be valued by the policy makers in government. Such a development would benefit all.

Generality and specificity

Every biological phenomenon is a special and unique event, but the description of unique events cannot constitute predictive science. There are too many phenomena and, to the extent that a phenomenon is unique, we can have no prior knowledge about it. Instead, we must use our knowledge of general patterns to set the particular case in the context of scientific theories, operating at various levels of generality from the specific to the general.

Most scientists prize generality. They admire and seek to build theories which apply across broad areas of our experience. General theories allow us to make and to judge statements outside of our narrow specialties, they place our studies in perspective, and they provide a basis for communication about common problems among scientists working with different organisms, sites, or processes. Popper (1985) stresses generality because general theories are subject to more tests and are therefore more 'risky' than specific ones. Ecological managers prize general relations because they can be accessed and used by those who do not have the skills, experience and information of the specialist. Finally, teachers and students need general theories because they cannot accommodate all the specific information available. Instead they depend on mastery of general theories to provide the paradigms for specific interpretations in the light of later specialization and experience. In short, generality is desirable because general theories provide more information than specific ones.

There are costs involved in the search for generality (Wiens 1983). General theories ignore many differences among ecological entities. They may sacrifice precision for breadth of application and are therefore unappealing within the domain of one's specialty. Moreover, each case presents some specific characteristics which, when taken into account, could improve predictive power. The problem confronting ecological prediction is therefore not whether one should prefer general or specific theories, but what balance should be sought between the two and what costs are involved in favouring one aspect over the other.

Ecologists frequently favour specificity over generality. Every volume of the primary ecological literature contains papers in which some obscure herb or process or ecosystem has been studied, apparently for its own sake – because 'it was there' – or because it was seen as a 'model system' for the study of some process. Examples are unnecessary. A recent trend in the critical literature regards this specificity as worthy of emulation, for the critics of theoretical ecology (e.g. Dayton 1979; Strong 1984; Simberloff 1980a; Wiens 1983) stress the need for specific studies of specific processes and organisms.

I cannot share this enthusiasm for specificity: the argument for specific studies can lead to an infinite regression to still more detail and the need for this information can excuse ecological inaction while more information is gathered (Chapter 5). The first part of this section points out that some general theories provide predictions which are as good as or better than those offered by highly specific alternatives, but that, even if imprecise, general theories may serve to validate and direct specific research. The second part of the section argues that model systems in ecology are a treacherous basis for important decisions, because they can be inappropriate. When highly specific information is required, specific studies are prescribed, but these must be specific studies of the object of interest, not of a model.

General relations and specific information

The question of whether a general relation is sufficient for a given purpose or if more specific information is needed has no single answer. The decision rests on the nature of the question and the quality of the competing theories. This can be illustrated by a series of examples in which the general relations of allometry overlap areas where I can claim some degree of 'special knowledge'.

A case in which a general allometric relation (Peters & Wassenberg 1983) fails to predict particular cases was mentioned earlier in this chapter with respect to bird densities (Juanes 1986) and need not be re-examined. The opposite case, in which general relations are more accurate and precise than a specific relation, is provided by comparisons between general models of nutrient release by zooplankton and some specific estimates (Rigler 1973; Peters & Rigler 1973; Peters 1987), in which the absence of a general model allowed erroneously high and low values to persist undetected in the literature because no one recognized them as implausible. That general models themselves sometimes do not apply (Olsen & Østgaard 1985) indicates that such comparisons are always relative.

In other cases, the general relations perform just as well as more specific information. Except in some special cases when 'foods' are toxic or of inappropriate size, general relations for the feeding rate of zooplankton estimate adequately the measured values for *Daphnia* (Lampert 1987). Similarly, the error associated with general curves predicting the metabolic rates of poikilotherms is not larger than the variation among estimates for either the genus *Daphnia* or for individual species of that genus (Peters 1987). In these cases, the knowledge of the specialist does not allow more precise prediction and therefore offers no advantage.

In most cases, general allometric relations are less effective than specific ones (Peters 1988a), but still allow approximations which may be sufficient for some purposes, but not for others. General relations describing the allometry of life history apply to both *Daphnia* (de Bernardi & Peters 1987) and primates (Peters 1988a), but in both cases, more precise and accurate estimates can be made by reference to specific relations for these groups. Which relations to use will depend on the question and our command of a diverse literature. Parenthetically it should be noted that this command depends on the ease of access to specific information. Eventually, computer-assisted retrieval ('expert systems') may shift the balance in favour of the most appropriate relations by minimizing the problems of availability and accessibility.

Model systems

One rationale for the intensive examination of particular ecosystems, species or processes is that the object of study may be considered a model for the behaviour of other more interesting, but less tractable systems. For example, Rigler (1978) hoped that the ecosystem of Char Lake in the high Arctic might be easier to treat than those in southern lakes because the fauna is depauperate. Wiegert (1975) chose the fly–algal system of hot springs for similar reasons. A variety of microcosms (Elliot *et al.* 1983) and mesocosms (Pilson, Oviatt & Nixon 1980; Bloesch 1988) have also been used as simple analogues of more complete, more complex and more difficult natural systems. For example, microcosms have been used to study bioaccumulation (Metcalf 1971), succession, competition (Frank 1952; Park 1962), and nutrient cycling (Whittaker 1961; Confer 1972). Similarly, some studies of specific organisms are undertaken because the species involved lends itself to investigation. The giant axon of squids has made them popular experimental animals in neurophysiology; rapid generation times and giant chromosomes have made *Drosophila* important for genetics; and the visibility of birds has made them attractive for the study of foraging behaviour (MacArthur 1958; Stephens & Krebs 1986). The rationale for specific studies is therefore popular. Unfortunately it courts irrelevance and inconsistency, for it is necessarily based on untested analogy.

Hume's argument against induction is also an effective argument against analogy (Ollason 1987). There is no logical reason that because two objects share characteristics A, B and C, that they should also share further characteristics D and E. To take Ollason's amusing example, the many physical similarities of two people, like Margaret Thatcher and

Mother Theresa of Calcutta, provide no basis for the contention that both have similar characters. Thus an analogy based on Mother Theresa's actions may not extend perfectly to Mrs Thatcher. To provide a more ecological example, Pianka (1988) suggests that the Australian wombat (*Vombatus*) is an ecological equivalent to the North American groundhog (*Marmota*). Unfortunately, we cannot be sure in what way this information can be used since, despite some similarities, the two animals differ in so many particulars (size and its correlates, life history, climate) that we are at least as likely to be wrong in divining the character of one animal from specifics of the other as we are likely to be right. This reservation likely holds for most cases of ecological convergence.

If the completeness of ecological convergence is suspect, it is even harder to accept that one species, community, or ecosystem can serve as a model for another when evidence for parallel ecology is weaker. Thus toxicologists recognize 'the myth' that there is a 'most sensitive species' (Cairns 1986), or other model system (Kimball & Levin 1985; Baudo 1987) which could serve as the basis of the definitive test in ecotoxicology: no single model, whether species or experimental system, can tell us everything we need to know about a potential environmental contaminant. Similarly, each set of animals potentially differs in so many ways that no single community can serve as a model for all others. For example, intensive studies of bird communities (e.g. MacArthur 1957; Fretwell 1972; Cody 1974) can only provide examples of how communities might be arranged. Before they can be used as models of other communities, it will be necessary to separate the aspects of the community behaviour of birds that reflect avian idiosyncracies (e.g., high visibility, low clutch sizes, high metabolic rates, diurnality, low population densities, migration, flight, etc.) from the aspects that can legitimately be expected in other groups. In practice, this means that community behaviour cannot be extrapolated from intensive studies of particular groups, but must be established by studies across the range of existing communities.

Model systems are analogies, and analogies are too undependable to serve as theories. Prediction differs from analogy because the prediction is justified by a theory that identifies distinct phenomena as instances of the same type. Consequently, some aspect of these phenomena can be deduced by reference to the theory. By definition, analogy deals with phenomena that are of different types, and therefore no such justification is possible. Analogy and metaphor may serve as a basis for conjecture, but until confirmed by test on the scale of interest, these conjectures must not substitute for theory. As a result, experiments with microcosms, mesocosms, or other models can only claim to be heuristic. Their extension to systems of real interest must be established by a further round of studies of those systems (Heath 1980; Schindler 1988). Given that limitation, it is always more desirable to work at the scale of real interest and to ignore the models.

This weakness of model systems is implicit in the rationale for their use. The model systems are used because they are significantly different from the real system in some way that renders their important characteristics easier to study. Thus the experimenter begins research knowing that the process of interest in the model system differs significantly from that in the real system of interest. It seems self-delusory to believe that this difference holds no other implications for the interpretation of the results. At most, model systems may help us select a few variables from a larger range of possibilities for study in the larger system of interest, but even then we must accept that winnowing of the variables in a model system may be irrelevant in the system of interest.

Economy

Scientific theories are needed to deliver us from doubt and to help manage the world around us. Particularly useful theories are economical in the sense that they provide relatively large amounts of information for relatively little effort. Thus economical theories are those that, on the basis of only a few, easily measured, predictor variables and boundary conditions, provide precise and accurate information about a phenomenon which would be comparatively very difficult to measure.

Economy of effort is a rather difficult property to address because the degree of economy depends on the context of the question. For example, the great effort and expense of weather forecasting is warranted because it gives us precious information about tomorrow's weather today and because, without this effort, the information would be unavailable. However, it would be very uneconomical to determine today's weather on the basis of yesterday's data plus the same forecasting tools, because a much simpler alternative, to look outside, is available. It is also uneconomical to use a meteorological approach to forecast weather more than a week in advance, because after seven days, meteorological forecasts are less effective than climatological averages (Tribbia & Antheis 1987). Similarly, the effort of palynology is warranted when it allows us to glimpse the species composition of long-lost plant assemblages, but would be wasted in determinations of the composition of contemporary forests.

Nevertheless, some ecological theories require information which is extremely hard to obtain. At best, such theories are needed only very rarely, because it will almost always be much easier to measure the predicted variable directly than to use the theory. Such theories engender research programs in which the main goal is to make the relevant measurements so that the theory may be applied, rather than to make or test a prediction.

Uneconomical constructs are widespread in ecology, probably as a reflection of our disinterest in the application of ecological theories. For example, Hutchinson (1959b) noted that coexisting, closely related competitors often differ in body weight by a factor of about two. He proposed that this may represent a limiting similarity beyond which competition leads to exclusion. Since this seminal work, the identification of Hutchinsonian ratios in guilds of species has become a centre of interest in ecology (Roughgarden 1983; Schoener 1984). Although the constant ratio now appears to be an artifact (Eadie, Broekhoven & Colgan 1987), the approach is also subject to the equally serious flaw that it is tremendously uneconomical. A generation of ecologists have busied themselves with a theory that predicts something as easy to measure as the ratio of body sizes, from something as difficult to measure as membership in the same guild (Connor & Simberloff 1984a). I can imagine no situation in which two or more species could be supposed to be members of the same guild, without any estimate of the ratio of their body sizes being available.

Large simulation models are also likely to prove uneconomical (Watt 1975). Clark et al. (1979) point out that highly complex models are likely to founder in unending data requirements and instead recommend ruthless parsimony in model construction. Beck (1981) similarly recommends small models as easier to use, less expensive to run, and more readily tested.

Finally, the controversial use of null models to identify alternate mechanisms underlying observed biogeographic patterns would represent a special case of an uneconomical model. This reservation is unconnected to the use of null models as points of contrast to see if observed distributions differ from those generated by a few widely accepted 'random' models. Biogeographic null models normally require extensive investigation of the fauna about which the 'casual', null distributions are to be generated. For example, the null model for island avifauna of the Bismarck islands discussed with respect to qualitative, verbal models, requires site visits to each of the islands included in the simulation. Therefore the null models require the same quantitative information about species co-occurrences as did the assembly rules, but because they make fewer assumptions about the patterns underlying the distributions, they actually make fewer predictions (Roughgarden 1983). These null models are primarily means of generating alternate historical explanations to demonstrate that point-by-point explanations are unnecessary. They may forestall the waste of studying a putative process which does not actually occur, but they do not offer more effective predictions. As predictors of the overall pattern, they are as uneconomical and uninteresting as the community assembly rules (Table 7.2) themselves.

Appeal

Scientific research is never finished until it is published and published research is never effective until it is read and has influenced the scientific community. Thus good science must have appeal, both so that other scientists will read the work and so that they will remember it. Appeal is a difficult criterion to examine because it is an amalgam of strong science, good presentation, commonsense, insight, and inspiration into a communication which is particularly effective for a given audience. Appeal includes the ideas of elegance, consistency, and beauty which Fagerström (1987) espouses. It also includes clarity and coherence that Stearns & Schmid-Hempel (1987) see as desirable for theoretical models. Most importantly, appealing theories are fruitful and productive because they lead researchers to form ideas they would not otherwise have had (Kuhn 1977; Fagerström 1987; Stearns & Schimd-Hempel 1987).

Although appeal is an important element in the success of particular scientific constructs, the criterion of appeal seems the most easily abused and most desirable to suppress of all the criteria cited in this book. Since beauty is in the eye of the beholder, Fagerström's (1987) elegance and beauty are subjective. Consistency and coherence, where they can be distinguished from accuracy, are too likely to justify uncritical common-sense notions about how the world should work. Clear and effective presentation is desirable, but care must be taken so that stylistic abilities remain handmaidens of science, not replacements (Symanski 1976). Fruitfulness and inspirational productivity are precious, but they must be measured in terms of the scientific theories that are produced, not in terms of busy-work inspired (Dayton 1979; Gray 1987). The weakness of ecological theories that is the theme of this book makes any appeal to fruitfulness of ecological writings dubious, if not contradictory.

I would not have included appeal as a criterion in this chapter were it not so frequently abused in ecology. Ecologists study mating in dung-flies (*Scatophaga*) because their competition for mates on the dung is a tractable system to study mate searching under intra-sexual competition (Parker 1978). Ecologists follow scorpion-flies (Mecoptera) because the mating gift of the male provides a measurable index of reproductive investment (Thornhill 1984). And ecologists study the Florida scrub-jay (Woolfenden & Fitzpatrick 1984) because of the prevalence of helpers at its nest. These are fascinating biological processes which may warrant extensive investigation; they have captured the imagination of ecologists as natural historians, and they appeal to our non-scientific sides. But these appealing areas are not models or examples for tests of general phenomena and should not be given that status. Ecologists must guard continuously against the temptation to allow the appeal of particular subject material to overrule more concrete and less disputable criteria in judging ecological theory.

Summary – Practicality and appeal

The predictive powers of all theories depend on a series of characteristics. Important theories are those that are relevant to the question in hand, and to the major questions of the discipline or to society; predictively powerful theories are those that have proven accurate, general, precise, quantitative and economical in application. In ecology, these virtues are often ignored in favour of theories which appear attractive, even though they are inaccurate, restrictively specific, imprecise, qualitative and uneconomical. This mistaken emphasis on appeal rather than effect continues to erode the standards of the science.

8 · Checklist of problems

Most of this book deals with weaknesses in the central constructs of ecology. In contrast, most ecological research addresses smaller, more tractable questions which relate only obliquely to the central tenets of the discipline. Many ecologists may feel that the faults discussed so far have limited bearing for their ecological specialties, and specialists may be tempted to ignore criticisms of general ecology. The same conclusion is encouraged by the exceptional character of the general constructs that are the focus of most ecological discussion and criticism. Lessons learned from the likes of Darwin, Elton, Hutchinson, MacArthur and May are hard to apply to the bulk of the scientific literature, simply because there are so few contributions of that stature, scope and form. Right or wrong, ecological syntheses rarely follow the standard format of a scientific paper; they rarely consider methods, and they rarely address the details of a particular case. In short, because seminal works in ecology are rare and exceptional, both their virtues and their faults misrepresent the field.

Dismissal of general ecological criticisms as irrelevant to the bulk of working ecologists would be unjustified. Intellectual leadership in a field is awarded by consensus of the field, and flaws in leadership reflect on the entire field. Those who read ecological journals and especially those who review manuscripts or research proposals know the problems of normal ecology and the many ways in which the frailty of the science is manifest in its primary literature. No critique of ecology would be complete if it confined itself to the problems of our leaders and ignored those of normal ecologists.

This chapter outlines common flaws in normal ecological articles, under the premises that the weakness of general ecology reflects weaknesses at its scientific base, and that the lack of an effective standard for scientific achievement has encouraged laxity at all levels. To a large extent, the chapter will cover familiar ground because the difficulties it addresses are those of writing or delivering a good scientific paper. This topic is covered by books, articles and the 'Instructions to Authors' found in most journals, and that large part of university education involving laboratory reports, term papers, and thesis evaluations that seeks to transfer acceptable standards of scientific reporting to the next generation.

Writing a scientific paper is not a so much literary exercise as a scientific one. The hallmarks of a good paper – logic, clarity, precision – are those of good science and the goal of scientific exposition, like that of research in general, is to provide a good piece of science. Like good science, good writing is straightforward, concrete, exact, rigorous, clear-headed and concise (Woodford 1968). When this is not so, more vigilant criticism is needed.

To facilitate application, this outline follows the classical format of a scientific paper – Introduction, Methods, Results, and Discussion – pointing out common deficiencies in each section like a checklist or field-guide. This device provides a structure for the chapter, but should not be taken rigidly, for ecological problems are rarely confined to single habitats. Since the weaknesses of larger ecological constructs are repeated in these smaller studies, there is necessarily some overlap between the criticisms of this chapter and those that went before, but this has been minimized.

A listing like this can never be complete, so the chapter is only a beginning for criticism of particular contributions. The reader is encouraged to expand this list, to modify it and to illustrate it with appropriate examples in the light of personal experience and particular application.

The Introduction

The Introduction of any scientific paper must do three things: It must establish the relevance of the topic for an informed reader; it must provide a context for the study by describing the status and shortcomings of present knowledge; and it must relate the study's goal to this context. For example, a paper introducing a model of the reacidification of lakes after neutralization with lime (Sverdrup & Warfvinge 1985) might begin by relating lake acidification and liming to the problem of conservation:

Anthropogenic acid precipitation threatens the biota of many lakes in granitic basins, but symptomatic relief may be had by liming.

Next there should be a concise statement of the present state of knowledge about liming, specifying the lacuna in this knowledge that the present paper will address:

Models exist to calculate the dose and form of calcium carbonate required to neutralize acidified lakes (Sverdrup 1983), but not the dose which optimizes the schedule for future treatments following the inevitable reacidification. Instead it is recommended simply that excess carbonate be added, although reacidification must depend on annual acid load to the lake, the flushing of the neutralized water from the lake, and the solubilization of sediment carbonate.

Finally, the introduction must explain how the paper fills this lacuna:

This paper (a) describes a model of reacidification that is based on flushing time, acid load, and the amount and distribution of carbonate on the sediment and (b) tests this model against rates of reacidification observed in limed lakes.

In general terms, good introductions present the questions being asked as hypotheses under test and show where these questions are relevant. Such introductions are easy to write if the study has been directed by explicit, testable hypotheses identifying the goals of the research. These goals should have been established when the research was proposed, not after the research is complete. Given a sound proposal, writing a good introduction is relatively simple, because its elements have been clear since the beginning of the project.

Many ecological papers fail to meet the elementary requirement of a clear hypothesis. In review of 87 publications about optimal diet theory, Gray (1987) was unable to determine what was being tested in 30% of the cases. A decade earlier, Fretwell (1975) reported that the proportion of papers in *Ecology* which tested explicit or implicit hypotheses had risen from about 5% of the total in 1950 to 1955 to almost 50%. This great improvement should not obscure the fact that, in 1975, over half of the papers in *Ecology* apparently did not address hypotheses. Such papers are extremely hard to judge or read because their authors' intentions are never clear.

A clear introduction is the key to a successful paper because it defines the hypothesis under test and therefore the question on which the paper depends. When this is done, the methods can be judged as to whether they are adequate to the test, the relevance of the results is clear, and the direction of the discussion is determined. However, this is possible only if the proposed question is clear and testable, if the methods chosen to make the test were appropriate, and if the results provide a clear answer to the question. These are requirements for both good papers and good science, but they are routinely ignored in ecology where the vagueness of questions addressed in single rescarch programs reflects a vagueness in the discipline. The uncertainty of ecology results in vague purpose and ultimately in vague papers. This is only compounded by the perceived personal need to publish at all costs so that even inconclusive tests now clog the literature.

Selection of hypotheses

The first problem for the scientist is to identify a relevant question or hypothesis to test as the goal of the research. This is likely also the most difficult problem, because it is the creative step. Once this step is taken, the rest of science involves deduction to determine what the hypothesis predicts, followed by the methodical application of relatively conventional intellectual and technical tools, like the appropriate instrumentation and statistics, to determine if the deduction holds. Most scientists consequently feel competent to test relevant hypotheses once these have been identified, but only a few are as confident of their continuing ability to form such hypotheses.

The challenge of selecting relevant topics for research explains the difficulty and frequent failure of impact assessments (Schindler 1976; Larkin 1984; Hecky *et al.* 1984). Valid assessment depends on what is assessed. Whether it is the agency requesting the assessment or the firm conducting it, someone must determine what environmental features are relevant in light of the proposed development for the site. Often this question is evaded with unthinking, routine observations about the presence and absence of different species and standard measurements of a few variables. However, because such variables are not always obviously pertinent to the problems posed by development, effective assessments require decisions about what, and in what way, observations are relevant.

The selection of variables to study requires fundamentally scientific decisions. Since identification of a single relevant variable tests the best minds of the science, it is scarcely surprising if ecologists in industry, who may lack the appropriate tools, surroundings and temperatment, sometimes find themselves stymied. As a result, poor impact assessments are so common that good ones stand out (Alexander & Van Cleve 1983; Berger 1977) as significant ecological achievements.

The difficulties in determining relevance were considered in the last chapter, so this section instead considers some of the ways ecologists have circumvented these problems of relevancy.

Justification by authority

Scientists idolize their heroes. The Newtons, Galileos and Eisteins are role models, even though very few researchers can achieve this stature (Glaser 1964). In ecology, this hero–worship is reflected in the zeal with which modern writers legitimize their position with quotes from

Darwin, Hutchinson or MacArthur. Some leading ecologists – Dayton (1979) cites Hutchinson, MacArthur, the Odums, and Watt – have had a Messianic appeal for their followers, so the views and interests of those men have defined much of modern ecology. Regrettably, the tentativeness and self-doubt that intellectual leaders commonly bear towards their own work are often lost when the approach is appropriated by acolytes whose admiration of the leader is so intense that they harden the great figure's ideas to an inflexible dogma.

One effect of charismatic authority is that the concerns of the master may become central to a field, without asking whether these are otherwise relevant or if they are scientifically tractable. Thus Hutchinson's famous question 'Why are there so many kinds of animals?' is still a central issue of ecology (Tilman 1982; May 1986), even though the form of the question begs an explanation which science can never provide. A similar situation arose when ecological opinion divided over the issue of whether the mechanism of population control was density-dependent or independent. This intractable question so fascinated leading ecologists, like Lack and Nicholson, that it overshadowed tractable problems associated with predicting population densities (Murdoch 1970; Strong 1983).

Ex cathedra statements

Ex cathedra statement represents a further aspect of authority often found in the introductions to ecological papers. Such assertions appear to be authoritative statements of fact, but are really only opinions that must be treated with circumspection. For example, the claim of Cates & Orians (1975) that slugs are generalist herbivores is central to their study of plant palatability, but the claim is not supported by rigorous studies of the diet of slugs relative to other organisms, but by unreferenced common knowledge that slugs eat a wide variety of plants. Similarly, the introduction of Macevicz & Oster (1976) declares that the social insects conform to a basic assumption in optimal foraging models that the systems be evolutionarily static. This claim is justified because 'the basic colony structure and strategies have remained largely unchanged for many millennia', but Macewicz & Oster (1976) cite no basis for such a seemingly substantive claim. Janzen & Martin (1982) base their argument for the coevolution of the trees and extinct megafauna of tropical America on anecdote, and Lehman (1986) suspects that 'applied ecology will be best served by continued basic inquiries into process and mechanism', but cites no evidence for that opinion. The critical reader should be suspicious of any unsupported assertions. Ex cathedra pronouncements may be an effective ploy in debate (e.g. Levins & Lewontin 1980), but unsupported opinion and hand-waving are treacherous foundations for intensive and expensive scientific research.

Emulation of authority

The attraction of authority has the further disadvantage that followers imitate the style of the great scientist as well. Thus MacArthur's iconoclasm freed his followers from stodgy rules of contemporary scholarship and breathed fresh air into the top ranks of American ecology (Fretwell 1975). This freedom is not wholly laudable, for MacArthur's style also seemed to sanction spotty and selective citation (Schoener 1972) and the evaluation of loosely formulated models with weak data (Fretwell 1975). Hutchinson's restrictive and idiosyncratic interpretation of the hypothetico-deductive method, whereby the hypothesis consists of an analogy between a syllogism and nature (Hutchinson 1978; Kingsland 1985), became the model for a generation of the brightest minds in the field, sometimes directing them to study syllogisms instead of nature.

In a field where general constructs are scientifically weak, the behaviour of those who built the constructs can be a poor example. Emulation tends to perpetuate and obscure the problems at the top by repeating them at the bottom. Thus mathematical, logical and verbal analogies or metaphors, inspirational writings based on vague and undefined concepts, why-questions about cause or mechanism, weak models, poor data, and unsupported opinion have become common elements at all levels of the ecological literature. However, the blame in this state of affairs does not belong to the leader, but to the followers, for science does not oblige us to follow anyone.

Faddism

The task of defining the hypotheses under test can be avoided by simply collecting data that are only 'related' to contemporary interests in some unspecified way. Unfortunately, refusal to specify relevant goals fosters research which is so unconnected to the rest of the field that it may wax and wane without affecting the science. Although such ephemeral research areas are easily identified as fads in retrospect, the failure of a research program to define its goals operationally and the circumlocutions that legitimize undefined research as 'providing insight' or 'furthering our understanding' are among the characteristics that should raise the critic's suspicions about faddism.

Dayton (1979) and Simberloff (1982) term 'competition' a fad, but

Abrahamson, Whitman & Price (1989) were unable to identify fads in the publication rate of papers within such large fields. If fads exist in science, they occur within narrow topics and the term should be reserved for lower elements in the hierarchy of ecological constructs. Thus competitive exclusion, limiting similarity, Hutchinsonian ratios, diffuse competition, and ecological divergence seem closer to the everyday use of the word 'fad' intended here, because interest in these areas is less enduring and more variable than the sum of all interests in the many aspects of competition, predation or life history.

Because fads are restricted in both the extent of their application and the duration of their interest, they are usually easier to identify within one's speciality. For example, a list of fads in freshwater biology (Peters 1988a) might include the measurement of primary production using $^{14}C - CO_2$ (Vollenweider 1971, and personal communication), nutrient turnover (Rigler 1975a), the size efficiency hypothesis (Hall *et al.* 1976), the sequence of studies of increasingly smaller organisms as we moved from studies of fish and zooplankton in the nineteenth centuries, to net plankton early in this century then on to nannoplankton, bacterioplankton, ultraplankton and picoplankton (Reynolds 1984; Stockner 1988). Other limnologists study the trophic cascade (Carpenter *et al.* 1985, 1987), nutrient responses and loadings (Peters 1986), top-down versus bottom-up control (McQueen, Post & Mills 1986), biomanipulation (Shapiro & Wright 1984) and food web manipulation (Benndorf 1987). In streams, the concepts of nutrient spiralling and the river continuum (Vannote *et al.* 1980; Statzner & Higler 1985; Minshall *et al.* 1985) are currently 'hot topics'.

These areas need not be weak. A great deal of good science can be done within them. Problems arise when researchers forget that their role is to test hypotheses and instead make observations that only 'pertain' to a topic in some unspecified way. Rigler (1982b) parodied this approach in an amusing model of the normal introduction to an ecological paper (Table 8.1). Such alternatives are especially tempting if it is easier and more enjoyable to make observations than to build and test hypotheses (Keddy 1987). In ecology, this choice is biased towards data collection by the pleasures of field work and away from hypothesis testing by the effort of intellectual rigour.

The trap of originality
Because many of the central issues of ecology are confused or untestable, ecologists are necessarily uncertain about what information the science

Table 8.1 *A substitute for hypothesis formation and a model for the Introduction of many ecological papers.*

X is a very
(*abundant, unusual, economically important*)
type of
(*species, community, ecosystem*)
Although other similar
(*species, communities, ecosystems*)
have been exhaustively studied, and some aspects of X have been studied
(references inserted here to demonstrate a knowledge of the literature)
X's
(*feeding, production, physiology, lead content, etc.*)
has not. Therefore I decided to study
(*one of the last set*).

Note:
Unimaginative researchers need only select specific terms from between the brackets to create the outline for a normal Introduction (from Rigler 1982b).

needs. Nevertheless, as good researchers, ecologists recognize clever and engaging research, independent of the problem under study or its place in science. When decisions are made about future research, funding or publication, such originality in research design is often rewarded at the expense of less technically brilliant and inventive research that might yield more relevant and useful information.

In science, ingenuity should be the handmaiden of wisdom. Wisdom leads to significant questions about nature, and ingenuity to the most effective solutions to those questions. Uncertainties about what is relevant to ecology have encouraged ecologists to skip the first step and to judge research in terms of an inadequate criterion of ingenuity alone. This substitution is a perversion of science.

Change and novelty have no scientific value in themselves. Theories and facts neither wear out with repeated use, nor should they become stale and unappetizing with age. Nevertheless, research fads and bandwagons, thoughtless reverence for new machines and techniques, disdain for the older literature, and evident preference for cunningly conceived research into trivial ecological hypotheses on the part of major journals, especially *Science* and *Nature*, show that contemporary ecology has placed too high a premium on novelty and originality. This is particularly problematic in America where government funding agencies prefer 'neat and nifty' ideas over painstaking scholarship and useful data, to the detriment of both science and society.

Verification and falsification

The imprecision of many ecological constructs makes them easy to confirm and difficult to falsify, because they are inconsistent with so few possible observations. As a result, it is easy to design research programs that will verify weak hypotheses, so verification must not be the goal of research (Murray 1986). As Dayton (1979) puts it, if ecologists are only interested in verifying a given thesis, this can be achieved most easily by designing experiments which maximize the chance of finding information that confirms the theory, and minimize any possibility of rejection. Such tests confirm the prejudices of the scientific community and are therefore likely to meet a sympathetic audience in both the reviewer and future readers. Introductions to such papers are pernicious because, if the research was undertaken to support an hypothesis, one purpose of the Introduction must be to obscure the intentional weaknesses and bias of the test.

Other introductory flaws

Logical lapses

The Introduction should form a logically ordered whole. Instead, relevance and context are sometimes 'demonstrated' by a bald list of other works in the area, with little attempt to describe the essence of this work, much less show how this precedent, thoughtfully considered, necessitates the present study. Sometimes, the relevance, context and lacunae described are vast, but the study is so narrow that the paper can only disappoint. Sometimes, the test is relevant to only part of the general hypothesis discussed in the Introduction so the reader feels cheated by the bit of information the paper holds. Sometimes the general hypothesis is so redefined by the test, that the test has little meaning for the hypothesis as it appears in the statement of relevance. The best defense against these non sequiturs is careful attention to the logical connections between each sentence and each paragraph (Woodford 1968).

Stock problems

Some pitfalls for Introductions to scientific papers need no discussion. The description of the paper's place in science should be brief, so that its point is not lost in a welter of references and detail. Since a research report is not a literature review, the point of the Introduction is to provide a logical context for the research and support for substantive claims by citing key references defining the hypothesis under test, rather than to prove one has read widely. This does not excuse a poor grasp of the literature. The lacuna in scientific knowledge identified by the paper must be real and the approach subsequently used to address that lacuna must be sound, not reflections of insufficient reading or selective interpretation.

Methods

The sections of a scientific paper are read in a different order from that in which they are presented (Woodford 1968), and doubtless with a different frequency, enthusiasm and impact. The actual reading order is Title, Abstract, Figures, Tables, Introduction, Discussion, Results, and Methods. This is understandable, but regrettable, for proper evaluation of a scientific paper requires evaluation of its techniques. Failure to read the Methods places more trust in the referees and authors than is extended to the remainder of the paper. This places a special onus on the authors as they select, describe and consider the methods.

The purpose of the Methods is to describe techniques in sufficient detail that an informed reader would be able to judge and repeat the work. The methods are more easily described if the researcher uses standard procedures, wherever these are appropriate, and is consistent in the application of whatever techniques are used. If modified or new methods are introduced, the novelties should be described, justified and tested. Explanation of the purpose of the major techniques, relating them briefly to the hypothesis under test, helps order and demystify methodological description. Many descriptions fall short of these standards.

Technical problems

Many of the shortcomings of the Methods involve highly technical and specific points that lie beyond the responsibilities of general criticism. One general weakness associated with these technical aspects involves the search for a 'technological fix' in the belief that better or just newer instrumentation by itself can resolve the problems of ecology. This can produce a succession of newer, larger, more expensive devices to monitor ecological systems at ever finer scales in space and time. It some areas, like physiological plant ecology, biogeochemical cycling, and remote sensing, this has led to a technological race, like the arms race, as different laboratories vie for the latest in analytical power.

Better analytical capacity is essential to test hypothesis that require better measurements, like those involving trace quantities of toxicants or minor nutrients. However more-sophisticated measurement must not become an end in itself, justified on the grounds that it somehow gives a truer picture of nature. Such a claim cannot be substantiated, because there are no independent estimates of what true pictures of nature might be. Our knowledge of the world consists of theories about how observations hang together (Hutchinson 1953) and the most sophisticated measurement has meaning only if incorporated in a theory relating that measurement to other variables of interest. The introduction of new instrumentation to a field should therefore illustrate the greater utility of the new measurements to theory.

New technology should not be applied for its own sake, and the pressure to justify the expense of new technologies by publication must be resisted. Papers that simply describe new methods seem to have little impact because they are among the least cited (Jumars 1987). This contrasts sharply with the success of papers that describe well evaluated methods with a clear role in the science; the latter are the most cited of all (Garfield 1988).

Citation analysis suggests that new technologies initially experience a phase of expansion, as they are examined and adopted by laboratories world-wide. Subsequently, citation rates may slow as doubts and second thoughts accumulate, and finally a steady, low level of citation reflects use of the technique in a few specific tasks (Peters 1989a). In aquatic sciences, this description seems to fit the evolutions of the ^{14}C-bicarbonate estimates of primary production in the phytoplankton (Vollenweider 1971) and of electronic particle size analyses in zooplankton feeding experiments (Peters 1984). No doubt technologies in many parts of ecology share a similar history.

'Mathematistry'

An almost identical situation occurs with respect to new mathematical and statistical tools. Periodically, the ecological literature is swept by claims for new calculating devices, like the logistic (Kingsland 1982, 1985), the simulation models of the 1970s (Patten 1975; Rigler 1976), catastrophe theory (Thom 1975), fractals (Mandelbrot 1982), and chaos (Gleick 1987). These may all be useful tools in certain situations, but initial propagandizing often oversells their importance and utility whereas initial 'applications' seem little better than self-justified abstractions (Hall & DeAngelis 1985) that may only repeat scientific common-places (Gray 1987). Box (1976) called this concern with mathematical abstractions 'mathematistry'. Critical readers in ecology should be wary of papers which serve only as vehicles for mathematical abstractions (Slobodkin 1975; Levin 1981b).

Statistical considerations

The most common weaknesses of the Methods are likely to be statistical. Proper tests invariably invoke statistical comparisons and these comparisons are generally stronger if the study is designed with a particular test in mind. Ecological statistics is too vast a field to be summarized here and much lies well beyond any claim of competence on my part. Statistics are better learned from direct applications of the statistics in the context of one's own research, supplemented whenever possible with appropriate readings, texts and courses. In the absence of such experience, Green's ten rules (Table 8.2) provide a concise summary of statistical advice for biological research, and will usually be a sound basis for critical assessments of the literature. Those rules related to the necessity of estimating the uncertainty in measurements merit the emphasis of reiteration here.

The difficulty of effective controls

Green's rules must be applied with the knowledge that real data only approximate his desiderata, and that too rigorous an interpretation may prove counterproductive. For example, it is unlikely that 'true' controls, those that differ only in one characteristic, are ever achieved in biology. This is particularly evident at the ecosystem level and Likens (1985) suggests that the term 'reference system' be used instead of 'control' for such comparisons.

Whatever term is used, the role of a reference or control is to determine the behaviour of systems not subject to treatment. This requires an estimate of both mean and variation associated with untreated systems which in turn requires replicated controls. Since the temporal variation within a single system can differ from spatial variation (Knowlton, Hoyer & Jones 1984; Likens 1985; Schindler 1987, 1988; Marshall, Morin & Peters 1988), reference systems must be either run parallel to the treatments or replicated in time. They must be replicated in space in any case. Without replication, the results will only support untested speculations, not valid conclusions. For example, the whole lake experiments of Carpenter et al. (1987) involve comparisons

Table 8.2 *Ten statistical principles for ecological research (Green 1979).*

(1) Be able to state concisely to someone else what question you are asking. Your results will be as coherent and as comprehensible as your initial conception of the problem.

(2) Take replicate samples within each combination of time, location, and any other controlled variable. Differences among can only be demonstrated by comparison to differences within.

(3) Take an equal number of randomly allocated replicate samples for each combination of controlled variables. Putting samples in 'representative' or 'typical' places is *not* random sampling.

(4) To test whether a condition has an effect, collect samples both where the condition is present and where the condition is absent but all else is the same. An effect can only be demonstrated by comparison with a control.

(5) Carry out some preliminary sampling to provide a basis for evaluation of sampling design and statistical analysis options. Those who skip this step because they do not have enough time usually end up losing time.

(6) Verify that your sampling device or method is sampling the population you think you are sampling, and with equal and adequate efficiency over the entire range of sampling conditions to be encountered. Variation in efficiency of sampling from area to area biases among-area comparisons.

(7) If the area to be sampled has a large-scale environmental pattern, break the area up into relatively homogeneous subareas and allocate samples to each in proportion to the size of the subarea. If it is an estimate of total abundance over the entire area that is desired, make the allocation proportional to the number of organisms in the subarea.

(8) Verify that your sample unit size is appropriate to the size, densities, and spatial distributions of the organisms you are sampling, then estimate the number of replicate samples required to obtain the precision you want.

(9) Test your data to determine whether the error variation is homogeneous, normally distributed, and independent of the mean. If it is not, as will be the case for most field data, then (a) appropriately transform the data, (b) use a distribution-free (non-parametric) procedure, (c) use an appropriate sequential sampling design, or (d) test against simulated H_0 data.

(10) Having chosen the best statistical method to test your hypothesis, stick with the result. An unexpected or undesired result is *not* a valid reason for rejecting the method and hunting for a 'better' one.

among three lakes representing two different treatments and one control; those of Schindler (1978) only one treatment and one control. Because the designs preclude any estimate of the variation associated with either treatment or control and because the number of degrees of freedom in any comparison is zero, no statistical comparison of the lakes or treatments is valid, and treatment effects cannot be distinguished from casual differences unassociated with the manipulations. Such distinctions can only be provided by lumping these treatment comparisons with other similar data from our experience outside the experiment, not from the limited evidence of the experiment alone.

Ineffective controls are not restricted to the ecosystem level where the scale of the study may put proper control beyond the budget of most ecological researchers. They are a common, but even less acceptable, feature of tests at all scales in ecology.

The necessity of replication

To determine if treatments have had a significant effect, ecologists often compare the means of different treatments or they correlate and regress responses against treatment levels. In all cases, effects can be distinguished only if they are large relative to the inherent variability of replicates and the number of samples that can be taken. Thus, variability is best estimated in a pilot study and can then be used to estimate the number of samples required to test for effects of a given magnitude (Elliott 1977). This is part of 'power analysis' (Toft & Shea 1983; Peterman 1990), for it indicates the statistical strength or power of the comparisons.

If the expected magnitude of the effects is so small that it is unlikely to be detected with realistic sampling effort, the study should be abandoned. If pursued, it will prove an inconclusive waste of valuable scientific resources. Insufficient sampling will incur a 'Type II' error in which no significant difference is found, even though such a difference may exist. For example, Emlen (1986) was unable to accept his own findings that Hawaiian bird densities were unrelated to site productivity, partly because his censuses were sufficiently erratic to obliterate any effect of productivity. The tests were invalid due to Type II error reflecting the poor design of the study.

The dangers of pseudo-replication

Hurlbert (1984) has identified a further pitfall associated with the analysis of controls and treatments: the biasing of the estimates by 'pseudo-replication'. Pseudo-replicates are repeated samples which misrepresent the range of possible values that treatments or controls might take because the replicates are not statistically independent and the number of independent measurements overestimated. For example, Fowler & Lawton (1985) criticize experiments purporting to show that herbivore damage to one tree induces a defense reaction in others, because of pseudo-replication. In this case, replicates did not involve the susceptibility of a series of different trees, but the susceptibility of one tree to a series of different larvae. Variations in herbivory estimated in such an experiment reflect variation in larval feeding rate, not variation in tree susceptibility, and any shift in the mean degree of herbivory on this tree might reflect any characteristics that distinguish this tree from the control, not just the effect of treatment. Even different trees in the same incubator may be pseudo-replicates because their response could as easily reflect an incubator effect as a treatment effect. A similar doubt hangs over the work of Hassell, Southwood & Reader (1987) who analyzed the population dynamics of white fly (*Aleurotrachelus jelinekii*) larvae on individual leaves of viburnum (*Viburnun tinus*). Since all the leaves were on a single bush, the supposedly independent populations could have instead been entrained and correlated by the behaviour of their single host.

Pseudo-replication is related to the older requirement that the sample represent the population. It differs because it does not emphasize the 'true' nature of the population being sampled, but the undesirable effects that interdependence of the samples can have for statistical description of the larger population. The result of this bias is that the parameters need not apply to the population as a whole or to other populations. Any statistical tests invoking parameters derived from pseudo-replicated studies could be invalid.

Problems of representative sampling

The problems of taking representative samples and of its converse, sampling bias, have had a long history in ecology (Southwood 1966; Elliott 1977; Downing 1979; Green 1979; Morin 1985). The problems endure because they are insoluble and they are insoluble because we can never know the reality beyond our samples. We are therefore limited to comparisons among different samplers and protocols to determine which comes closest to the device and technique we consider most reliable. Reliability is usually subjective.

A more effective criterion for the best sampling protocol would be the utility of the estimate, where utility refers to the function of the estimate as a variable or boundary condition in one or more theories. The best sampling strategy results in estimates that lead to the best predictions within logistic constraints. Such estimates are 'representative', because they provide more information. Critical scrutiny of the methods should therefore consider the potential utility of the estimates in available theory and ensure that the methods represent the variables as defined in the theory under test.

It is usually assumed that the most reliable samples are also the most useful. That this is not always the case is demonstrated by the coexistence of pure and applied science (Cartwright 1983). Applied science exists as a separate field because pure science does not work well in application; presumably pure science exists because it is considered more reliable (i.e. realistic) notwithstanding. Comparisons of the utility and reliability of planktonic chlorophyll a concentrations estimated with sophisticated high pressure liquid chromatographic (HPLC) and with standard spectrophotometric analysis of crude pigments in an organic extractant provide an ecological illustration. HPLC estimates are surely more reliable, but crude extractions are far more informative, because they appear in predictive relationships for transparency (Carlson 1977), fish production (Jones & Hoyer 1982), zooplankton biomass (McCauley & Kalff 1981) and primary production (de Lafontaine & Peters 1986). The HPLC estimates cannot be introduced to such equations, because these reliable estimates bear no consistent relation to the crude extractions on which the informative regressions are based (Jacobsen 1975; Sartory 1985); HPLC estimates may be up to an order of magnitude below the crude estimates and would therefore underestimate the response variable in these equations. It is even doubtful that relations built on reliable HPLC chlorophyll a values would perform as well as the existing relations. The crude extracts likely reflect the range of biologically important properties of the phytoplankton community better than the sophisticated measurement. This situation will be resolved only when freshwater biologists give up the fiction that traditional techniques actually measure 'chlorophyll a' and adopt a different term for the object of the crude extractions.

Results

'The Results' describe the observations, establishing what differences were significant and what were not. Since the study should be directed towards an hypothesis, these findings are easier to grasp if they are related

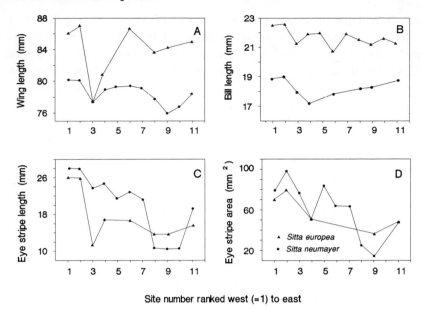

Fig. 8.1. Peak analysis of the covariance between four morphological characteristics of two sympatric species of rock nuthatches in 11 areas lying along a cline from Yugoslavia to Iran. Although the original author felt that these show 'good correspondence between the species', his own correlation analyses revealed significant correlations only in the case of eye stripe area and bill width. The latter relation was not plotted here because it was omitted in the original (From Grant 1975).

to the hypothesis under test, leaving comparison of the results with the literature for the Discussion.

Results and Discussion are frequently fused to avoid needless repetition. Although these papers may be easier to write and read, the merger may obscure the distinction between observation and conjecture. Writers who fuse Results and Discussion should therefore be especially careful to separate the two trains of thought with appropriate paragraph structure and verb tense.

Description of the Results

One of the most vexing problems in assessing papers in contemporary ecology is to discover just what the results were. High publication costs, editorial precepts, competition for journal space, and the awesome capacities of modern computers encourage extensive reduction of the data before publication. Although some reduction of the raw data is essential to presentation, this can be carried so far that the observations are obscured. Even when familiar data-reduction techniques are applied, the results are often so digested that no check on the accuracy of the author's calculations is possible (Gray 1987). This might be acceptable if ecologists made few calculation errors, but that premise would strain the credulity of the most casual critic. Full disclosure of the data is an elementary precaution against both honest mistakes and charlatans. In any case, the financial support provided by society is normally given so that the results can be used by the community and scientists should therefore feel obligated to make their results public.

The problems of data access could be easily circumvented if authors accepted that publication of the research entails the availability of the data to the community. Extensive data may be placed in inexpensive data banks and repositories dedicated to this purpose (Prothero 1986; Downing 1979), but most scientists seem unaware of such services. In my experience, direct requests for the data behind published reports are usually either refused or ignored. Many scientists are too busy or lazy to prepare data in accessible form, some are too jealous or defensive to relinquish their proprietary rights, but most simply cannot find the old data. In all cases, their behaviour is antithetical to the open commerce of knowledge upon which science depends.

Analysis of the results

Once the results are established, they must be analyzed to determine whether patterns predicted by the hypothesis under test occur or not. Attendant problems are normally associated with inappropriate techniques for pattern analysis. Once the analysis is complete, it should be a simple process to decide if the hypothesis has been supported or not.

Bias in peak analysis

A preliminary step in the analysis of some data arrays is to plot different variables along a single axis. The plotted arrays are then examined for co-occurring peaks and troughs in the data set, so the technique can be called peak analysis (Fig. 8.1). This approach is often used to compare biological response and environmental factors along geographical or temporal gradients, but it is equally effective in any ordered display of the data.

Peak analyses are only appropriate for casual explorations of the data, because they are biased. Outliers are highly influential in identifying peaks and troughs, whereas the bulk of the data receives little or no weight. In addition, the correspondence between peaks and troughs in different variables is often hard to judge because the data are presented in different panels. This uncertainty is sometimes abused by implicit or ad hoc time lags to allow for imperfect matches in the data. Data may be ordered in this way if the analysis examines the relation between the response variables (e.g., phenological patterns or geographical gradients in biological response) and the ordering variable (e.g. time, site). Such ordinations are also effective in isolating eccentric data. However, if the point of the analysis is to relate the biological response to some other correlate of space or time, the appropriate visual tool is a scatter diagram relating biological response to the environmental correlate. The appropriate statistical analysis is a regression or correlation, not a verbal summary of the fancied correspondence between high and low values in space and time.

Misapplication of statistical models

Parametric statistics depend on the fit of the data to an abstract statistical model, like the normal or binomial distribution. This is inevitably an approximation (Box 1976), but many statistical tests are robust to such abuse. Nevertheless, there is a real possibility that lack of fit between the data and the assumed statistical model will bias any statistical test. For example, the frequency distributions of many biological data are closer to log-normal than normal (Koch 1966, 1969). When this is the case, the data should be analyzed after logarithmic transformation, for parametric analyses of the raw data would be misleading. This is the contention of Eadie *et al.* (1987) who explain the regularity of 'Hutchinsonian ratios' as an artifact of comparing the arithmetic means of log-normally distributed data. Such confusions can often be avoided by scrutinizing the data to see that it agrees with the results of statistical tests. When this is not so, transformation or non-parametric analyses (Conover 1971) may be required.

Negative evidence

Statistical tests are always probabilistic and therefore they are never certain, even if applied well. The possibility that Type II error might obscure real differences (and therefore lead us to 'accept the null hypothesis of no difference as true when it is false') was considered briefly in discussing experimental design. Type II error is so common in ecology that ecologists are wary of any 'negative evidence' that purports to show the absence of an effect. Such results are less interesting to consider and are generally hard to publish (Connell 1983; Toft & Shea 1983; Rotenbury & Wiens 1985; Peterman 1990). These difficulties can be overcome if the study includes a careful consideration of the possibility of Type II error and if the results of the test are in some sense unexpected. In this case, evidence that no difference was found is fully as informative as 'positive' evidence that differences exist.

Type I error

Type I error refers to the complementary problem of falsely identifying a significant difference where none exists. If the statistical models used are appropriate to the data, the probability of a Type I error is the level of significance (the α level) of the test. Thus when there is a 5% probability that a certain observation occurred by chance, the probability of Type I error is also 5%. Because the fit of the data to the model is always approximate, such probability levels should not be interpreted too strictly; nevertheless they are usually the best available indicators of the chance of a Type I error.

The likelihood of Type I error can be increased if an extensive data set is analyzed for many possible patterns. This might be the case in 'all subset regression' which regresses every variable against every other or in multiple pairwise comparisons. In 100 regressions or comparisons, one would expect to find that five of these were significant at the 5% level, on the basis of chance alone. An example is available in the key factor analysis of *Viburnum* white fly mortality by Hassell *et al.* (1987). This study considered 108 regressions between different life stages of the fly in different years and corresponding population densities, and identified 16 significant relations. Since about a third of these can be expected to represent Type I error, the significance of all 16 must be suspect. Because similar ambiguities result whenever extensive multiple comparisons are made, the critical level of significance must be lowered. A Bonferroni correction designates the new critical level to be $5/n$ % where n is the number of regressions considered (P. Legendre, personal communication). In such cases, significance at the typical levels of 5% or 1% can only be a tentative indication that a relation may exist.

Discussion

The Discussion should further relate the results of the study to the hypothesis, acknowledge weaknesses that may compromise this interpretation, point out the consistency of the results with existing knowledge, and suggest some of the implications of the new results for scientific theory. Like the Introduction, the Discussion is usually one of the most interesting sections to write because it allows the writer some scope. As a consequence, it is also likely to share debilities with the Introduction, including loss of focus with attendant confusion of the reader. Such difficulties frequently result because the author's views are so skewed by precept that the paper is distorted to support these views, rendering the order and selection of material idiosyncratic and illogical. Those who read Discussions, and those who write them, should be aware of the many problems that can be encountered there.

Flew's flaws and fallacies

Many of the problems in scientific discussions occur in most human discourse and are only specific examples of the general difficulty of clear thinking. These everyday problems have been described by Antony Flew (1975) following an old suggestion of the philosopher, Schopenhauer, to give each of many fallacies and logical errors a short and obvious name 'so that when a man used this or that particular trick, he could at once be reproached for it.' Although this chapter can no more do justice to applied logic than it can to statistics, some good comes from a simple list defining Flew's terms with ecological examples. If nothing else, this exercise may encourage others to read his short and witty book, but it should also serve to warn of some of the problems that may be found in the Discussion.

Denying the antecedent and affirming the consequent

In logic, propositions are given the form 'if P, then Q', but there is nothing intrinsic to logic in the form. Because scientific theories are also propositions, they fit the same model: If an animal is a 6-kg North American omnivorous mammal, then its population density will be between 0.2 and 100 km^{-2} (Fig. 2.1). Science must be concerned with both the truth (i.e. the predictive or factual value, which is usually addressed in Results) of the proposition and the logical validity of inferences drawn from the proposition (which is primarily the concern of the Introduction and Discussion). Logic, the topic of this section, deals only with logical validity.

Two valid inferences can be drawn from the proposition 'if P, then Q'. Assuming the proposition holds, then wherever P obtains, Q will also obtain, and wherever Q does not obtain, then P will not obtain either. Thus, if the proposition holds, one can deduce that the density of a 6-kg mammalian omnivore lies between 0.2 and 100km^{-2} and that an animal whose density is 10,000 km^{-2} is unlikely to be a 6-kg omnivorous mammal. Two other inferences would be fallacious: if P does not obtain, there is no reason to assume that Q also does not obtain; and if Q is found, one cannot infer that P also occurred. In the concrete terms of the theory in Fig. 2.1, just because an animal is not a 6-kg North American omnivorous mammal, we cannot infer that its density is not between 0.2 and 100 km^{-2}; and a density between these limits is not enough to identify an animal as a 6-kg North American omnivore.

Reiterating the explanation in Chapter 7, the traditional name for the fallacy of asserting P when Q is observed is 'affirming the consequent', because one correctly affirms that the consequent (Q) was observed in order to assert fallaciously that the antecedent (P) also occurred. The second fallacy is called 'denying the antecendent' because it correctly states that the antecedent (P) was not found but fallaciously concludes that the consequent (Q) could not hold. Both fallacies occur widely in ecology. For example, it may be proposed that if two mice species are strong competitors, then they will be allopatric, but it would be fallacious to conclude from that proposition that allopatric mice are competitors, because this affirms the consequent (allopatry) in a vain ploy to affirm the antecedent (competition). The fallacious claim that mice should be sympatric simply because they are not competitors instead denies the antecendent (competition) to deny the consequent (allopatry).

Similar fallacies are often involved in mechanistic arguments. For example, it might be proposed that if a large composite process, like primary productivity, has a Q_{10} of 2, then the rates of its constituent processes double with each 10°C in temperature. This is a testable scientific proposition. However, one cannot affirm that the Q_{10} of the whole will be 2 because that value was determined by laboratory study of one or more constituent processes, nor can one presume that Q_{10} of the constituents is not 2 simply because the whole responds differently to temperature change. In ecology, these seeming inconsistencies may be explained away with multiple causality and scale effects, but as logical fallacies, they need no such effort.

The Un-American fallacy and the no-true-Scotsman move

The 'Un-American fallacy' is named to recall Senator Joseph McCarthy and the House Committee on Un-American Activities who used the fallacy to identify communists as all those who shared some belief, pacifism or atheism for example, with communists, even though the belief was in no way particularly or necessarily communist. The traditional name is the 'fallacy of the undistributed middle', and this fallacy imputes all characteristics of a given class to any object having one of these characteristics. As a result, any 'average' characteristic applies to all members of the class.

This fallacy can be used to smear opponents in ecology, too. For example, because Christian fundamentalists are opposed to the principle of evolution by natural selection, defenders of evolution may be tempted to protect their position by implying that all evolutionary critics are fundamentalists (Ruse 1982). Lehman (1986) associates the empirical work of his limnological opponents with the fraudulent, but putatively empirical, work of the psychologist, Cyril Burt. The fallacy may find less-emotional application in bolstering typologies: for example, if K-selected species tend to lay large eggs or live long, then it is tempting to class any organism that has large eggs or a long life as K-selected (Pianka 1970).

A second and related fallacy excludes undesirable cases as unrepresentative of the class. Thus a proud Scot might save his nationalist self-esteem simply by reclassifying some deplorably behaved highlander as 'no-true-Scotsman'. In ecology, one might accept both the evidence in Fig. 4.3 that animals which eat both plants and animals are common and Pimm's (1982) contention that omnivores are rare (Chapter 7), by defining omnivores as organisms which take at least 20% of their diet from non-adjacent trophic levels (Pimm 1980). In other words, the omnivores in Fig. 4.3 are 'no-true-omnivores'.

Redefinition is not fallacious or reprehensible in itself. What must be avoided are the logical inconsistencies that result if familiar words are used in unfamiliar senses without first warning the audience of this change in meaning, or if different definitions are confounded, for example, by proposing a new definition but then shifting ground to use the term in its older and more widely accepted sense. In other words, Pimm's assertion has no bearing on the scarcity of omnivores in the English sense of animals that eat both plants and animals.

The logically-black-is-white slide, the genetic fallacy, and the heaper

Definitional problems often arise where distinctions are so vague that a more or less arbitrary line is drawn between classes. 'The logically-black-is-white slide' exploits the resulting uncertainty to argue that all distinctions based on that division are in reality non-existent or unimportant. This argument denies significance to differences in degree and stresses the importance of quality over quantity. This fallacy encourages and is encouraged by widespread acceptance of types, categories and individuals as ecologically important, and by mistrust of the mathematical arguments that often accompany quantitative treatments, even though quantitative treatments lend themselves better to more rigorous and logically transparent analyses than typologies permit.

The arbitrary use of critical significance levels provides a concrete example of the problem. Because any critical level is arbitrary, one may be tempted to assume that the distinction between significance and non-significance is illusory and subject to reinterpretation. Such a refusal to play by the accepted rules and policies of science can represent a significant rupture of the implicit contract between the scientific expositor and his or her audience.

'The genetic fallacy' is a special case of the logically-black-is-white slide. The genetic fallacy holds that whatever developed or evolved from something is still really the same as its antecedent. A chicken is always an egg. Flew (1975) uses the example of Desmond Morris' (1968) contention that man is only a naked ape on the grounds that man evolved from the apes. Leaving aside the issue of which end of the evolutionary continuum should be more offended by this assertion, the naked ape remains an example of refusing to consider differences in degree as important.

The 'heaper' or 'sorites' is the related fallacy that no matter how many small sequential changes occur, there is never a point where one thing changes into another. Thus if a heaper (in Greek, *Sorites*) adds grains of sand to a small pile one-by-one, there is never a point at which this pile changes into a heap.

Three fallacious premises

The 'truth-is-always-in-the-middle damper' refers to doctrinal adherence to the principle of the golden mean. Although the truth may often lie between opposing viewpoints, this cannot be taken for granted and Flew demonstrates that, in practice, the position is incoherent and absurd. If the truth in any debate lies at the mid-point between the two debaters, then the truth will vary with the extremity of the positions taken. A Machiavellian opponent could coerce the truth to any position simply by taking extreme enough positions for the sake of argument.

This might work in bargaining for public opinion, but it ought not sway scientists for whom the value of a construct depends on its predictive value in dealing with nature, not on a gimmick of exposition.

It is regrettable that Flew redefines all of his fallacies in contemporary English, for the long title of 'the whatever-follows-must-be-the-consequence fallacy' is less euphonious than the traditional 'Post hoc ergo propter hoc'. Both English and Latin phrases refer to the fallacy of attributing causal force to whatever antecendent conditions may have been observed or remembered. In ecology, this fallacy frequently occurs as casual explanations used to explain away outliers with ad hoc devices, in the causal attribution following insufficiently replicated treatments, and in discussion of dichotomous natural experiments that relate observed changes to whatever variable the writer feels is important. The problem recurs in the broader stricture that correlation is not causation and reflects the general difficulties and ambiguities involved in causal attribution (Chapter 5).

The 'first maxim for Balliol men' is that even a truism may be true. Flew (1975) introduces this aphorism to counteract the insidious and fallacious premise that only new propositions need be entertained or that propositions may be dismissed simply because they are old, boring or hackneyed. This premise also encourages the pathological emphasis on originality noted in discussing the Introduction earlier in this chapter.

The subject/motive shift and four derivatives

One of the difficulties in writing or reading any scientific prose is to keep one's mind on the subject. All too easily, discussion can shift to related but different problems and a convincing argument can be elaborated which has nothing to do with the topic under scrutiny. The best defense against such shifts is to stress what substantive thing one wants to know and how that knowledge may be achieved, instead of stressing who 'wins' the argument. Flew (1975) discusses these problems in relation to a change from the subject of discussion to the motives behind the various positions: the 'subject/motive' shift.

'The but-you-can-understand-why evasion' shifts the debate from a consideration of whether a given position is scientifically correct because the theories it espouses are better predictors to a justification of the position in terms of scientifically extraneous criteria that explain why an erroneous position is held. Valid motives for maintaining a position are confused with valid positions. Thus we may understand if capitalists stress competition and socialists stress mutualisms as key ecological interactions, but the scientific strength of those positions is entirely independent of their supporters' politics.

The 'but–they–never–will–agree diversion' confounds adequacy of support for a position with the effectiveness of that position in convincing the opponent. Scientific questions must be decided on the merits of different views as evidenced by information value and not on the extent of opposition or approbation offered by others. The opinion of others may have no basis at all in fact, but depend entirely on precept and misinformation. Scientific debate must focus on which position has the strongest theoretical and factual support, not which has proven most appealing to other scientists.

The grounds for discussion can also be shifted by simply taking the subject of the discussion for granted, by 'begging the question'. Thus the debate between theoretical and empirical approaches to an ecological question may lead to the dismissal of each side by the other as 'too theoretical' or 'only empirical'. Such entrenchments beg the question, but do not resolve disagreement.

Begging the question is a defence similar to 'the fallacy of the pseudo-refuting description' whereby the opponents' positions are dismissed by the simple act of giving them a name. For example, 'reductionism' may be used falsely to refute mechanistic studies or 'naive falsificationism' can be used to decry the criterion of predictive ability. In these cases and many others, the critic must explain what is fallacious in those positions, and not simply tar them with a vaguely condemnatory name.

Two debating ploys

The 'pathetic fallacy' is better termed a misconception than a fallacy, for it is not a logical error but an unsupported proposition based solely on analogy. Nevertheless, Flew (1975) calls attribution of human emotions, motives, and desires to non-human entities 'the pathetic fallacy' because it requires us to experience a sense of pathos or compassion for the object of our research. Since there is rarely evidence for the feelings of such entities, this is scarcely a strong position, yet ecologists repeatedly employ the pathetic fallacy when they discuss 'ecological strategies'. This may be an acceptable shorthand, but the critic must be wary that the strength of the argument depends on the power of the theory to which the analogy leads and not the appeal of the analogy.

The 'fallacy of many questions' is also not a fallacy, but simply a trick to set the audience on a desired track. This is achieved by posing questions that build false assumptions into any answer. The stock example is 'When did you stop beating your wife?' A number of ecological examples were collected in Table 1.1.

Comparisons of data and hypothesis

In science, instruments fail, sampling schedules are disrupted, unexpected behaviours appear, statistical assumptions are only approximated, and unusual events occur, so the results of even the best-planned experiments are less clearcut than intended. In these cases, the empirical support for the hypothesis depends on interpretation of the data. In ecology, this is especially necessary because hypotheses are poorly phrased, methods uncertain, and data irrelevant. Consequently, discussions frequently begin by interpreting the data in light of the hypothesis and critical readers must be alert to the possibility of bias in this interpretation.

Superficial and partial tests

Superficial analyses take the agreement between some general aspect of the phenomenon and some implications of a mechanistically or hierarchically structured hypothesis as a confirmation of that hypothetical structure at all levels. This was discussed with reference to mechanism in Chapter 5 and as the fallacy of affirming the consequent in this chapter. Since different processes can induce the same gross phenomenon, such tests do not effectively discriminate among alternate mechanisms. Loehle (1987a) holds that this is a common error in the evaluation of computer models. Dayton (1973) addressed this problem in his complaint that models might be right for the wrong reason, because analyses of the hypothesis at lower hierarchical levels were not upheld. Belsky (1987) has termed this a confusion of scale. Kalff (1989–MS) details the problems associated with making ecological predictions from physiological studies.

Partial tests present a similar problem, because the support for one aspect of a complex hypothesis is held to confirm other aspects, but is distinguished from superficiality in that the other phenomena are at similar scales. Simberloff (1978) holds that this is the case in island biogeographic theory where the existence of the species–area curves is taken to confirm the existence of a stable equilibrium between unmeasured rates of colonization and extinction.

The confusion of correlation and causation is another example: simply establishing a correlation between two variables does not imply that manipulation of one will induce variation in another. Manipulability must be established in separate experiments. For example, evidence from dogs suggests that the correlation between mammalian size and gestation time is not manipulable, because dogs of all sizes gestate in about 63 days. In contrast, a similar correlation between size and food intake is manipulable, for big dogs eat more than small ones. Both partial and superficial tests are sometimes called inferential because their confirmation is used to infer qualitatively different attributes of the system from those which were observed or are necessitated by the observations. Such inferences may make good hypotheses for future studies, but they are not deducible from the evidence available in partial or superficial tests.

Special pleading

Many studies generate eccentric values that contradict the hypothesis under examination, but are not considered falsifications of the hypothesis under examination. Such data are normally explained away by invoking bias, or artifact, or some other ad hoc explanatory device. This is entirely acceptable if it is not overused by claiming unusual circumstance more frequently than necessary (Weatherhead 1986). Nevertheless, the ploy protects the hypothesis from falsification and may hide particularly interesting data that do not fit conventional theories. The reader must be informed when data are excluded and told how much such exclusions influence the interpretation. Only then can the critical reader decide whether the exclusion was acceptable or not.

The quagmire of plausibility

In 1887, Stephen Forbes could legitimately claim that, given the virtual absence of appropriate ecological information and the effort required to collect such information, it might be more effective to use the 'a priori road' of speculation and plausible argument. A century later, the same argument has a hollow ring, but plausibility still plays a large role in contemporary ecology. Plausible arguments are an amalgam of clever writing, commonsense and dogma which should be regarded with considerable suspicion. As Howe (1985) writes in his critique of the view that tropical American plants coevolved with now extinct gomphotherian mammals as agents of dispersal: 'It is unsettling that the idea is most plausible when applied to fruits we know virtually nothing about'. The a priori road may be valid for erecting hypotheses, but it is a poor tool for judging them.

Literature comparisons

Any ecological measurement could be biased by factors which the author was unable to consider. As a result, discussions seek to reassure the reader that the data collected were appropriate and representative of the phenomenon under study. One way to do this is to cite other measurements from the literature which indicate that similar values have been observed elsewhere and that the results of this test are consistent with the expectations based on an accurate reading of the literature. If this is not the case, the difference between the literature and the observations should be explained with a testable hypothesis.

Selective comparison and self-delusion

Where the literature is extensive, it is easy to flatter one's point of view by unconsciously selecting references and information. This selection might be achieved by restricting comparisons to trends rather than absolute amounts. In the early 1970s, the literature on phosphorus excretion by zooplankton claimed agreement because all studies showed that specific excretion rate declined with increasing size. This agreement virtually excluded the further observation that published estimates of excretion for animals of the same size differed by more than 1000-fold (Rigler 1973) or that the rates of decline with increasing size differed.

Comparisons are therefore best made quantitatively and statistically, but this is effective only if the data are an unbiased sample from the literature. If this is not the case, comparisons with the literature may degrade to meaningless polemics (Schoener 1972) or simply to self-delusion whereby we support our biases by unconciously selecting data that are most similar to our own.

Over-interpretation

Given sufficient good will and a sufficiently loose hypothesis, aspects of many different phenomena will seem consistent with our hypothesis and beliefs. One flaw in discussion may therefore consist of an accumulation of diverse observations which are interpreted as consistent with the hypothesis, even though they are only incidental. For example, Howe (1985) sees many of the points raised to demonstrate the coevolution of gomphotheres and neotropical trees (Janzen & Martin 1982) as too subjective, qualitative, and irrelevant to test the hypothesis adequately. As a result, using these points to support the hypothesis is over-interpretation. Such over-interpretations are easy to make when the hypothesis they appear to support is particularly engaging (as in the theory of megafaunal dispersal), or particularly widely accepted (as in the case of competition theory; Simberloff 1982) or particularly loose (as in the case of adaptationist arguments; Gould & Lewontin 1979; Ollason 1987). One litmus test against misinterpretations is to ask if the hypothesis would have been falsified by observing the converse of what was actually observed. When this is not the case, the evidence has been over-interpreted.

Over-interpretation is a common vice in ecology, for many ecologists see that their purpose is to organize the diversity of ecological observations, rather than to predict then. This clogs the literature with intellectual jig-saws that exclude and overshadow testable hypotheses.

Significance and magnitude

If the sample size is large enough, even very small differences are highly significant. It is therefore appropriate to ask if very small, but statistically significant, effects are also biologically important. Fowler & Lawton (1985) adopted the position that small differences obtained in laboratory tests may not be biologically important in the field because the highly defined, invariant conditions of laboratory testing may prove unrepresentative on extrapolation. For this reason, isolated statements of probability are insufficient; the magnitude of the difference and its associated variation must also be considered.

Vacuous contrast

A frequent trick in favorably comparing one's results to the literature is to make a vacuous contrast, claiming that the hypothesis of choice has some special virtue not contained in an unspecified, and actually non-existent, alternative. For example, Janzen & Martin (1982) cite the rotting of seeds and fruit under the parent canopy of certain South American trees as evidence in favour of their hypothesis that the extinct megaherbivores once ate these fruits and so dispersed the seed; This would carry weight only if the seeds of other trees, whose dispersal agents are not extinct, did not rot under trees. Since they do (Howe 1985), this is a case of vacuous contrast.

Critics often use vacuous contrasts to create and criticize straw men. This step may be necessary in ecology because many ecological constructs have to be defined before they can be addressed. However, this opens the criticism to the charge of irrelevance, because some aspect of the ecological concept is almost inevitably excluded by the process of definition (see Chapter 4), and vacuous contrast, because the model being criticized is a figment of the critic.

Since it seems unlikely that any scientific paper is flawless, this discussion of all the ways a scientific paper could be improved may also seem a vacuous contrast. This is not so because the argument is not that ecology should consist of perfect papers (thereby invoking a non-existent perfect science), but rather that the science will progress faster if existing flaws are recognized. The contrast between the literature and perfection would be vacuous, that between better and worse elements in the existing literature is not.

Extensions and hypotheses

Perhaps the most important function of a scientific paper is to describe where we go from here as scientists: the Discussion is therefore important because it draws out the implications of the hypothesis. It explains what we now can predict that might not have been possible before and how we can build on this with future research. This should be one of the most exciting parts of the paper, but often it falls short because the authors are so unaware of the limitations of their work that they feel all questions have been resolved, or because they do not recognize the importance of new hypotheses, or because they have become so confused by their own rapportage that the hypotheses they offer are either unrelated to the paper or untestable as stated. The critical reader should be aware of these difficulties and be alert to any hypotheses which the author has made, but not recognized, or recognized but not made.

Tunnel vision

Most scientists work with a limited set of ideas and these limitations restrict their breadth of view on their work. As a result, the self-evaluations represented by discussions are likely to favour prevailing dogma and personal beliefs, and alternative approaches are less likely to be considered thoroughly.

For example, fisheries biologists are so firmly wedded to the concept of density-dependent population control that they find it difficult to conceive of alternatives, even though the evidence for density-dependence in fish stocks is virtually non-existent (Beverton *et al.* 1984). Beverton *et al* (1984) claimed that 'it is difficult, if not impossible, to conceive that the upper extreme of population size is not limited ultimately by density dependent processes, which is one manifestation of self–regulation (homeostasis) . . . ' and Rothschild (1986) can protest that, despite all the evidence to the contrary, 'there must be some relation of recruitment to stock, otherwise stocks would not be persistent . . . '

The historical record of any persistent stock should show a tendency to decline from high populations and to rise from low ones. With enough data, this pattern, which is inherent in the definition of an historical mean population size, will generate negative relationships between the size and growth rate of the population and will result in regression coefficients of less than unity when the logarithm of population at time t is regressed against that of the population at time $t+1$. Neither of these common tests for density-dependence entails a density-dependent, causal process (Eberhardt 1970; Maelzer 1970; St Amant 1970). Murdoch (1970) recommends that the words be used simply to indicate relationships between density and the change in numbers, rather than some underlying homeostasis. Fisheries biologists have mistaken precepts and logical models for the data, even though these precepts and models do not apply to the data in hand. It would be surprising if this confusion led to particularly effective fisheries management policy.

In the context of the Discussion, this bias is likely to lead to one–sided arguments in which the preferences of the author usually carry the day. For example, supporters of optimal foraging theory find that it predicts effectively (Stephens & Krebs 1986; Stearns & Schmid-Hempel 1987), but its critics conclude just the opposite (Gray 1987; Pierce & Ollason 1987). Similarly, Gilpin & Diamond (1984b) are frustrated at the inability of Connor & Simberloff (1984a,b) to see the virtues of competition theory. No doubt, Connor & Simberloff (1984b) are just as puzzled that the vices of competition theory are not equally apparent to Gilpin & Diamond. It is important therefore to recognize the existence of scientific bias. Writers of Discussions should strive to free themselves from this handicap by self-criticism; but since they are likely to fail, critical readers should be alert to the possibility of bias.

Gould (1981) has demonstrated the vulnerability of science to bias in his analysis of the evidence supporting sexist and racist preconceptions of human intellegence. Gould (1986) divides these into three groups: the frauds, like Cyril Burt, who manufacture evidence; the finaglers, like Morton, who subtly and perhaps unconsciously bias their results by selective application of otherwise acceptable corrections; and those, like Broca, who suffer such disabling bias that their interpretations of the data are warped and untrustworthy.

Ad hoc hypotheses
Many papers introduce ad hoc hypotheses to explain away eccentric data from the literature or from their own experiments. This is a normal process and is not too damaging if the ad hoc hypothesis is testable (Fretwell 1987). Ideally, this test is performed in the context of the paper, but if this is not possible, it should be tested in some future study. If this is not done, the test offered in the current paper is so compromised that it is meaningless, because whatever conclusion was reached rests on a crucial, untested assumption. In practice many ad hoc assumptions are not intended to be tested and therefore should be discounted as special pleading.

Field relevance
Theories in ecology are relevant only if they can be applied in the field. Consequently, one of the most interesting elements in the Discussions of ecological papers describes the implications of the paper for organisms in nature. To the extent that a study has been removed from nature, this consideration is pivotal, and the reader should expect to find some evidence that such an extension is warranted. Alternatively, there should be warnings that no effective field testing has been done, specifying the sorts of information which would constitute such a test. This need not be extensive, but it is a useful protection against unwarranted claims for field relevance.

Extraneous material
Of all the parts of a paper, the Discussion is most likely to contain material that is not essential to the evaluation of the hypothesis under test, and the inclusion of such extraneous material is one of the most confusing aspects of many Discussions. For example, ecologists often confuse the precise use of words to present data and theories with their vague use to motivate and suggest ideas (MacArthur 1972b). Since only the former use is appropriate in the Discussion of a scientific report, this confusion can bewilder the reader. The same result is achieved when the Discussion is used to introduce arguments, facts and theories into the literature, not because they are germane to the hypothesis under test but because they concern the author. Woodford's dry advice to the author who wishes to write a good Discussion is 'avoid megalomania'. No reader wants to know all of the author's thoughts about something, so the author should decide which points are most germane, make these points and keep the rest to himself or herself.

Minims

Mentis (1988) identified a special problem of megalomania in ecology as the production of trivial hypothesis or minims. This term is an analogue of a maxim and is defined as a statement of proverbial form having no general application or practical use. Any test of the proposed hypothesis would require so much effort and yield so little information that no one would wish to do so. It is an aphorism which appears to offer wisdom but actually says very little. The competitive exclusion principle (Hardin 1961) might be one example.

Extrapolation

Extrapolation in its broadest sense refers to the extension of an hypothesis beyond the original domain over which the hypothesis was built and tested. In one sense, this is inevitable. Because every situation is different in some way, each application of the hypothesis represents an extension of the original domain of application. Such extrapolations present no new problem for science.

The charge of extrapolation is usually limited to cases where one of the boundary conditions or one of the predictor variables takes a value outside the domain of application specified in the theory. The commonest case is likely the extension of a regression line beyond the range of the data set (Fig. 8.2). Such extrapolations are considered risky because they presume that the effect of a variable is unaffected by scale. Sometimes this is so: general allometric relations for rates of metabolism seem to extrapolate relatively well to very large organisms. But this is not always the case: relations relating speed to size overestimates the maximum speeds of large animals (Bonner 1965).

One widely discussed extrapolation involves the growth of the human population. Von Foerster, Mora & Amiot (1960) fit a set of past estimates for the human population of the Earth as a function of date (in years AD) such that:

$$\text{Population} = 1.79 \times 10^{11}/(2026.87 - \text{Date})^{0.99}$$

This leads to the expectation that world population will reach infinity on Friday, 13 November, 2026 – Doomsday. This relation has done surprisingly well at predicting human growth (Caswell 1976), although it now underestimates the world population (Umpleby 1987). However, since we can be certain that the population will never reach infinity, we know that it cannot extrapolate to doomsday. Indeed, even before doomsday, the equation must prove false for it would require doubling

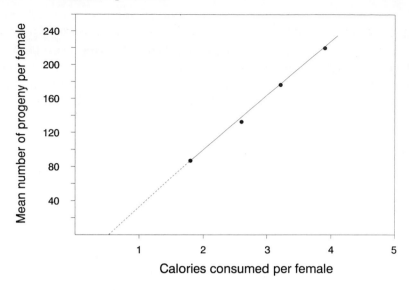

Fig. 8.2. An extrapolation beyond the domain of the data on which the regression, describing reproductive output as a function of ration for *Daphnia*, is based. (Modified from Hassell 1978.)

rates of the human population that are beyond the biological capability of our reproductive system (Deevey 1987).

Extrapolation does not always introduce a fatal flaw. The extrapolation in Fig 8.2 is particularly interesting, because it does not accord with the expectations of most zooplankton ecologists (Porter, Orcutt & Gerritsen 1983; Frost 1985; Geller 1985; Stemberger & Gilbert 1985) who prefer to fit such data as a concave downward plot. However, the available data suggest that in this case linear extrapolation is more likely to be consistent with the evidence than a more complex curvilinear relation (Condrey 1982; Condrey & Fuller 1985).

Summary – The challenge of good science

Science is best when it is straightforward, logical and concise, but achieving these simple goals is not easy. Science requires a judicious blend of wisdom and knowledge in choosing an appropriate question, single-minded dedication to that question in selecting and pursuing the appropriate techniques, followed by a perceptive and unbiased judgement about the importance and relevance of the findings. This development requires sound ethics and broad knowledge to select a relevant, soluble, unsolved problem, a practical grasp of epistemology to address the problem so that it will yield meaningful results, appropriate statistical and technical tools so that meaningful results can be properly evaluated, and basic but powerful logic to pursue the implications of the analyses, all bound together in varying proportions by effective writing. Strong science is difficult to achieve, but that is no excuse to try for less.

9 · Putting it together – competition

This book concentrates on describing the kinds of criticisms that could be applied in ecology, rather than specific instances of weak ecological constructs. As a result, the text is arranged as a sequence of criteria illustrated with a variety of ecological examples. The alternative topical approach would select a few important themes in ecology and criticize each in turn with whatever criticisms are appropriate. For example, different chapters could have addressed evolutionary theory, island biogeography, fisheries models, simulation and so on.

Several factors make the latter structure less attractive. Topical arrangement stresses examples rather than critical principles thus confusing a major purpose of the book, which is to arm ecological readers with the criteria needed to judge whatever material they encounter. Topical treatment would restrict the scope of the examples used, whereas an ancillary goal of the book is to argue that the flaws in ecology are common and widespread. Topical treatment would also require frequent repetition of those common problems that recur throughout the discipline. Finally, the ecological literature already contains many topical critiques (e.g., Lewontin 1978; Gould & Lewontin 1979; Gilbert 1980; Hurlbert 1984; Strong 1986a,b; Ollason 1987; Gray 1987; Paine 1988) so there is less need for a topical text.

Categorical criticisms, like those used to this point, can be less effective than topical ones because the latter show better how different weaknesses in ecology interact to foster one another and to obscure their central failing, the inability to predict. Treatment of ecological criticisms as a series of separate classes also risks the introduction of yet another ecological typology, further compounding the problems the critique should curtail. As partial protection against those possibilities, this chapter summarizes and integrates many of the foregoing criticisms by applying them to a single area in ecology: competition. The status of competition has been addressed frequently (Connell 1980; Schoener 1982; Simberloff 1982; Keddy 1989a), so that there is no need to develop a comprehensive critique. Instead, I will concentrate on how the multiple shortcomings of the subdiscipline interact, thereby bringing a substantial proportion of the criticisms outlined in the preceding pages to bear on a single subject.

256

The chapter begins by acknowledging that competition occurs. It then summarizes the operational difficulties surrounding the use of competition in ecological theories and contends that these difficulties have encouraged mathematical and verbal tautologies whenever competition is discussed. The remainder of the chapter establishes the ramifications of these problems for competition theory. It argues that the availability of a tautological model eases the incorporation of competition into simple historical explanations of ecological observations, but does not provide predictive power. Another line of ecological development tries to determine the competitive 'mechanisms' and 'causes' behind competition, in the belief that a knowledge of the mechanisms, rather than of the phenomenon, will permit predictions. This chapter argues that such a research program is likely to founder in infinite regress, reductionistic analysis, and unwieldy complexity. The chapter concludes by comparing the format of the major elements in competition with ideals for predictive theory. The components of 'competition theory' are nebulous, imprecise, specific, qualitative, and uneconomical. That this has not impeded the development of a substantial literature dealing with competition apparently reflects the charisma of the topic and its leaders, more than any information about nature that the research in the topic has provided.

The prevalence of competition

Competition has long been one of the most popular areas of study in ecology. Jackson (1981) has shown that papers on competition have constituted 5 – 6% of the ecological literature for over 60 years. May & Seger (1986) examined the relative emphasis placed on various inter-organismic interactions in the ecological literature around 1970. They found that the ratio of the number of papers on competition : predation : mutualism in ecological journals was 4 : 4 : 1, and the ratio of pages devoted to these topics in ecological textbooks was 5 : 6 : 1. Together, these three interactions represented the focus of 12% of the papers published and 16% of the pages in textbooks. Keddy (1989a) analyzed the indices of 12 recent texts and essentially confirmed the findings of May & Seger (1989): the ratio of index entries for competition : predation : mutualism occur in the ratio 8 : 5 : 1. The higher proportion of competition in the last study reflects the emphasis on competition in texts by Hutchinson (1978) and Odum (1971) which were not used by May & Seger (1986).

The continuing popularity of competition is surprising since competition is much less apparent than predation and at least as hard to observe as mutualism. This paradoxical emphasis on a process that is difficult to observe may reflect subtle, extra-scientific factors that encourage the study of competition despite its difficulty (Keddy 1989a): cultural biases in western thought may favour application of the metaphor of competition in human affairs to nature (e.g. Diamond 1978); gender bias in contemporary science may encourage a male view of life as a contest, like a football game; taxonomic biases towards birds and other vertebrates may stress highly competitive organisms; and the ecological community may itself be biased towards competition by the ecologists who have studied competition. For example, competition is indexed more than 10 times as frequently as predation by both Odum and Hutchinson (Keddy 1989a) and these authors have had a tremendous impact on ecology both through their own writings and through those whom they have influenced. As McIntosh (1987) comments on the refutation by May & Seger (1986) of the view that 'competition was the only game in town' (Lewin 1983a), 'competition may not have been the only game in town, but it certainly was the noisiest'.

There is no question that competition occurs. Generations of plant ecologists have shown that competition is a regular element in their discipline (McIntosh 1970; Jackson 1981). Connell (1983) and Schoener (1983) reviewed the literature and found that competition occurred in the majority of instances where it was sought. Because studies of competition are presumably biased in favour of finding competition, these surveys cannot establish the frequency with which competition occurs (Keddy 1989a), just as Endler's (1986) survey of field studies of natural selection could not establish the frequency of that process (Chapter 3). For similar reasons, such comparisons are unable to determine which interactions among or within species are more important in nature. Any question about the occurrence of competition probably reflects a tradition of studying competition where it is most elusive: between such similar competitors that the competitive outcome is variable and uncertain (Keddy 1989b). Comparisons of big and small animals (Grant 1972b; Peters 1983) or plants (Keddy & Shipley 1989; Keddy 1989) repeatedly and convincingly show that competition is a common element in the economy of nature.

The important question is not 'does competition occur'? This has been answered in the affirmative. Nor can we ask about how important competition is or whether competition is more important than predation and mutualism. These questions are impossible to answer without a random sample of all interactions. Instead, ecologists must ask how competition serves in predictive theories that link competition to other variables and thereby inform us about the world around us. This question is rare because the answer is obvious. Competition has proven so weakly informative that it is premature even to speak of 'competition theory' (Keddy 1989b). The remainder of this chapter examines the basis of the failure of competition to develop theories worthy of the name.

Operationalization

Because competition is difficult to observe, because there are no meters or probes or litmus papers to measure it, operational definition of the term has proven elusive. Milne (1961) considers 12 definitions and stresses the confusion and dangers associated with the thoughtless proliferation of meanings. He ends by defining competition as 'the endeavour of two or more animals to gain the same particular thing, or to gain the measure each wants from the supply of a thing when that supply is not sufficient for both (or all)'. Keddy (1989a) suggests 'The negative effects which one organism has upon another by consuming, or controlling access to a resource that is limited in availability'. Both warn against too close an interest in definition itself, stressing that the value of the concept lies in its relation to nature, rather than to words.

Both definitions carry the seeds of operational difficulty. Neither definition embeds competition in a theory, so there is no indication of what we might do with a measure of competition or why we would measure it. In the spirit of Chapter 4, variables are defined better by their theoretical relations than by verbal fiat, so that the goal of the definition and the definition itself will remain nebulous, until these definitions become part of a relationship between entities. Competition only represents a potential variable that will not play a role in theory until we are told how to measure that variable and how to use it in prediction.

The definitions of Milne (1961) and Keddy (1989a) define competition relativistically with respect to the needs and requirements of two or more organisms and their mutual effects. To be invoked, one must simultaneously determine what the competitors are endeavouring to gain, what their rates of gain are, how the rates affect competitor performance, and how the presence or absence of the competitor affects these determinations. The complexity of relative measures is not fatal, but it renders any empirical treatment of competition tedious and unattractive, because it involves simultaneous, repeated estimates of an organism's response to at least one other organism's effect over an entire range of resource states.

Just demonstrating the occurrence of competition is enough to discourage most field biologists. Reynoldson & Bellamy (1971; cited in Wise 1984) propose five criteria as sufficient to establish interspecific competition beyond a reasonable doubt. These include casual observations that suggest the possibility of competition: (1) indirect evidence for resource limitation, and (2) indirect evidence for competition. Presumably these are rather easy to collect, but provide only unconvincing, circumstantial evidence that would not withstand close scrutiny. (3) More formal pattern analysis is the next simplest step, but the resulting correlations do not suffice to demonstrate competition (Keddy 1989a), so this should be followed by experimental manipulations that demonstrate (4) intraspecific competition through changes in population density in response to varying resource levels, and (5) interspecific competition as appropriate responses to experimental removal and addition of potential competitors.

Keddy (1989a) notes that process can never be inferred from pattern and therefore he too recommends experimental, field manipulations of competitor and, if possible, resource levels as part of the demonstration of competition, but he recognizes that no set of experiments could remove 'all reasonable doubt'. Underwood & Denley (1984), discussing presumed instances of competition in marine communities, have shown that plausible alternatives always exist if our minds are open enough to look beyond the first explanation that is offered. This wealth of choice indicates that explanation of existing observations by reference to competition is so facile as to be uninteresting.

The interest and utility of a scientific explanation depend instead on its predictive implications. Unfortunately, the protocol to identify competition involves so massive an experimental effort that most studies exhaust available scientific resources just to establish that the process occurs. This leaves little reserve to develop or test any implications of this discovery.

Still other operational difficulties dog any application of competition. There is every likelihood that limiting resources, the competing organisms, the degree of competition and the mechanisms involved in competition will change spatially and temporally, so the environmental context and scale of the interaction are extremely important. Because this is so, the definitions associated with competition – resource, competitor, negative effects – are open-ended in the sense that they can

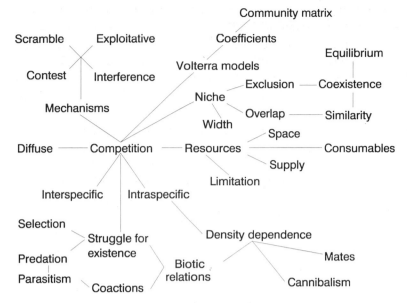

Fig. 9.1. Some of the conceptual connections and associations of competition.

represent a potentially infinite series of entities and phenomena. Thus predictive failure of an operationalized theory of competition from one study does not falsify the general approach but only forces the scientist to consider more and more resources, organisms and responses with changing scales. Although this is not a fatal objection (because the possibility of an effective alternate operationalization always exists), it does suggest that empirically justified or calibrated theories of competition are likely limited to a given scale and situation.

These operational problems have encouraged ecologists to use the term 'competition' without the constraints imposed by questions of applicability. As a result, the term is used in so many incommensurate ways that the reader can no longer be certain what meaning is intended. Depending on the author, competition may include the struggle for existence, density dependence, biotic interactions, parasitism, and predation. 'Competition' remains an omnibus term (Milne 1961).

Competition has benefited from the flexibility of its definition to assume a central position in the conceptual literature of ecology, linking a number of key concepts which are themselves uncertainly defined (Chapter 4). Thus the term is at the centre of a broadening superstructure of concepts (Fig. 9.1).

Several authors have tried to re-establish order and meaning within the proliferating definitions of competition. Suggested classifications have been based on the characteristics of the resource or the mechanism of competition: for example, competition for resources and for space, interference and exploitative competition, or scramble and contest competition. Tilman (1982) developed an elaborate model differentiating among the different resource supply rates and strategies which condition competition. Any such reorganization presents a suite of difficulties.

In one overview of the role of competition in ecology, Schoener (1982) points to six main propositions that Table 9.1 treats as a summary of competition theory. He noted that most of these statements contained vague qualifiers: For example, competitive exclusion will only occur if 'sufficiently' similar competitors coexist for 'long enough'. These qualifiers must be defined if the theory is to be applied (Murray 1986). Schoener (1982) does not treat the difficulty in defining other elements in these propositions – coexistence, resource, similarity, niche, adaptation, competitive pressure, or overlap – but since each of these variables is open to the same operational problems as competition, the operationalization of competition as part of a predictive ecological theory will require considerable thought and effort. Operationalization is not impossible, but the ecological community has generally shown itself unwilling to make the investment required.

Recognizing that a complex series of operations, like those required to demonstrate competition, discourages empirical study, Keddy (1989a) recommends against measuring the traditional components of competition models and instead commends a 'phytometer' (Gaudet & Keddy 1988) to measure the performance of plant competitors in habitats with competition, relative to their performance in similar habitats without competition. The depression of performance could then be used as an index of competition pressure and related to other aspects of the environment in an ecological theory of competition. For example, Wilson & Keddy (1986a,b) developed a relation between this measure of competitive pressure and total biomass in communities of lakeshore plants. This may seem impractically laborious to those accustomed to the rapid algebra of competition, but it is unavoidable. Indeed, the use of an organismic titre of competition is implicit in the competition coefficients (a_{ij}'s) of the Lotka–Volterra equations, since these measure the depressive effect of each member of population j on the growth of population i, relative to the effect of each i on the growth rate of its own population. If

Table 9.1 *Six main propositions of interspecific competition.*

(1) Species 'too' similar in the resources they use cannot coexist 'for long'. This is Gause's law.
(2) Species that coexist in nature do so by virtue of 'sufficient' differences in ecological niche or, equivalently, in resource use. This is the theory of limiting similarity.
(3) Interspecific competition is a powerful evolutionary force, selecting for adaptations that result in species differing in use of resources.
(4) Species 'too' similar ecologically have disjunct ranges and given enough time, competitive pressures therefore determine how many and which species coexist in a community. Propositions 3 and 4 represent the ghost of competition past.
(5) Species may compete by interference, as well as by depletion of resources, but interference is unlikely to evolve if resources are not 'sufficiently' scarce. This interest in the way that organisms compete has led to studies of the mechanisms of competition.
(6) Interspecific competition should be detectable by experiments performed on species with 'substantial' overlap in their use of resources. This proposition has since been questioned (Pianka 1981), but there is still a substantial effort aimed at detecting competition.

Note:
Quotation marks indicate vague qualifiers (Schoener 1982).

Keddy's solution is appropriate only for some systems, then the traditional models of competition are at least equally restricted. Any operationalization of a concept cluster will exclude some current nuances from the operational definition (Chapter 3).

Tautology

The difficulty of empirical studies of competition has not dissuaded ecologists from modeling the process algebraically and graphically. Because operational problems prohibit the full specification of the models with respect to any natural situation, the models cannot be applied to nature and remain purely deductive constructs. These models are therefore ecological tautologies (Chapter 3). Because parts of the model logic can be associated with observable phenomena, for example with population density, the models can be used in the non-rigorous, verbal interpretation of observations, even though these could not be predicted. Moreover, because tautologies can accommodate any possibility, available observations always match some aspect of the model.

Limiting similarity

Some of the basic components of 'competition theory', like the Lotka-Volterra models of competition (Chapter 3) and the competitive exclusion principle (Chapter 4), have been addressed earlier in this book and elsewhere (Slobodkin 1961, 1986b; Colinvaux 1973; Keddy 1989a). Abrams (1983) examines another aspect of competition in his discussion of the 'theory of limiting similarity' which states 'There is some limiting similarity (some maximal use of a set of resources in short supply) between competing species that will allow coexistence'.

This proposition is usually represented graphically (Fig. 9.2) or algebraically (MacArthur 1972a). Consider two competing species, 1 and 2, with differing 'resource utilization functions', u_1 and u_2, that describe the response (resource use, population size, growth rate, etc.) of the populations as a function of the level of resource j. For the sake of simplicity, these functions are assumed to be normal curves with the same standard deviation, σ, lying at different points along the resource gradient, such that the midpoints of the two curves are separated by a distance d. The question addressed by the theory of limiting similarity is 'How large must d be, measured in terms of σ, for coexistence of both populations'? Several assumptions later, it can be shown that the limiting similarity occurs when d is of the order of σ. The exact value depends on the assumptions (May 1981c; Schoener 1982).

If we do not consider the necessary measurements too closely, it is easy to imagine that the relative effectiveness of utilization of a limiting resource supply would determine the size of coexisting populations, and that common use (overlap) of the resource would inhibit both populations. These assumptions make it possible to equate resource use at stable equilibrium, as given by either the curves in Fig. 9.2 or the resource utilization functions, to the product of population size and the competition coefficient used in Lotka-Volterra models of competition. Unfortunately, this approach is extremely difficult in practice because its constituent concepts are fuzzy (Abrams 1983) and the relevant measurement would be almost impossible to make in nature (Pianka 1981). As a result, the 'theory of limiting similarity' can explain away observations that are already in hand or serve in logical extensions to the Lotka-Volterra models, but it cannot predict the extent of similarity or overlap.

The abandonment of empirical or predictive pretensions in favour of logical development by ecological modelers has had a profound effect (Strong 1986a,b). Because the models are not expected to address the

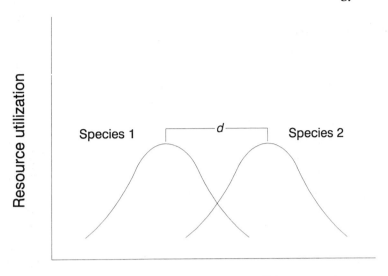

Fig. 9.2. A graphical representation of the central problem of the 'theory of limiting similarity'. The curves represent resource utilization functions, u_1 and u_2, of two species, 1 and 2, competing for resource j which is distributed as a gradient represented by the X-axis. For convenience, the curves are assumed to be normal and to have the same variance σ. The limiting similarity is the value of d, the intermodal distance measured in terms of σ, that will allow coexistence of both species.

vagaries and inconsistencies of nature, they may stress more purely intellectual virtues of elegance, simplicity, and beauty, rather than practicality, applicability, predictive power and accuracy. Once free of empirical constraints, modelers can develop simplifying assumptions, axiomatic frameworks, and deterministic relationships which, if they relate to anything, must relate to other concepts because they cannot relate to predictive theory or observable nature. Thus competition theory posits homogeneous environments, identical organisms, equilibrium populations and characteristic constants, less because these are thought to represent nature than because they render the models logically and mathematically tractable. Strong (1986b) has denigrated this as 'prim' science (Table 3.1).

Since tautologies do not exclude any logical possibility, they can be matched rather easily with (and in that sense, 'explain') particular observations. In fact, the indeterminacy of the coefficients and parameters produces a looseness of argument that allows several explanations for any observation. For example, since no one expects a single interaction to explain the ecological world (Schoener 1982), the absence of a shared resource may be dismissed as irrelevant because competition is absent. Alternatively, since 'resources' represent an open-ended set (e.g. Tilman 1982), lack of overlap may be explained as a case where the proper resource was not examined. Or, since competition need not be continuous in time or space (MacArthur 1972a), lack of overlap may represent a period in which competitive pressures are relaxed, due to some biotic or abiotic disturbance. At the other extreme, complete overlap on one resource axis by supposed competitors might be interpreted as showing that the limiting resource lies on another, unmeasured, axis, or that competition is not occurring, or that competition is intense but equilibrium unachieved. Any intermediate degree of overlap can be fitted similarly to the model.

Because most of the parameters in the theory of limiting similarity are unmeasurable, the theory can only be used qualitatively. Any difference in resource use may be taken as a confirming instance of limiting similarity, or the observation can be explained in the same terms as complete overlap, and still confirm the model. Moreover, since resource utilization functions are not in fact measured but replaced with observable surrogates, like geographical distribution and temporal displacements, the model can be applied to any observed pattern of distribution.

Historical explanation

Because the components and principles of competition theory are tautological, they tell us nothing about the world around us that we should not already know. However tautologies in general and models of competition in particular do provide tools for post hoc rationalizations of observations that are already in hand. It is always possible to interpret observations relating to resource use, adaptation, or geographic distribution in terms of some competitive scenario and to use the logic of the deductive model to organize observations.

The fit of observations to a pre-existing explanatory scheme must not be confused with a successful test or instance of the theory. Instead, one should always ask if theories, 'confirmed' by existing observations and proposed explanations, would be inconsistent with different results. Explanations which could be consistent with both what was and what was not observed (e.g. with any degree of resource overlap) are not scientific explanations at all. In competition 'theory', such confirmation has replaced predictive tests so often and so successfully that the fraud is rarely noticed.

Even if historical explanations do not rule out results different from those observed, they are acceptable in some contexts, provided that their distinct character and limitations are recognized (Chapter 6). Such explanations are judged in terms of elegance, parsimony, familiarity, plausibility and heuristic content. Competitive explanations meet these criteria so well that competition may have become a panchreston, a device that explains virtually anything and everything (Simberloff & Boecklen 1981) – disjunctive distributions, clinal gradients in abundance, temporal and successional changes, zonations, as well as patterns of adaptation, behaviour, and habitat preference.

Competitive explanations are so routine that weakness of the evidence is often ignored. Frequently, competition is inferred from patterns, an example of confusing cause with correlation (Keddy 1989a). Even experimental evidence may be ambiguous in its support for competition (Underwood & Denley 1984), although this may go unappreciated because the competitive explanation is the only one considered (Keddy 1989a). Explanations of many evolutionary and biogeographic patterns may be entirely untestable because they refer to unique conditions and events set in remote prehistory. This is the ghost of competition past (Connell 1980). Nevertheless, competition gains by repeated application because the scientific audience inevitably becomes less critical and more accepting.

Competitive explanations often combine different sorts of historical explanations. They may be called normic explanations because the more frequently these explanations are invoked, the more they are seen as norms for ecological explanation, and the more likely they are to be invoked again. They may be termed genetic explanations because they relate the present observations to earlier situations through a plausible sequence of events; this sequence is itself unpredictable because many essential characteristics of the sequence can only be surmised after the fact. Elements of these sequences can serve as norms for further normic explanations and, when assembled together, the weight and extent of these genetic and normic explanations may give greater legitimacy to the individual parts. This recalls colligative explanation which explains the diverse components of our experience by subsuming them in a single process, in this case competition. Competitive explanations that invoke unique combinations of events and places, like the evolution of Darwin's finches, may represent narrative explanations which explain by combining as many factors as possible into a unique story of past events. Certainly, competitive arguments that explain massive displacements or colonization by reference to the competitive abilities of brown rats (*Rattus norvegicus*) or starlings (*Sternus vulgaris*) contain aspects of individual explanation, which explains historical events by reference to individual characteristics, in this case large coefficients of competition in some individual organisms or situations.

The challenge of historical explanation is to blend all the elements of competition so skilfully and convincingly that the resultant composite explanation will leave the audience intellectually satisfied. Thus the audience can claim to understand a process they could never predict, because they have achieved a complex, empathetic, and intensely personal feeling of understanding about what went on. The prevalence of competitive explanations, in the face of relatively weak evidence and in the absence of predictive ability, is a measure of the importance of this feeling. Nevertheless, the measurement of scientific success by the degree of intellectual satisfaction is a case of Flew's (1975) subject/motive shift (Chapter 8) in which the substantive issue under debate is supplanted by consideration of the motivation of the debaters.

Mechanisms of competition

Competition is not a mechanism of organismic interaction. Instead, it is a black box, relaying only the information that species j reduces some index of the vigour of species i. This competition coefficient of the Lotka–Volterra equations may seem an inappropriate and arbitrary end-point for scientific enquiry and considerable effort has been made to reveal the mechanisms underlying these coefficients (Tilman 1987).

Researchers into this aspect of competition accept that mechanisms and causes cannot be revealed by pattern analysis, so they depend on experiment. In addition, because this research is directed at what are believed to be fundamental processes and mechanisms of organismic interaction, the results of field experiments seem ambiguous because they confound the effects of many uncontrolled variables (Bender *et al.* 1984; Diamond 1986). The mechanisms of competition are best revealed by carefully controlled laboratory experiments. Predictions in nature would therefore require the reassembly of the unit mechanisms investigated in these experiments into a model of the functioning whole (e.g. Tilman 1988).

Experimental analysis of the mechanisms of competition may seem to test the proposal of competition, but in practice the results from such studies can only render competition more plausible, by identifying potential mechanisms that could have induced the observed competitive interaction. For example, an experimental test of the hypothesis that two plants compete for water can only have ambiguous results. If watering the plants encourages both competitors, one could not assert that the plants compete for water, since it is possible that water is only one of several causal controls and that water acts indirectly, for example, by liberating nutrients. To suggest that this experiment shows that competition for water limits the plants is to affirm the consequent. Failure to demonstrate an effect of water would not show that the plants are not competitors, nor even that the amount of water is unimportant to the interaction, since the amount of competition in the unmanipulated case is contingent upon initial conditions and these are changed by the addition of water. To insist that, because additional water did not affect them, the plants were not competing is to deny the antecedent. In either case, researchers committed to competitive mechanism could salvage their beliefs whatever the experiment showed. At most, if the proposed mechanisms prove ineffective in experiment, only the details of the competitive model need change and the hypothesis of competitive interaction in nature would be unaffected.

Laboratory analysis of ecological competition in nature presupposes a mechanistic view of nature (Chapter 5) which has yet to prove effective. There is certainly as much reason to believe that the processes which result in observed competition coefficients are not deterministic units, but the stochastic summaries of a series of interactions, the precise mix and expression of which is determined by a complex of environmental effects and factors. If this is the case, some mechanisms may evaporate under analysis as their composite structure is destroyed. In these cases, dissection or analysis simply dismembers the composite which appeared to be a unit mechanism at some grosser level. Other mechanisms simply will not pertain in the laboratory or will be so grossly perturbed by study as to be inapplicable in nature.

The search for mechanism also presumes that there will be some convenient and obvious stopping point, when the mechanism of one interaction is known and the researcher can presumably switch attention to another interaction. Experience suggests that this is improbable. The researcher is more likely to be sucked into the maelstrom of reductionism. Chapter 5 termed this the problem of infinite regress whereby each layer of the mechanism is peeled back to reveal more and deeper sets of causal processes.

Finally, the reassembly of these unit studies into a functionally predictive model of a community seems Utopian. Simulation modelers have abandoned the hope of a realistic composite model as unworkably complex (Watt 1975; Joergensen 1986). Such a model would be impossible to specify (Clark *et al.* 1979), the necessarily massive calculations will inflate error until the predictions are rendered meaninglessly uncertain, and the purpose of the model will be obscured by its own complexity (Chapter 5).

Research into the processes underlying competitive patterns may merit study. For example, crop science has extensively investigated competition in terms of planting density, although this remains an empirical science with little effective connection to the classic models of competition. This work extends easily to multiple crop systems where the total number of species is small. Perhaps, specific examination of the elements of diffuse competition will eventually prove useful for some forms of range or pest management. However, the success of competition relative to predation, parasitism, and mutualism in agriculture suggests that in practice, competition is likely the least controllable and least promising of the coactions.

Mechanism and mechanistic approaches to competition reflect a desire for causal explanation of the world around us. Except in the restricted sense of cause as a regular or manipulable pattern, this is a false hope. Science cannot give deeper explanations, so ecologists should adopt a more mature attitude to what they can legitimately provide and make this realistic measure the goal of their enterprise.

Science provides tools to help predict and control the world around us. If we ask science for something else, we can only be disappointed. Moreover if we seek something besides scientific tools, we are likely to fail on two counts, finding neither the tool we need (because we do not look) nor the deeper understanding we seek (because science cannot provide it).

The theoretical status of 'competition theory'

This book began with a basic form for scientific theories whereby the investigator is instructed to make some measurements in order to achieve a desired prediction (Fig. 2.3). Thus, scientific theories are only invoked when their predicted variables are of interest and any scientific theory can only be applied if a set of predictor variables is identified. The 'theories' of competition (Table 9.1) can be forced to fit this model, but the theories that result are very weak, because they exclude so very little.

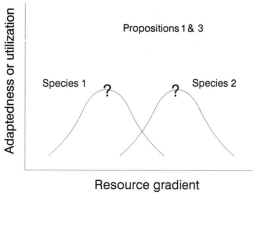

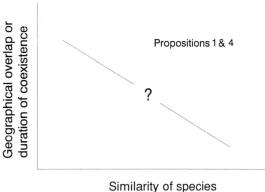

Fig. 9.3. The theoretical implications of the propositions in Table 9.1. These theories are extremely weak because the variables are virtually undefined and because they are irretrievably qualitative. The question marks on the curves are included to stress the uncertainty about the precise shape of these relations.

Figure 9.3 purposely discounts the immense difficulty of operationalizing variables and instead addresses the question of what sort of theories we would have if the operational problems were solved. Until empirical studies generate coefficients for the relationships and error estimates for the predictions, all the relations would be qualitative and imprecise. We are presently limited to crude statements of trend about the effects of resource availability and ecological similarity on coexistence and resource use by competitors in space and time.

Although only empirical study can correct this vagueness, there is no assurance that empirical study would actually do so. A priori, it is as likely that the empirical relationships which emerge from extensive research will differ greatly for different resources, environments, and competitors. If the theories must be cast in terms of particular species and particular resources, they would be debilitatingly specific.

Like many explanatory ecological constructs, the purpose of considering competition is not always explicit. Presumably it forms part of the resolution of Hutchinson's (1959b) question 'Why are there so many kinds of animals'? (Brown 1981), and a fully elaborated theory of competition would therefore predict species number and abundance. However, the relevance of the propositions in Table 9.1 and the relations in Fig. 9.3 to this question are not obvious. The predictions in Fig. 9.3 seem almost unrelated in this regard because, even if they were operational, quantitative and precise, the researcher would still be at a loss about how to use these rules to predict species diversity. Moreover, it seems extremely unlikely that any of these theories could achieve the status whereby we would find it easier to predict co-occurrence from resource supplies and similarities than to measure co-occurrence directly. The proposed theories are simply too indirect, too impractical and too uneconomical to merit use.

None of these arguments can show that competition could never lead to successful ecological theory. New researchers with new ideas, using new techniques and new approaches may be able to erect different theories of competition with greater generality and precision with less effort than now seems possible. Some of Keddy's (1989a) ideas suggest this possibility, but scientific intransigence towards change and refusal to recognize the evident failure of competition theory make improvement unlikely. The potential of competition theory has been explored by some of the brightest minds in ecology for decades without notable success. The rest of us might do well to look elsewhere.

The search for alternatives will be encouraged by fuller recognition of the limitations of contemporary ecology. Criticism lays bare such limits, but is rendered difficult by the intricacy and complexity of the problems of ecology. We have misapplied our ingenuity to turn defeat into victory through intellectual machination, and succeeded only in deluding ourselves about the extent of our muddle. Ecology now stands in desperate need of ruthless criticism to liberate itself from the accumulated weight of intellectual baggage, so that the brightest researchers will once again be free to engage the tremendous scientific and practical difficulties that confront humanity.

Summary – The muddles of ecology

The weakness of the central constructs of contemporary ecology results because ecology compounds its single failings. Operational impossibilities spawn tautological discussions that replace predictive theories with historical explanations, testable hypotheses with the infinite research of mechanistic analysis, and clear goals for prediction with vague models of reality. The resulting melange obscures appropriate research and attainable goals with sloppy, ineffective activity. As a result, the central constructs in ecology yield predictions with difficulty and these are often so qualitative, imprecise and specific that they are of little interest and less utility. The complexity of contemporary ecology makes criticism difficult, because the critic scarcely knows where to begin. This predicament protects ecological constructs from all but sustained critical scrutiny.

10 · Predictive ecology

Like most critiques, this book is largely censure. It presents a negative view of contemporary ecology. The critic also has the smaller, but essential, duty to provide a positive vision of what the subject might and should be like. Such a vision is the goal of this chapter which, as its title suggests, describes the characteristics of an explicitly predictive, and therefore more rigorously scientific, more informative and more useful ecology.

The chapter begins with examples from eight classes of predictive ecological theories or models. Most of these theories are well known and are referred to elsewhere in this book. Here, they are discussed in enough depth that their range, similarities and differences can be appreciated. The examples are purposely diverse. They deal with allometry (Peters 1983; Calder 1984), with responses to different levels of specified resources (Coe et al. 1976; East 1984; Peters 1986), with lake nutrient loading (Vollenweider 1968; Dillon & Rigler 1975; OECD 1982), with the interrelationships among physiological processes like animal respiration, plant water stress and productivity (Humphreys 1979; Turner 1988), with the prediction of qualitative variables like plant form (Box 1981), with aspects of a major problem of environmental concern, the ecotoxicology of organic chemicals (Kaiser 1984), and with variables of longstanding ecological interest, like the abundance of a given narrow taxon (Morin & Peters 1988) and species diversity (Strong, McCoy & Rey 1977; Currie & Paquin 1987). These examples then illustrate discussion of the many facets of two attractive qualities of predictive ecology: pluralism and practicality. Pluralism is reflected in hierarchical structure, complementarity, and the maintenance of multiple working hypotheses, practicality in the use of simplifications, abstractions, instrumentalism, empiricism, holism, sound variables and adaptive management strategies.

Eight classes of model theories in predictive ecology

The work described in this section necessarily meets the criteria for scientific theories described in Chapter 2 and reiterated throughout the book. All the examples make falsifiable predictions by identifying some logically possible phenomena as improbable in practice. All use operationally defined variables and, in all cases, the accuracy of the predictions has been tested against observation. Most of these predictions could be represented on a Cartesian co-ordinate system, like Fig. 2.1 and 2.3. Many ecologists would find these models disappointing if advanced as explanations and none of the theories purports to describe the detailed mechanisms of ecological systems, but all identify patterns in nature. Most of the relations are general, quantitative, and probabilistic, but the imprecision of all indicates the need for further research. Since most are relevant to both science and society, these areas are likely growing points in ecology. All these characteristics relate to the formal requirements for scientific theory developed in Chapter 2 and used throughout the book in evaluating ecological constructs.

Allometry

An example of an allometric approach to ecological problems has already been detailed in the discussion of population density and size (Chapter 2), and need not be redescribed here. This relationship, like many others in allometry, was developed through a literature search for estimates of population density and body weight for a wide range of mammals. Regression analysis described the relation between these variables empirically, quantitatively, and stochastically. The relation is imprecise, simplistic, and highly abstract in the sense that it represents very little of the animals or their environment. In compensation, it is likely the best general predictor of density. This body-size relation is only one of a very large number of relations that predict aspects of the morphology, physiology, and ecology of organisms from a simple estimate of size. These have been detailed elsewhere (Peters 1983) to serve as an extended example of predictive ecology.

Resource response

A number of different models predict standing stocks or biological activities on the basis of some measure of what may loosely be termed the 'resource level', although the meaning of this phrase must be operationally determined for each relationship. I am most familiar with the suite of limnological regressions based on total phosphorus concentration (Table 10.1), but since these are discussed elsewhere (Peters 1986; Seip & Ibrekk

Table 10.1 *Selected regressions describing the relationships between total phosphorus concentration (TP, mg m^{-3}) and other lake characteristics; n refers to the number of lakes* (From Peters 1986).

Dependent variable	Units	Equation	r^2	n
[Chlorophyll]	mg m^{-3}	$Y = 0.73\,TP^{1.4}$	0.96	77
Transparency	m	$Y = 9.8\,TP^{-0.28}$	0.22	87
[Phytoplankton]	mg wet wt m^{-3}	$Y = 30\,TP^{1.4}$	0.88	27
[Nanoplankton]	mg wet wt m^{-3}	$Y = 17\,TP^{1.3}$	0.93	23
[Net plankton]	mg wet wt m^{-3}	$Y = 8.7\,TP^{1.7}$	0.82	23
[Blue-greens]	mg wet wt m^{-3}	$Y = 43\,TP^{0.98}$	0.71	29
[Bacteria]	millions ml^{-1}	$Y = 0.90\,TP^{0.66}$	0.83	12
[Crustacean plankton]	mg dry wt m^{-3}	$Y = 5.7\,TP^{0.91}$	0.72	49
[Zooplankton]	mg wet wt m^{-3}	$Y = 38\,TP^{0.64}$	0.86	12
[Microzooplankton]	mg wet wt m^{-3}	$Y = 17\,TP^{0.71}$	0.72	12
[Macrozooplankton]	mg wet wt m^{-3}	$Y = 20\,TP^{0.65}$	0.86	12
[Benthos]	mg wet wt m^{-2}	$Y = 810\,TP^{0.71}$	0.48	38
[Fish]	mg wet wt m^{-2}	$Y = 590\,TP^{0.71}$	0.75	18
Avg prim. prod.	mg C m^{-3} d^{-1}	$Y = 10\,TP - 79$	0.94	38
Max. prim. prod.	mg C m^{-3} d^{-1}	$Y = 20\,TP - 71$	0.95	38
Fish yield	mg wet wt m^{-2} y^{-1}	$Y = 7.1\,TP^{1.0}$	0.87	21

1988), I will not pursue them at length here. There are, however, similar relationships for terrestrial systems that also demonstrate the powerful effect of resource levels and which should be better known.

In grasslands, annual net production and therefore maximum standing stock of plants increases with rainfall (Walter 1973; Sinclair 1975; Coupland 1979; Sims & Coupland 1979; McNaughton 1985) or some other measure of available water (Singh *et al.* 1980; Sala *et al.* 1988). The biomasses of both vertebrate (Coe *et al.* 1976; East 1984) and invertebrate (Kajak 1980) herbivores are in turn correlated with water availability or plant biomass. This pattern is repeated for predators and their prey (Kajak 1980; East 1984; Fig. 10.1).

These relations are typically described by simple regressions and scatter diagrams between standing stock and some measure of a consumable benefit or resource in the ecosystem. They differ from traditional ecological discussions of 'resource limitations' and 'limiting nutrients' because the resource in question is explicitly and operationally defined. In lakes, the concentrations of total phosphorus and nitrogen in

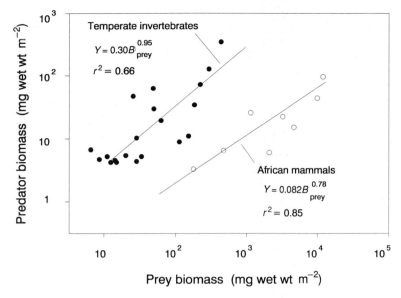

Fig. 10.1. The relation between predator and prey biomass (data from East 1984; Kajak 1980).

the water are usually powerful predictors. In grasslands, available water or rainfall seems to work. Appropriate resource measures for marine systems and forests are more debatable, but nitrogen (Ryther & Dunstan 1971; Caraco *et al.* 1987) and available water (Lieth & Whittaker 1975) respectively are likely choices. No such prime resource has yet been identified for running waters. All these relations are empirical and probabilistic. Little effort has been directed to define the 'resource' relative to particular consumers. As a result, many causal steps and ancillary factors that might distinguish among systems or organisms are necessarily ignored. This excessive simplicity increases the uncertainty of the relationship which in turn underscores the need for improvement.

Lake loading models

Loading models for lake nutrients are an essential counterpart to the nutrient response models in Table 10.1. These models allow one to predict the phosphorus concentration in a lake on the basis of the annual 'loads' or budgets of water and phosphorus entering the lake from the watershed and the air. These values are in turn estimated from crude models based on land use, geology and runoff.

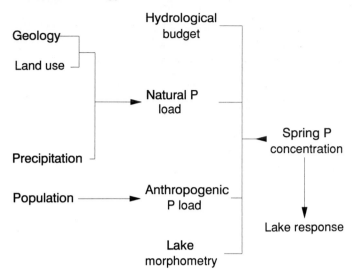

Fig. 10.2. Flow diagram indicating the steps involved in predicting lake phosphorus concentration (from Dillon & Rigler 1975).

Loading models are relatively complex in that they involve several steps (Fig. 10.2). First the volume of water flowing through a lake (Its hydrological load, $m^3 y^{-1}$) is determined. This requires measurement of the lake and watershed areas (m^2) as they appear on topographic maps, and estimates of precipitation, lake evaporation and watershed runoff per unit area ($m^3/m^2 = m$), which are obtained as long-term averages from meterological records.

Hydrological load = lake area × (precipitation − evaporation)
+ watershed area × runoff   (10.1)

Natural phosphorus load is calculated from average values for the phosphorus falling on the lake in rain and dust ($P_{rain} = 30$ mg $m^{-2} y^{-1}$) and the phosphorus entering the lake from the watershed. The latter is expressed as a P export coefficient (mg $m^{-2} y^{-1}$) and the averages used reflect differences associated with land use and geology (Table 10.2):

Natural P load = (watershed area) × (P export coefficient) +
(Lake area) × P_{rain}.   (10.2)

This natural load is further increased by about 0.5 kg P y^{-1} per person for each human inhabitant of the basin:

Anthropogenic load = population × 0.5   (10.3)

Table 10.2 *Phosphorus export coefficients (mg P m⁻² watershed γ⁻¹) as affected by land use and geology.*

	Geological classification		
Land use	Igneous	Sedimentary	Unspecified
Forest	4.7	12	22
	0.7–8.8	6.7–18	2–45
Forest +	10.2	23.3	—
pasture	5.9–16	11–37	—
Agriculture	—	—	45
			10–300
Urban	—	—	190
			50–500

Note:
Listed values are the means and ranges (Dillon & Kirchner 1975) or, for the last column, midpoints (Reckhow & Simpson 1980), and ranges of reported values.

The per capita P release is about twice as high in areas still using high phosphate detergents (Reckhow & Simpson 1980). From these two figures the weighted average phosphorus concentration of the inflowing water may be calculated as:

$$[P]_{load} = (\text{natural load} + \text{anthropogenic load})/\text{hydrological load} \quad (10.4)$$

Some portion of the phosphorus entering the lake is lost to the sediments and so the phosphorus concentration in the lake must be adjusted for this retention (R, expressed as a fraction):

$$[P]_{lake} = [P]_{load} \times (1 - R) \quad (10.5)$$

A number of formulae provide estimates of R (Canfield & Bachmann 1981; Nürnberg 1984). Ostrofsky (1978) gives an empirical estimate of R which has the advantage that it can be made from lake surface area and therefore topographic maps alone:

$$R = 0.201 \exp(-0.0425 q_s) + 0.574 \exp(-0.00949 q_s) \quad (10.6)$$

Where q_s = hydrological load/lake area. Another estimate of R can be made if estimates of lake volume are available from bathymetric surveys:

$$R = 1/\rho^{0.5} + 1) \quad (10.7)$$

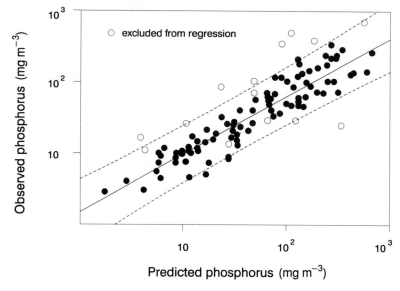

Fig. 10.3. The relation between predicted and observed phosphorus concentrations in lakes. Points excluded from the regression represent lakes where internal loads are so substantial that the annual phosphorus load from rain and inflows (P_i) are thought to underestimate total load or these points represent reservoirs where sedimentation loss are reduced by rapid flushing (from OECD 1982).

where ρ is the annual flushing rate of the lake (ρ = lake volume/ hydrological load, y^{-1}).

This formulation is derived from a different algebraic treatment, based on the assumption that retention depends on lake phosphorus concentration, rather than the loaded phosphorus concentration. The derivation leads to a more widely used, but apparently less effective (Prairie 1987) expression favoured by OECD (1982) which uses the inverse of flushing rate, the water residence time, τ in years:

$$[P]_{lake} = (P\ load/water\ load)/(1 + \tau^{0.5}) \qquad (10.8)$$

This model is remarkably successful in predicting phosphorus levels in lakes (Fig. 10.3) and has become pivotal to phosphorus abatement and eutrophication control. When coupled with nutrient response models (Table 10.1), it allows one to estimate the biological impact of increased use on recreational lakes and the probable benefit of phosphorus abatement.

This set of relations differs from the models of allometry in that it is not based on isolated and independent regressions but on an articulated series of models. The heart of the model is the gross simplification that the lake and the phosphorus cycle can be treated as a single 'continuously stirred tank reactor', but this engineer's abstraction is blended with empirical relations for retention and crude averages for the runoff coefficients to yield the inelegant and jerry-rigged sequence in Fig. 10.2. Nevertheless, this model and its various competitors have the great advantages that they have been tested in relevant field settings, they provide a tool to manipulate our environment, and they seem to work.

The scatter in Fig. 10.3 shows that the model is still uncertain, but it is far better than the models it superseded. For example, in Ontario, where much of the early work on loading models was conducted, the capacity of a lake to accept cottage development was previously estimated from the area required to operate a boat safely and the average number of boats per cottage. Loading models are now the basis of a worldwide program of phosphorus abatement, an application which Schindler (1987) argues is the greatest single success of ecological knowledge over environmental problems. It is an achievement of which the profession can be proud.

Physiological relations

The covariation among many physiological processes has been used to develop a variety of physiological relationships with ecological implications. At the level of the individual, there are examples that relate rates of growth to ingestion in aquatic animals (Condrey 1982), contaminant uptake to metabolism in fish (Neely 1979), and net growth to transpiration in plants (Turner 1988). Other authors have used essentially physiological approaches to predict characteristics of populations or communities. For example, Humphreys (1979) has shown that annual population production can be estimated from population respiration rates; Kajak (1980) developed regressions that relate assimilation and production by communities of grassland invertebrates to consumption by the same community. Presumably respiration and elimination are similarly related.

One of the most versatile of these basically physiological variables is evapotranspiration, which was originally developed by Thornwaite (Thornthwaite & Mather 1957; Mather 1978) to predict phenological events in agriculture but has since been exploited in many areas of hydrology, agriculture, and ecology. Although others have addressed the related problem of water balance (van der Beken & Hermann 1985) and alternatives for some aspects of the model exist (Penman 1964), Thornthwaite's comprehensive model remains one model of choice (Box 1981). The model is a cumbersome tabulation – a FORTRAN version was developed by Willmott (1977) – summarizing the empirical relations and approximations that determine evapotranspiration. Basically, the model compares the potential for evapotranspiration, calculated from day length and measured mean monthly temperature, with the water available from rainfall. These two variables are then modified to allow for differences in runoff, soil water capacity, and rooting depth. Actual evapotranspiration equals potential evapotranspiration so long as there is enough rain, but where rain is insufficient, the model allows calculation of actual evapotranspiration, change in the store of soil water, and the reduction in transpiration induced by the unavailability of water. The amount of water in runoff and snow melt is also estimated. The model has been usefully applied in calculating runoff (Calvo 1986), agricultural production, irrigation schedules, soil moisture losses (Mather 1978), and primary production (Lieth & Whittaker 1975). Evapotranspiration is also a component in models for litter breakdown (Meentenmeyer 1978) and plant life form (Box 1981).

The original formulation has recognized flaws. For example, it underestimates potential evapotranspiration in extremely dry, windy or cold environments and its tables seem anachronistic in the computer age. A tremendous amount of information exists which could be used to update and improve the model, but this has not been done. In part, the sweep of Thornthwaite's concept was so great and successful that it impeded change in those global models. Instead, a number of specific changes have been proposed for various sites and applications. Models, like this one, now await replacement by a generation of models that make better use of modern computational power and the information available through remote sensing (Monteith 1973; Waring 1983; van der Beken & Hermann 1985).

Thornwaite's models differ from some of the other examples of predictive ecology, because they do not consist of regressions or scatter diagrams. They are only semi-empirical and are not phrased probabilistically. They are most like the lake loading models in that they seem mechanistic, while omitting a host of details. Also like loading models that predict phosphorus concentration, the main thrust of Thornthwaite's model is to estimate an important independent variable, actual evapotranspiration that is related to other variables by simple regressions. Nevertheless, some of the steps in both developments, like runoff and phosphorus export, are interesting variables in their own right.

Plant life form

The foregoing models are all quantitative. Although quantification is desirable in science (Chapter 2), Box's (1981) macro-climatic model to predict plant life form shows that qualitative models can also be successful. Box took floral descriptions from a large ($n = 1151$) sample of sites around the world and used his botanical intuitions to describe these floras as mixtures of up to 86 types of plant form. Each form was identified by a series of 25 characteristics describing plant size (tall, normal and dwarf), type (tree, shrub, herb or grass, and cactus-like 'stem-succulent'), leaf size (large, normal, small, or very small), shape (broad, narrow, grass-like, or none), structure (herbaceous, leathery, schlerophyllous, succulent, woody stems, pubescent), seasonal habit (evergreen, semi-evergreen, rain green, summer-green, suffrutescent – those with a permanent woody base and annual herbaceous shoots, marescent – persistent grasses and herbaceous shoots, and ephemerals). He then turned to the world climatic literature to determine the likely climate at each sample site in terms of regularly reported variables and finally settled on eight of these: the highest monthly mean temperature, lowest monthly mean temperature, annual range of monthly mean temperatures, average annual precipitation, annual potential evapotranspiration, highest average monthly precipitation, lowest average monthly precipitation, and the average precipitation in the warmest month. These data were then used to describe climatic niches for each plant form as the hyperspace enclosed by the maximum and minimum limits of all climatic variables. These descriptions, and some alternatives, were iteratively fitted to his sample floras until he obtained the best possible description of the climatic niches of the floral types in the model building data set. Additional criteria were then introduced to determine the extent of cover of the ground by plants (which rises with the ratio of rainfall to potential evapotranspiration) and dominance (which is largely a function of plant size) in the sample, and to eliminate forms that were close to their limits of tolerance (i.e. he eliminated all forms that were within 10% of either extreme of any variable, where the range between maximum and minimum for each climate variable represents 100%).

After identifying a 'best' model in terms of its description of the original data, Box applied that model to 74 'validation' sites, also representing a world–wide data set, that had not been used in the original calibration. This comparison could not use standard statistics because the data are qualitative. However, Box concluded that the model was successful in 92% of these cases, predicting the correct mix of forms and correct dominants in 37 sites and the correct mix of forms (but not the correct dominants) in 31. The remaining six sites were remote areas in South America and Asia where it is conceivable that weaker meterological records increased the disparity between prediction and observation.

This model is a stunning achievement. Like the others in this section, it is empirical, abstract, simplifying, and general. It ignores traditional divisions and has stood up to testing. Moreover, although it is a relatively complex model, Box (1981) conscientiously described its basis and its faults so that the model can be used by other researchers.

QSARs

There are now well over 10 million organic chemicals known to science. Over 60 thousand of these are in common use and a thousand or more are added to the market each year (Stumm, Schwarzenbach & Sigg 1983). Some of these are highly dangerous carcinogens, teratogens, and toxics whose control is a major challenge to mankind. Because there are so many chemicals that contaminate so many kinds of organisms in so many kinds of habitats, it is impossible to treat 'the ecology' of each combination. Indeed, testing even one of these chemicals for the simplest of ecological effects, bioconcentration and acute toxicity, costs more than $10,000, comprehensive testing would cost 10 to 100 times more (Nirmalakhandan & Speece 1988). Given the magnitude of the costs, the tasks, and the time frame in which we need answers, we must find general rules to identify potentially dangerous products and predict their effects.

Quantitative structure activity relationships (QSARs) describe the relations between the behaviour of a chemical and some quantitative index of chemical structure. Such relationships have long been employed by pharmacologists (Kier & Hall 1976) and QSARs are increasingly used in the ecotoxicology of organic pollutants in aquatic systems (Stumm et al. 1983; Connell 1987, 1988a,b).

In ecotoxicology, the relevant aspects we wish to consider include the sorption of the contaminant to non-living materials, like the sediments and soils, uptake and bioaccumulation by organisms, persistence and detoxification of the chemicals by living and non-living process, and the toxic effects of the material (Connell 1987). All these processes are affected by the nature of the contaminant, and this in turn reflects its molecular structure. Simple as molecules are, they are complex enough that characterization involves a host of potential descriptors. These include some familiar descriptors, like molecular weight, aqueous solubility, boiling point, and number of carbon atoms per molecule, and some less familiar, like connectivity (a numerical measure of the branching of the chemical skeleton), molecular surface area (the two-dimensional space available for a water molecule in contact with the solute) and parachor (a composite of surface tension, density and molecular weight).

The most frequently used measure of chemical structure is the 'octanol–water partition coefficient'. This is measured from the equilibrium distribution of the chemical dissolved in a two-phrase system of water and octanol:

$$\log P_{ow} = \log_{10}(P_o/P_w) \tag{10.9}$$

where P_o and P_w are the proportions of the total solute in the octanol and water phases, respectively. This property is in turn related to a number of other characteristics including all of those listed above and more (Cramer 1980; Verschueren 1983; Mailhot & Peters 1988). This has led to a series of investigations to determine if another index, possibly one which is easier to determine than P_{ow}, could be used instead (Chiou et al. 1977). Further details of this expanding field need not be pursued here.

P_{ow} has proven to be a powerful tool in the prediction of ecotoxicological effects. Compounds which have high values of P_{ow} tend to be fat-soluble, toxic, carcinogenic, bioaccumulable, and persistent; those with low P_{ow} tend to be water-soluble and relatively innocuous. For example the value of $\log P_{ow}$ for DDT is 5.7, that for ethanol is -0.3 (Mailhot & Peters 1988). Early work (Neely, Branson & Blauu 1974; Moriarty 1983) showed that $\log P_{ow}$ was positively related to bioaccumulation and now a suite of relations exists between the two variables (Fig. 10.4). It is noteworthy that bioaccumulation in these plots seems relatively little affected by food-chain effects, for similar levels are obtained from systems in which there is no food, in microecosystems in which the food chain is drastically simplified (Metcalf et al. 1975), and in nature (Bysshe 1982). More recent work has related rates of uptake (Mailhot 1987) and release (Oliver & Niimi 1983) to $\log P_{ow}$, to sorption by sediments, soils and decontaminating agents like activated charcoal (Verschueren 1983) and still another section of ecotoxicology has related $\log P_{ow}$ to toxicity

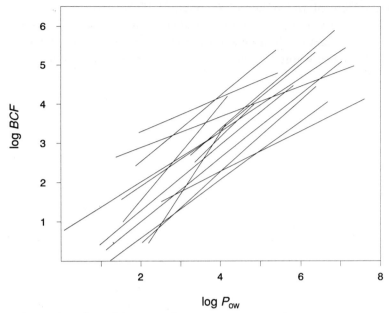

Fig. 10.4. The effect of the octanol–water partition coefficient (P_{ow}) on bioconcentration factor of organic compounds ($BCF = $ [body]/[water]) by aquatic organisms. Each line represents a separate regression from the literature (Mailhot 1987).

(Veith, Call & Brooke 1983; Friant & Henry 1985; McCarty 1986). This area is still small (Kaiser 1984), but it too promises to develop rapidly.

Like most other relations in this section, QSARs are empirical, statistical regression equations. They have been 'explained' after the fact as the result of relative solubilities in fat and water, and as the interplay of fugacity and Henry's law (Mackay 1982; Mackay & Patterson 1982), but their chief attraction lies in the utility of the patterns they describe.

Biomass of larval black flies

The previous examples have all dealt with general phenomena but predictive ecology also works to predict the properties of particular, narrow taxa. This should not be surprising because applied sciences, like agriculture and public health, have always used such theories. This section simply adds an ecological example to demonstrate that predictive ecology is not limited to community level phenomena or to common properties of animals distinguished only by their average weight. The example is a model to predict the abundance of a small group of

Table 10.3 *A multiple regression to predict the microhabitat distribution of larval black flies living in the outlet rivers of lakes.*

Response	Predictors	Units
log biomass =	2.92	mg dry mass m^{-2}
	-0.62 log(distance)	m
	$+0.030$ velocity	cm s^{-1}
	-0.00012 velocity^2	cm s^{-1}
	-0.019 depth	cm
	-0.0094 periphyton	mg m^{-2}
	$+1.03$ log[chlorophyll]	mg m^{-3}
	-0.42 log[chlorophyll] × log(distance)	
	-0.007 log[chlorophyll] × velocity	
	$R^2 = 0.55$ Residual mean square $= 0.36$	

Note:
Distance refers to distance of the sample site from the lake outlet, velocity is the mean current velocity of the stream above the sample site, depth is depth of water above the sampled substrate, [chlorophyll] is the concentration of chlorophyll in the water, and periphyton is the mean standing stock of chlorophyll on the substrate where the larvae were collected (From Morin & Peters 1988).

organisms using more information than would be appropriate for more general analyses.

Black flies (Simuliidae) are a group of small biting dipterans which infest the forests of Canada and many other parts of the world. In Africa, these insects are the vector for 'river blindness' a scourge of human communities on the rivers of West Africa. In the North, they are usually only a pest, but the nuisance of black flies is great enough to effect vacation schedules, the price of land, and the balance of tourism. The larvae of these flies live in the fast waters of streams, especially at lake outlets, where they attach to the substrate and filter particles (seston) from the stream. This stage may be important as fish food, but it is indubitably important as the target for biological control using the bacterium *Bacillus thuringinensis*.

Antoine Morin has developed two sets of multiple regression models to predict the abundance of these larvae in winter and early spring. His microdistribution model (Table 10.3) describes the distribution of flies in a single river as a function of water depth and flow, seston levels, and temperature measured at each sampling station. Such a model is limited

Table 10.4 *Two multiple regression models to describe the macrodistribution of the biomass (mg dry weight m^{-2}) of larval black flies in streams exiting lakes in Quebec.*

Response	Predicators	Units
Model I log biomass =	2.11	
	−0.52 distance	m
	+0.04 velocity	cm s^{-1}
	−0.0021 velocity^2	
	+0.54 log[chlorophyll]	mg m^{-3}
	−0.033 periphyton	mg Chl m^{-2}
	$R^2 = 0.71$	Residual mean square = 0.17
Model II log biomass =	0.55	
	−0.37 log distance	m
	+0.045 velocity	cm s^{-1}
	−0.00024 velocity^2	
	+0.13 phytoplankton	g C m^{-3}
	$R^2 = 0.62$	Residual mean square = 0.22

Note:
Model I is based on in situ measurements, whereas the variables in model II can be estimated from maps, meteorological data and loading models (from Morin & Peters 1988).

in its applications because the effort in measuring the independent variables is high, relative to direct measurement of animal biomass. Morin therefore developed two large-scale models which predict the abundance in different streams (Table 10.4). The first of these is based entirely on directly measured properties – stream flow, distance from the lake, and average concentrations of periphyton and suspended chlorophyll. Morin recognized that even this model would be unlikely to be of use, because of the effort involved in measuring chlorophyll and periphyton levels. He therefore erected a third model based on properties that could be estimated from a topographic map and an adaptation of Vollenweider's loading models to estimate stream flow and seston levels. This last model could be used, for example, by scientific advisers in government or industry to determine the suitability of remote sites for cottage development or black fly control programs (Morin & Peters 1988).

Diversity

The study of biological diversity has long been a part of ecology and, before that, of natural history. As part of this interest, researchers have identified correlates that can be used to predict some measure of biotic diversity.

The utility of these relations is weakened somewhat by the measures used (Hurlbert 1971; Peet 1974). Which measure is 'best' cannot be decided without first knowing why diversity should be of interest, but the literature is surprisingly unhelpful with this issue. The long-standing connection between diversity and stability is now largely discredited (Goodman 1975) and the role of species diversity as an indicator of stress is questionable (Schindler 1987). Strong *et al.* (1977) show that the diversity of crop pests is related to the area under cultivation and this may have implications for pest control. Sometimes, diversity seems to have attracted attention simply because some patterns could be identified. For the most part, however, diversity seems to be a naturalist's measure that reflects the likelihood of encounter between a naturalist or researcher and a new species.

Hurlbert (1971) suggested a statistic to measure this last aspect of diversity, but the information required to use his statistic is often unavailable. Of the remaining choices, the least obscure is species number because it is least manipulated. This statistic has the disadvantages that it is affected by sampling effort, methods and sample size, but it has the advantages that it can be approximated from the sizable literature in natural history, a literature which is inappropriate for many other diversity measures, and that it is readily interpretable by the likely audience of naturalists. Moreover, for many of the most interesting groups (trees, birds, mammals, reptiles, butterflies), species numbers as reported by natural historians are likely closely related to the probability of encounter. In any case, without any stated purpose for an estimate of diversity, we cannot decide what the most relevant measure is going to be. Species number is at least a representative estimate of diversity for which a lot of data are available.

Patterns in species number have long fascinated ecologists (Pianka 1966; Lewin 1989). Some deal with general phenomena like the species–area curves (Connor & McCoy 1979) or island biogeography (MacArthur & Wilson 1967). A number of specific perturbations have been shown to reduce species number (Patrick 1963; Sanders 1969; Harvey 1975) and the relative numbers of species of different faunal elements are frequently correlated (Krebs 1985; Sprules 1975). Evapotranspiration has been correlated with the numbers of both trees (Currie & Paquin 1987) and birds (Brown 1981). If these relations can be viewed free of the trappings of theoretical ecology, they are simple, probabilistic, general, empirical patterns in the abundance of species. Like other predictive relations in ecology, these are relevant only to a specific set of questions, but in that area they can be effective. It is a measure of the confusion of ecology that diversity relations are so little appreciated.

The attractions of predictive ecology

As theories, each of these very different constructs must be judged independently on the basis of its ability to predict. As examples of a subdiscipline called 'predictive ecology', they share a characteristic style and vision. It is these common qualities lying beneath the differences, rather than the formal implications of predictive power, that are the focus of this discussion.

Pluralism

Schoener (1982) suggested that the ecology of the future would be characterized by a more pluralistic approach than that of the past, and his view has been seconded by McIntosh (1987). Reimers (1986) suggested the similar concept of a dualistic or multiple approach to ecosystem study. The diversity of the few examples listed above is one aspect of this pluralism. Pluralism is reflected in the hierarchical structure of contemporary ecology, in the realization that different problems require different approaches, variables and solutions, and in the maintenance of competing hypotheses, so these topics are discussed below. The complexity of pluralism and complementarity necessitates some device, other than explanatory causal or mechanistic narratives, to organize existing knowledge; the concept of predictive matrices is introduced as such a device.

The eight exemplary theories in this chapter address different aspects of very different ecological phenomena and the amount of interaction among them is small. The theories do not assume that a single approach, technique, variable or process will succeed in all systems, but rather show that different factors and different approaches may be needed for different systems and different questions. Modelers who built these theories have taken inspiration where they found it: from engineering models like the stirred reactor of Vollenweider, from applied chemistry for ecotoxicological QSARs, from existing scientific lore like the nutrient response models and even from the interests of theoretical ecology, like diversity. Although many of the models use regression, this is not an essential element of the approach, but only a useful tool to describe relations and summarize data. The regressions are drawn because our inspirations led us to suppose that there might be a relation. The source of that inspiration might be based on mechanisms, precepts, concepts, and whatever else we could use, but the source should be irrelevant to any judgement of the final product.

It is important that this pluralism be distinguished from the view that 'anything goes'. In the context of predictive ecology, pluralism is limited to sources of inspiration, types of models, choice of variables, modes of analysis and kinds of prediction. Pluralism should not extend to the criteria of validation; these criteria remain predictive power and its implications.

My doctrinaire position seems ungentlemanly and chauvinistic, compared to more even-handed and diplomatic views that variety is desirable in all parts of science and that any form of scientific doctrine would unfairly constrain intellectual freedom and creativity (Hutchinson 1978; May 1981b; May & Seger 1986; McIntosh 1987). However, value-free views about what constitutes science have eroded the standards of ecology so much that they precipitated the crisis described in Chapter 1. The only hope of developing a predictive, informative, useful body of ecological theory requires pluralism and inventiveness in theory building, coupled with the rigorous criticism of theory evaluation implicit in tests of their predictive power. Only the latter distinguishes science from other human endeavors (Chapter 2).

In a sense, ecologists must replace their previous rigid views about appropriate methods for creating theory with tolerant ones and their previous careless views about what constitutes theory testing with strict ones. Ecologists who are not dedicated to prediction, those who prefer the pleasures of academic discussions and theoretical ecology, will find this switch in emphasis difficult. However, those who find joy in science, in knowledge and in offering some measure of protection to humanity and to nature in these dangerous times, will find predictive ecology far more satisfying and involving.

Hierarchical structure
The call to recognize the hierarchical structure of the external world, or at least the science that describes it, is another valid aspect of the pluralism about what constitutes scientific method and theory. Contemporary

'hierarchies' are just the opposite to a military or bureaucratic hierarchy. There is no rigid, highly interactive chain of command in which each theory of lower generality confirms, but specifies, the trends in more general theories. A modern hierarchical view implies that any system may be viewed at a variety of scales as a multilayered composite. Moreover, different aspects of the system may require different hierarchies so that the system is simultaneously described by a series of virtually independent models (O'Neill *et al.* 1986). Thus phosphorus concentration is useful in addressing differences among lakes, but other, more biological variables may be required to address differences within lakes (Sommer *et al.* 1986; Benndorf 1987; Carpenter 1988). Some theories of limnology may invoke phosphorus concentration, others lake area or volume, others pH or zooplankton numbers. Similarly, there is no longer any need to expect that solutions at specific levels should be mirrored by more general relations. Size may be useful in predicting gross differences in abundance of animals (Peters 1983) but not differences within populations of larval black flies (Table 10.4). This hierarchical structure of science is also reflected in the limitation that model systems are no longer considered pictures of reality but only simplifications for a limited purpose (Chapter 8). Contemporary ecologists must be more willing to accept this variety, its possible inconsistencies, and its untidiness, than they were in the past.

Modern hierarchical concepts allow so much freedom for independent interpretation, invoking different variables or different effects of the same variables at different hierarchical levels and scales, that it is scarcely hierarchical in the older sense of stratified at all. The components of each level of the hierarchy, the different levels of the hierarchy, and even the entire hierarchy itself assume a modular construction which allows restructuring, rephrasing and replacement of various parts of our scientific world view without dramatically affecting the remainder. Hall & DeAngelis (1985) touch on the same view by suggesting that ecological science is a patchwork, not a monolith. The monolithic view of science is no longer a working hypothesis of science, even if it remains a philosophical or perhaps religious element in the thinking of many scientists. The rise of modern hierarchical thinking is reflected in the decline of the view that the universe or its parts should be interpreted as a machine (Chapter 5).

Complementarity and cohesion
The incommodity of a patchwork science is loss of cohesion. The narrowing of our interests to focus on prediction of defined variables

will almost certainly reduce interactions among ecologists working in very different areas who now interact through the central concepts of the science. Even within a subject area, a modern hierarchical structure is likely to loosen the connections among different scientists approaching different problems. This is not entirely regrettable, for this book has laboured to show that such cohesion may be grounded in misconception. Concepts, like niche breadth or community stability, may give limnologists and desert ecologists a basis of communication, but the resultant dialogue may be meaningless to all practical intents and purposes.

Cohesion also results from overlap in different research programs and the breadth of vision of the scientists in the area. In a predictive science, overlap and breadth are encouraged because theories define variables by their relations to other variables. Thus a variable is never a stopping point, but a connection to other theories that use the same variable to make other predictors (as phosphorus concentration is used in Table 10.1), theories that predict this variable (as Equations (10.1) to (10.8) predict phosphorus concentrations), theories that use elements from these predictions in further predictions of still other variables (as the phytoplankton concentration in model II of Table 10.4 is determined from maps, phosphorus loading models and biological response models). Cohesion is further encouraged by common technological problems, by the search for appropriate models in other parts of the literature, and by scientific curiosity about areas outside one's specialty.

Reimers (1986) suggests that we would encourage cohesion in the science by accepting that different theories may be complementary to one another and thereby allowing ourselves to address a greater variety of questions. This would allow us the intellectual freedom to work at different scales, using different variables and models, with less fear that our research will prove unacceptable because it does not fit some established view (Dayton 1979; Van Valen & Pitelka 1974). In addition, Reimers hopes a rationalization of the resulting plethora of different constructs will allow us to see the interrelations in our work. He terms his view 'complementarity', a reference to Niels Bohr who proposed that physical phenomena be addressed with complementary theories, as light may be treated as a particle or a wave, depending on the question being asked.

Reimer's complementarity could replace the false cohesion of contemporary ecology. It would be encouraged if we accept the lessons of both philosophy and history of science that all our models are erroneous, that they all need improvement, and therefore that they are all provisional and ephemeral. The opposite belief, that our constructs catch some element of the truth and therefore should endure, impedes interaction among scientists, because personal research must always seem closer to the truth than the alternatives chosen by others. Reimers' complementarity can be achieved only if we open our minds to accept the validity of those alternatives.

Multiple working hypotheses

Predictive ecology encourages the coexistence of complementary and potentially competing models. Different models can rarely be dismissed out of hand, because they all differ in some aspect of the domain or the data base that may limit their applicability in a new case. As a result, several competing theories must be maintained by the scientist until or unless later observations show which relations are most effective in what cases and which relations may be ignored. For example, Dillon (in Peters 1986) states that there are over 60 phosphorus chlorophyll relations; Canfield & Bachmann (1981) list over 20 formulations for the retention term in loading models; Fig. 10.4 shows 15 variations of the relation between bioaccumulation and P_{ow}, Peters (1983) list 63 allometric equations for standard and basal metabolism for homeotherms, and Connor & McCoy (1979) list over 100 species–area curves. Even when only one model exists, the models of predictive ecology are so simplistic that it is easy to imagine other variables and therefore other ways of predicting the phenomena.

One very salutary effect of this multiplicity of choice is that predictive ecology encourages 'multiple working hypotheses' (Chamberlain 1890) and tests among explicit hypotheses or 'strong inference' (Platt 1964). In both cases, the existence of alternative hypotheses heightens the importance of the data because new data may allow one to select among the competing hypotheses. Thus predictive ecology should encourage the growth of science by showing where disparities exist and fostering tests to select the best among existing alternatives.

Predictive matrices

One device to foster complementarity is to develop expert systems that allow access to matrices of predictions for different sciences (Fig. 10.5). In such a matrix, the cells are filled with the different models relating the response and predictor variables at the beginning of each row and column respectively. If the matrix is well constructed, the details in each cell should permit comparison of the precision and the domains of

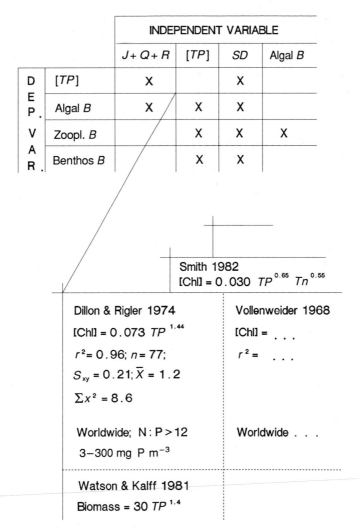

Fig. 10.5. A portion of a predictive matrix describing the eutrophication of lakes. An effective version of the matrix in the top half of the figure should contain details indicated in the expansion in the lower half, describing the relation between the variables beginning each row (measure of algal biomass) or column (total phosphorus concentration). The parallel entry for *TP* and total nitrogen (*TN*) implies that multiple regressions can be fitted into the same framework by employing more dimensions. This small section of the matrix deals with phosphorus loading estimated from hydrological load (*Q*), the phosphorus load (*J*), and retention (*R*), total phosphorus concentration (*TP*), Secchi disc transparency of the water column (*SD*), and the biomass (*B*) of different biotic components (from Peters 1986).

existing models to facilitate the appropriate choice by each user. In one sense, these theories are competitors, because they predict the same thing. They are also complementary in the important sense that they allow prediction of some desirable piece of information from different sorts of knowledge. Predictive matrices therefore encourage both the maintenance and the confrontation of multiple working hypotheses. Knowing the variable that one wishes to predict, one can read along the row to determine what variables and models permit this prediction. By reading down the column, one can determine what aspects of the system can be predicted given an estimate of one or more predictor variables. The matrix can also serve to decide further growing points in the science, for example by showing what relationships have not yet been investigated and what relations most need improvement by extending their domain, reducing their imprecision or developing alternatives.

The most complete example of such a matrix of which I am aware is given in appendices in Peters (1983) which list general allometric relations to predict various aspects of the autecology, physiology and morphology of individual organisms. This only represents a single column in an extensive matrix describing these predictions from other variables. Nevertheless, it should be possible to focus on aspects of this matrix, for example on relations for ingestion rate, or life history traits, or population density, to generate more restricted matrices involving alternative predictors. Some such smaller summaries appear in the *Biological Data Books* (Altman & Dittmer 1968). Extensive matrices are too large to be represented on a printed page, but could easily be accommodated by 'expert' computer systems (Seip & Ibrekk 1988).

Such matrices require extensive, quantitative reviews of the existing literature. These reviews differ from those now typical in ecology because they would summarize the actual theories and models of ecology. Present reviews, those of *Annual Reviews in Ecology and Systematics* for example, too often reduce whatever quantitative patterns exist in the literature to pallid, verbal summaries that do not lend themselves to prediction, test or evaluation.

Practicality and instrumentalism

For many, scientific theories like those of predictive ecology are uninteresting because they are too practical. They do not invoke the flights of high scholarship that have characterized ecology for the past generation. Moreover, they put a premium on the ability to measure some characteristic of the world, and the mathematical, statistical and logical skills they require are quite ordinary. Finally, the theories tell us something about what we may need to know to manage the Earth, so that the theories involve aspects of practice and application which have often been regarded as second-rate in ecology. Predictive ecology threatens the ivory towers of academic ecology and may force ecologists to compete with engineers and other scientists on the unforgiving ground of real life rather than in the meadows of arcane theory.

Chapter 5 contended that scientific theories are instruments, tools to help estimate and predict the likely configuration of the external world. The alternative position, that theories are approaches to the truth of the real world, does no harm so long as this position is accepted as entirely metaphysical. It may influence how we feel about science but it should have no effect on how science is done. Unfortunately, metaphysical views too often impinge on and distort practice.

The questions addressed by predictive ecology also reveal the extent of the break with traditional ecology. On the whole, these questions have not been the focus of general ecological interest, although they have interested some schools within ecology. They include the question of plant life form, a major area in plant ecology for a century but never of much interest to other ecologists; similarly, eutrophy has been a focus for limnology since the early days of the science, but one which failed to capture the interest of theoreticians, as theoreticians. Allometry has had an even longer career. These areas are growing not because they have recently been discovered. They are growing because we are finally asking resolvable questions about areas of concern to people dealing with ecosystems and landscapes, rather than with the artificial concepts that have grown out of a theoretical ecology.

Predictive ecology marks a step in a general movement in science away from grand questions that are almost impossible to treat and towards humble questions that can be answered. This movement is an implicit recognition of the wisdom of Medawar's (1967) thesis that science is the art of the soluble. We get nowhere by asking intractable questions like 'what is life?', 'what is the limiting similarity?', or 'how does stability change with stress?'. These questions are so vast that they cannot be dealt with. Only when scientists become content with little questions that can be answered, does science begin to progress (Jacob 1977; Weisskopf 1984; Slobodkin 1986b). This section details some of the attributes of predictive theories in ecology which make them practical and useful tools.

Authority, abstraction and simplicity

All of the relations described at the beginning of this chapter are simplistic abstractions. They describe one aspect of a phenomenon effectively and economically, and nothing else. As a result, they give a good estimate of a single variable, but a very incomplete picture of nature. If they represent an intellectual achievement, this consists of seeing a simplicity underlying the complexity of nature and ignoring the convolutions offered by traditional work in these areas. Since they are among the simplest possible ways of achieving prediction, they may be considered minimalist representations of the phenomenon (Slobodkin 1986b) and applications of Ockham's razor.

Following Ockham's dictum that there is no need to invoke more than is necessary, the constructs of predictive ecology identify explicitly what information they provide as the prediction and what information they need as the predictors and boundary conditions. Thus it is easy to determine whether the theory should be invoked, what one must measure to make the prediction, and what has been left out. Minimalism raises some special burdens for authority. Predictive ecology is attractive because it offers practical approaches to real problems. Its theories can therefore attract potential users in government, industry, and the population at large; moreover, their simplicity opens them to use by other scientists, engineers, planners and laymen. Researchers in predictive ecology must therefore take special care to explain the limitations of their theories, and non-expert users must consider these limitations in their applications, employing adaptive management strategies wherever possible. Regardless, widespread use will certainly result in some abuse and misapplication, because these inevitably result from application. If this has been a smaller problem in the past, it is because traditional ecology had so little to offer environmental practice.

The simple models of predictive ecology contain still another pitfall for authority because they are open to scrutiny and criticism from outside the science. Complicated models are a refuge for the expert, because they cannot be easily interpreted or understood (Dennis, Downton & Middleton 1984), whereas simple models encourage participation, even from the general public. Since the public is even less prepared than scientists to accept abstract, general, simple, empirical, probabilistic models, public involvement in predictive ecology is likely to expose predictive ecologists to heavy censure and hard questions. The solution to this problem lies in the daunting prospect of better public education in science, but the alternative, to exclude the public from the decision-making because they might understand the basis of the process, seems morally untenable, as well as politically untenable over the long terms that ecologically sound decisions require.

Adaptive management

This acceptance and encouragement of strong inference, of competition among theories, and of application should also encourage the development of adaptive management strategies (Walters 1986). If managers recognize both the unfinished nature of science and the uncertain basis of even our best ecological theories, they will also recognize the opportunities for improvement. Moreover, the existing models are so simple that they are likely to demand adjustment and improvement by managers with more particular knowledge of specific cases. If this proves to be so, there should develop a niche for scientific advisors who help plan, monitor and assess these adjustments, and who ultimately feed the results to a central agency for evaluation of the management experiment. Failure to do this has already resulted in the waste of thousands of potential experiments associated with forestry, agricultural practice, regulation of fishing, hunting and water levels, soil conservation and species introductions. Sverdrup & Warfinge (University of Lund, personal communication 1987) estimated that almost 1000 early trials of liming in Sweden gave results ranging from 'illustrious to disastrous', but were so poorly documented that few could be of any future use at all.

Sound variables

The variables used in the examples that began this chapter are characteristic of predictive ecology, for they tend to be simple, measurable, operationally defined variables which have long been the focus of research: diversity, phosphorus concentration, body weight, population density, etc. The exceptions are primarily technological advances representing the intrusion of technology into ecology (e.g. P_{ow}), or combinations of variables of long-standing interest (e.g. evapotranspiration; phosphorus load). None of these variables resulted from the introspections of abstract theory.

This gives the relations a certain homeliness because they deal with variables that have concerned field biologists for years. The variables also give a provincial quality to the science, enforcing the modular aspect of hierarchical structure, because such qualities do not always extend beyond the limits of a subdiscipline. For example, loading models and phosphorus response models are restricted to lakes, QSARs only have meaning in the context of ecotoxicology and pharmacology, diversity may interest naturalists but be incidental to foresters and toxicologists.

It is tempting to generalize these models to types, as I did in lumping phosphorus, available water and nitrogen as resources, and production and standing stocks of plants, algae and animals as 'responses' to discuss 'resource response models'. This aggregation involves a dramatic loss in resolution, information and operationalism, since neither resources nor responses are defined variables. This is the choice of Andrewartha & Birch (1984) who take resources as one of four basic aspects of the environment, the other three being mates, malentities, and predators. This is inoffensive as an expositional device, but these aggregates must not replace or obscure the variables that actually appear in theories or the other characteristics that distinguish the systems under study.

Schoener (1982) suggested that the growth of pluralism would see the decline of 'megaparameters' like niche breadth and resource overlap which attempt to represent large sections of ecological thought or many aspects of an organism's ecology in a single term. The same may be said of stability or limiting similarity and other concepts which have long been the hope for theoretical ecology. Predictive ecology is unlikely to invoke such variables.

A further characteristic of these relationships is that the predictor variables are only minimally affected by the predicted variable. Orians (1980) recognized that this independence of the variables was of primary importance in building useful relationships. This limitation figured in earlier discussions (Chapter 4) of relativistic definitions of the variables and of the possibility that some variables are so uneconomical and redundant that it would be more effective to measure the dependent variable directly than to use the theory (Chapter 7).

Empiricism

All of the examples discussed above are distinctive because they make extensive use of the existing data. All elements of predictive ecology use data because data distinguish promising theories from just bright ideas. Ecology has so long chased after untested thoughts and unconfirmed models with such unimpressive results that we should have learned that models unsupported by data merit little attention. As a result, theories of predictive ecology are invariably accompanied by the data that give them warrant and patterns in these data are identified by relatively simple statistical manipulations, like averaging and regression.

Statistical analysis of existing data is the fastest, easiest, least expensive and most effective way to determine a predictive relationship (Poole 1978). This effectiveness is achieved because it is the sole purpose of the statistical models used, and it is a matter of record that such models outperform their competitors in prediction. The drawback to these models is that they presume that the future will resemble the past and therefore they only work if the system is not changed dramatically with respect to the regression. This is indeed a drawback, but it is not one that is confined to regressions and other statistical models. It is the nature of hypothetical knowledge and applies to all models and theories. The pretention that this limitation can be escaped if the models are based on precept or understanding was addressed at length in Chapters 5 and 6, but the argument is of such importance that it is summarized again here.

The first contention, that the predictive power of empiricism depends on the absence of significant change is correct; but the second, that 'if we do not know why a particular relationship conforms to the data, we cannot guess when it will fail' (Lehman 1986) gives the false impression that we can escape the first. To argue that we can know the reason or mechanism better than data justify is an example of begging the question (Chapter 8). The argument assumes that some non-empirical way of knowing exists, for how else are we to know 'why' the relation holds, when that is the point under discussion. Empiricists argue that the only scientific tests are empirical, and thus that any claim to know 'why' a relation works ultimately rests on past observation of a regularity. Non-empiricists must therefore argue for some alternative to observation that can show why a relation holds. Yet, all researchers realize that many different mechanistic, historical, mathematical, and logical models can be proposed to explain any set of observations, so the mere existence of non-empirical alternatives cannot ensure that these alternatives are better theories or entitle them to claim to explain why a relation holds. The histories of bitter controversies both within science and without, shows that strength of belief in an explanation is also no guarantee of validity. Indeed, dearly held models of complex processes may give diametrically opposed forecasts (Lewin 1983b), so the only effective way to distinguish which is better is to determine which describes available data more effectively. When no data are available, logically sound, competing explanations and models should survive until they can be evaluated on the evidence in scientific tests. Thus both empirical and explanatory models ultimately rest on empirical support, both hold only if conditions do not change significantly, and both must be treated as hypothetical when they are applied in new situations. The contention that there is another, surer way of knowing, through non-empirical explanation, is a red herring, a case of vacuous contrast (Chapter 8).

Many models consist of little more than a regression line and its associated statistics (e.g. allometry, phosphorus response models). Reliance on regression and other statistical procedures has been made possible by the advent of the computer-assisted analyses, but this facility introduces a danger of widespread abuse. For example, regression might be applied to all possible combinations until something pops out, and the likelihood of collinearity often limits the number of variables that can be considered simultaneously. In practice, these particular problems are less important than might be suspected by those who have never tried such empirical analyses. The assembly of large data sets is time-consuming and expensive, and becomes more so with each additional variable. In the case of literature review, the likelihood of finding appropriate information in the literature for several variables is surprisingly small. All subset regressions or extensive multiple regressions are only rarely a realistic alternative in empirical analyses, so the problems they introduce have been small heretofore. The future development of the field will however require more careful attention to these problems, as more factors are considered.

Since all factors cannot be considered simultaneously, empirical models must be based on a few powerful available predictors, expressing the effects of unmeasured variables as residual variation. This is one of the great advantages of empiricism over controlled experiment or mathematical modeling: factors which could not be studied directly are still represented in the model as residual error. This representation can only be estimated empirically. As a result, all predictive models are probabilistic and ecologists must stretch their minds to include uncertainty at all levels of their interpretations (Dayton 1979). In some cases, for example in the early development of Vollenweider's (1968) loading model and in Box's (1981) model of plant form, the data were insufficient or inappropriate for statistical descriptions. Nevertheless, data can be used to establish the extent of variation in patterns proposed on the basis of precept.

There is a seeming inconsistency in my contention that empirical models based on existing values in the literature are somehow an alternative to the studies which generated the data. This paradox is more apparent than real, for in principle a focussed study of the data needed to build positive predictive relations would be faster, more cost-efficient and less subject to error. However, in the absence of the time and funding for such focussed studies, and given the availability of appropriate data, one can mine the literature to good effect. If appropriate data are not already available, the data will necessarily be sought with appropriate primary research.

The effectiveness of reinterpretations of the published literature may seem to suggest the importance of serendipity in scientific advance and argue against hypotheses-testing and goal-directed research. Such a position is absurd. It holds that we are more likely to find what we want by not looking for it. Certainly, we should be ready to take advantage of chance and serendipity, but we cannot abandon rational scientific research to do so.

Holism

'Holism' is at least as protean a term as 'niche' or 'competition'. For some, it implies that the entire system must be considered simultaneously, a view which the success of the various examples has shown to be false and which a moment's reflection shows to be impossible, because we cannot measure or know everything in a system. The theories discussed above could be termed 'holistic' in another sense, because they make no attempt to reduce the system to sets of interacting parts or to detailed mechanisms. They show that successful predictions in ecology can be developed from 'the top' (Brown 1981) using a 'black box' approach (Walters 1971), and need not sum individual processes and mechanisms (Clark *et al.* 1979; Orians 1980).

Rigler (1975a) has argued that it is a matter of record that science proceeds from the identification of pattern, to the construction of empirical theory, and then to dissections of this pattern to develop more explanatory theories. A similar model must underlie the emphasis of many different ecologists on the concept of pattern in ecology (Williams 1964; MacArthur 1972a; May 1986). Unfortunately, many ecologists take the alternative approach and try to explain patterns which had not yet been identified (Roughgarden 1983; Rigler 1975a). Predictive ecology has taken the promising approach of first identifying patterns, and then deciding whether to develop explanatory theories by dissecting the phenomena, to establish the patterns more precisely with improved models, or to address other questions by looking for complementary patterns. Given the extent of our environmental ignorance and our environmental problems, the use of precious public funds to seek explanation as disinct from prediction would seem unconscionable. Therefore, the thrust for predictive ecology is likely to remain holistic in the sense that it seeks patterns rather than explains them.

For the most part, this form of holism is in keeping with use of holism in hierarchy theory which refers to the construction of models which derive all variables from the same or adjacent levels in the hierarchy. This has been recommended by Brown (1981) as a top-down approach to ecological theory, rather than a more reductionistic bottom-up approach. It seems part of what Ulanowicz (1986) calls 'top down causality' and it is a working principle for systems modeling (Clark *et al.* 1979).

The theories of predictive ecology may also seem holistic in the sense that many cases (resource response models, loading models, diversity) deal with properties of the whole system. This is often so, but models like that for black fly density were discussed to show that this is not essential. The critical element is not so much whether the theory deals with the whole system, however that is defined, but rather that the theory be relevant, immediate, and sufficient (Chapter 7). All models of predictive ecology are sufficient and relatively immediate to their relevant goals.

Summary – A scientific alternative for ecology

Eight classes of examples show that ecology can produce useful, quantitative, general, predictions from simple models which, if drawn together, could represent a viable alternative to those aspects of contemporary ecology that have been criticized elsewhere in this book. Such a science would be distinguished by pluralism and practicality. Pluralism reflects the hierarchical structure of knowledge and the recognition of this structure will require the introduction of expert systems and predictive matrices to organize alternate hypotheses and make them readily available. The theories of predictive ecology tend to be empirical, because only empiricism allows realistic estimates of the uncertainty associated with unconsidered factors, holistic and simplistic because complex or mechanistic theories seem inapplicable, and practical, because the theories are often inspired by pressing questions about nature rather than the scholasticisms of academia.

References

Abbot, I. (1980). Theories dealing with the ecology of landbirds on islands. *Advances in Ecological Research*, **11**, 329–71.

Abrahamson, W. G., Whitman, T. G. & Price, P. W. (1989). Fads in ecology. *BioScience*, **39**, 321–5.

Abrams, P. (1983). The theory of limiting similarity. *Annual Review of Ecology and Systematics*, **14**, 359–76.

Alexander, R. D. (1977). Natural selection and the analysis of human sociality. In *Changing Scenes in Natural Sciences 1776–1976*, Special Publication 12, ed. C. E. Goulden, pp. 283–337. Philadelphia: Academy of Natural Sciences.

Alexander, R. McN. (1982). *Animal Mechanics*, 2nd edn Seattle: University of Washington Press.

Alexander, V. & Van Cleve, K. (1983). The Alaska pipeline: a success story. *Annual Review of Ecology and Systematics*, **14**, 443–63.

Allee, W. C., Emerson, A. E., Park, O., Park, T. & Schmidt, K. P. (1949). *Principles of Animal Ecology*. Philadelphia: Saunders.

Allen, T. F. H. & Starr, T. B. (1982). *Hierarchy: Perspectives for Ecological Theory*. Chicago: University of Chicago Press.

Altman, P. L. & Dittmer, D. S. (1968). *Metabolism*. Bethesda Md.: Federation of the American Society of Experimental Biologists.

Anderson, G., Berggren, H., Cronberg, G. & Gelin, C. (1978). Effects of planktivorous and benthivorous fish on organisms and water chemistry in eutrophic lakes. *Hydrobiologia*, **59**, 9–15.

Andrewartha, H. G. (1984). Ecology at the crossroads. *Australian Journal of Ecology*, **9**, 1–3.

Andrewartha, H. G. & Birch, L. C. (1954). *The Distribution and Abundance of Animals*. Chicago: University of Chicago Press.

(1984). *The Ecological Web*. Chicago: University of Chicago, Illinois, U.S.A.

Andrzejewska, L. & Gyllenberg, G. (1980). Small herbivore subsystem. In: *Grassland Systems Analysis and Man*, ed. A. I. Breymeyer & G. M. Van Dyne, pp. 201–67. Cambridge: Cambridge University Press.

Antonovics, J. (1987). The evolutionary dys-synthesis: which bottles for which wine? *American Naturalist*, **129**, 321–31.

Arthur, W. (1988). The ecological niche: Nexus or nonsense. *Endeavour*, **12**, 66–70.

Auerbach, M. J. (1984). Stability, probability, and the new topology of food webs. In *Ecological Communities: Conceptual Issues and the Evidence*, ed. D. R. Strong, D. Simberloff, L. Abele & A. B. Thistle, pp. 413–36. Princeton University Press.

Baird, D. & Ulanowicz, R. E. (1989). The seasonal dynamics of the Chesapeake Bay ecosystem. *Ecological Monographs*, **59**, 329–64.

Ball, I. R. (1975). Nature and formulation of biogeographical hypotheses. *Systematic Zoology*, **24**, 407–30.

Barker, A. D. (1969). An approach to the theory of natural selection. *Philosophy*, **64**, 271–90.

Barr, T. C. & Holsinger, J. R. (1985). Speciation in cave faunas. *Annual Review of Ecology and Systematics*, **16**, 313–37.

Barton, D. R. & Taylor, W. D. (1985). The relationship between riparian buffer strips and trout habitat in southern Ontario streams. *North American Journal of Fish Management*, **5**, 364–78.

Baudo, R. (1987). Ecotoxicological testing with *Daphnia*. *Memorie dell'Istituto Italiano di Idrobiologia*, **45**, 461–82.

Beatty, J. (1980). Optimal-design models and the strategy of model building in evolutionary biology. *Philosophy of Science*, **47**, 532–61.

Beauchamp, T. L., ed. (1974). *Philosophical Problems of Causation*. Encino, Calif.: Dickenson Publishing Company.

Beck, M. B. (1981). Hard or soft environmental systems. *Ecological Modelling*, **11**, 233–51.

Begon, M., Harper, J. L. & Townsend, C. R. (1986). *Ecology: Individuals, Populations, and Communities*. Sunderland, Mass.: Sinauer Associates.

Belovsky, G. E. (1986a) Generalist herbivore foraging and its role in competitive interactions. *American Zoologist*, **26**, 51–69.

(1986b) Optimal foraging and community structure: Implications for a guild of generalist herbivores. *Oecologia*, **70**: 35–52.

Belsky, A. J. (1987). The effects of grazing: Confounding of ecosystems, community, and organism scales. *American Naturalist*, **129**, 777–83.

Bender, E. A., Case, T. J. & Gilpin, M. E. (1984). Perturbation experiments in community ecology: Theory and practice. *Ecology*, **65**, 1–13.

Benndorf, J. (1987). Food web manipulation without nutrient control: a useful strategy in lake restoration? *Schweizerische Zeitschrift für Hydrologie*, **49**, 237–48.

(1988). Objectives and unsolved problems in ecotechnology and biomanipulation: A preface. *Limnologica*, **19**, 5–8.

Benndorf, J., Schultz, H., Benndorf, A., Unger, R., Penz, E., Kneschke, H., Kossatz, K., Dumke, R., Hornig, U., Kruspe, R. & Reichel, S. (1988). Food-web manipulation by enhancement of piscivorous fish stocks: long-term effects in the hypertrophic Bautzen reservoir. *Limnologica*, **19**, 97–110.

Berger, T. R. (1977). *Northern Frontier, Northern Homeland: The Report of the McKenzie Valley Pipeline Inquiry (Canada)*. Ottawa: Supply and Services.

Berkowitz, A. R., Kolasa, J., Peters, R. H. & Pickett, S. T. A. (1989). How far in space and time can the results from a single long-term study be extrapolated? In *Long-Term Studies in Ecology*, ed. G. E. Likens, pp. 192–9. New York: Springer.

Berlinski, D. (1976). *On Systems Analysis: An Essay Concerning the Limitations of some Mathematical Methods in the Social, Political, and Biological Sciences*. Cambridge, Mass.: MIT Press.

Bethel, T. (1976). Darwin's mistake. *Harper's Magazine*, **252**, 70–5.

Beverton, R. J. H., Cooke, J. G., Csirke, J. B., Doyle, R. W., Hempel, G., Holt, S. J., McCall, A. D., Policansky, D. J., Roughgarden, J., Shepherd, J.G., Sissenwine, M. P. & Wiebe, P. H. (1984). Dynamics of single species. In *Exploitation of Marine Communities*, ed. R. M. May, pp. 13–58. Dahlen Konferenzen. Berlin: Springer-Verlag.

Beverton, R. J. H. & Holt, S. J. (1957). On the dynamics of exploited fish populations. *Fishery Investigations, London* (2), **19**, 1–533.

Billings, W. D. (1952). The environmental complex in relation to plant growth and distribution. *Quarterly Review of Biology*, **27**, 251–65.

Birks, H. J. B. (1987). Recent methodological developments in quantitative descriptive biogeography. *Annales Zoologici Fennici*, **24**, 165–78.

Bishop, J. A. & Cook, L. M. (1980). Industrial melanism and the urban environment. *Advances in Ecological Research*, **11**, 373–404.

Biswas, A. K. (1975). Mathematical modeling and environmental decision-making. *Ecological Modelling*, **1**, 31–45.

Bliss, L. C. (1984). Ecologists need to increase their investment in society. *Bulletin of the Ecological Society of America*, **65**, 439–44.

Bloesch, J. (1988). Mesocosm studies. *Hydrobiologia*, **159**, 221–2.

Blueweiss, L., Fox, H., Kudzma, V., Nakashima, D., Peters, R. & Sams, S. (1978). Relationships between body size and some life history parameters. *Oecologia*, **37**, 257–72.

Bonacina, C., Bonomi, G. & Mosello, R. (1986). Notes on the present recovery of Lake Orta: an acid, industrially polluted, deep lake in North Italy. *Memorie dell'Istituto di Idrobiologia*, **44**, 97–115.

Bonner, J. T. (1965), *Size and Cycle. An Essay on the Structure of Biology*. Princeton, N. J.: Princeton University Press.

Botkin, D. B., Maguire, B., Moore, B., III, Morowitz, H. J. & Slobodkin, L. B. (1979). A foundation for ecological theory. *Memorie dell'Istituto Italiano di Idrobiologia*, **37** (supplement), 13–31.

Box, E. O. (1981). *Macroclimate and Plant Forms: an Introduction to Predictive Modeling in Phytogeography*. The Hague: Dr. W. Junk.

Box, G. E. P. (1976). Science and statistics. *Journal of the American Statistical Association*, **71**, 791–9.

Bradie, M. & Gromko, M. (1981). The status of the principle of natural selection. *Nature and System*, 3, 3–12.

Brady, R. H. (1979). Natural selection and the criteria by which a theory is judged. *Systematic Zoology*, **28**, 600–21.

(1982). Dogma and doubt. *Biological Journal of The Linnean Society*, **17**, 79–96.

Briand, F. (1983). Environmental control of food web structure. *Ecology*, **54**, 253–63.

Briand, F. & Cohen, J. E. (1987). Environmental correlates of food chain length. *Science*, **238**, 956–60.

Brocksen, R. W., Davis, G. E. & Warren, C. E. (1970). Analysis of trophic processes on the basis of density dependent functions. In *Marine Food Chains*, ed. J. H. Steele, pp. 468–98. Berkeley: University of California Press.

Brookhaven Symposium in Biology (1969). Diversity and Stability in Ecological Systems. No. 22. Upton, N.Y.: Brookhaven National Laboratory.

Brown, J. H. (1981). Two decades of homage to Santa Rosalia: toward a general theory of diversity. *American Zoologist*, **21**, 877–88.

Brown, J. H. & Bowers, M. A. (1984). Patterns and processes in three groups of terrestrial vertebrates. In *Ecological Communities: Conceptual Issues and the*

Evidence, ed. D. R. Strong, D. Simberloff, L. Abele & A. B. Thistle, pp. 281–96. Princeton: Princeton University Press.

Brown, J. H. & Maurer, B. A. (1986). Body size, ecological dominance, and Cope's rule. *Nature*, **324**, 248–50.

Brown, S. & Lugo, A. E. (1988). Exaggerating the rate and impact of deforestation. In Chiras, D. D., *Environmental Science: A framework for decision making*, 2nd edn, pp. 200–1. Menlo Park,. Calif.: Benjamin Cummings.

Brown, W. L. Jr. & Wilson, E. O. (1956). Character displacement. *Systematic Zoology*, **5**, 49–64.

Burkholder, P. R. (1952). Cooperation and conflict among primitive organisms. *American Scientist*, **40**, 601–31.

Busack, S. D. & Hedges, B. S. (1984). Is the peninsula effect a red herring? *American Naturalist*, **123**, 266–75.

Bysshe, S. E. (1982). Bioconcentration factor in aquatic organisms. In *Handbook of Chemical Property Estimation Methods: environmental behavior of organic compounds*, ed. W. J. Lyman, W. F. Reehl & D. H. Rosenblatt, pp. 5–30. New York: McGraw-Hill Book Company.

Cain, A. J. (1977). The efficacy of natural selection in wild populations. In *Changing Scenes in Natural Sciences 1776–1976*. Special Publication 12., ed. C. E. Goulden, pp. 111–33. Philadelphia: Academy of Natural Sciences.

Cain, S. A. (1947). Characteristics of natural areas and factors in their development. *Ecological Monographs*, **17**, 185–200.

Cairns, J., Jr. (1986). The myth of the most sensitive species. *BioScience*, **36**, 670–2.

Calder, W. A. (1984). *Size, Function, and Life History*. Cambridge, Mass.: Harvard University Press.

Calvo, J. C. (1986). An evaluation of Thornthwaite's water balance technique in predicting stream runoff in Costa Rica. *Hydrological Sciences Journal*, **31**, 51–60.

Canale, R. P. (1976). *Modelling Biochemical Processes in Aquatic Ecosystems*. Ann Arbor, Michigan: Ann Arbor Science.

Canfield, D. E., Jr. & Bachmann, R. W. (1981). Prediction of total phosphorus concentrations, chlorophyll a and secchi depths in natural and artificial lakes. *Canadian Journal of Fisheries and Aquatic Sciences*, **38**, 414–23.

Cantlon, J. E. (1981). *Productivity of Resources and Environments: A National Assessment of Research Trends and Needs*. Indianapolis, Indiana: Institute of Ecology.

Caplan, A. (1977). Tautology, circularity, and biological theory. *American Naturalist*, **111**, 390–3.

(1978). Testability, disreputability, and the structure of the modern synthetic theory of evolution. *Erkenntnis*, **13**, 261–78.

Caraco, N., Tamse, A., Boutros, O. & Valliela, I. (1987). Nutrient limitation of phytoplankton growth in brackish coastal ponds. *Canadian Journal of Fisheries and Aquatic Sciences*, **44**, 473–6.

Carlson, R. E. (1977). A trophic state index for lakes. *Limnology and Oceanography*, **22**, 361–9.

(1984). The trophic state concept: A lake management perspective. *Proceedings of the Annual Conference of the North American Lake Management Society*, **3**, 427–36.

Carpenter, S. R., ed. (1988). *Complex Interactions in Lake Communities*. New York: Springer-Verlag.

Carpenter, S. R. (1989). Replication and treatment strength in whole-lake experiments. *Ecology*, **70**, 453–63.

Carpenter, S. R. & Kitchell, J. F. (1984). Plankton community structure and limnetic primary production. *American Naturalist*, **124**, 159–72.

Carpenter, S. R., Kitchell, J. F. & Hodgson, J. R. (1985). Cascading trophic interactions and lake productivity. *BioScience*, **35**, 634–9.

Carpenter, S. R., Kitchell, J. F., Hodgson, J. R., Cochran, P. A., Elser, J. J., Elser, M. M., Lodge, D. M., Kretchmer, D., He, X. & von Ende, C. N. (1987). Regulation of lake primary productivity by food web structure. *Ecology*, **68**, 1863–76.

Cartwright, N. (1983). *How the laws of Physics Lie*. New York: Oxford University Press.

Case, J. T. (1978). On the evolution and adaptive significance of postnatal growth rates in the terrestrial vertebrates. *The Quarterly Review of Bioloogy*, **53**, 243–82.

Castrodeza, C. (1977). Tautologies, beliefs, and empirical knowledge in biology. *American Naturalist*, **111**, 393–4.

Caswell, H. (1976). The validation problem. In *Systems Analysis and Simulation in Ecology*, vol. 4, ed. B. C. Patten, pp. 313–25. New York: Academic Press.

(1978). Predator mediated coexistence: a non-equilibrium model. *American Naturalist*, **112**, 127–54.

(1988). Theory and models in ecology: a different perspective. *Ecological Modelling*, **43**, 33–44.

Cates, R. G. & Orians, G. H. (1975). Successional status and the palatabilty of plants to generalized herbivores. *Ecology*, **56**, 410–18.

Caughley, G. (1970). Eruption of ungulate populations, with emphasis on Himalayan Thar in New Zealand. *Ecology*, **51**, 53–72.

Caughley, G. & Lawton, J. H. (1981). Plant–herbivore systems. In *Theoretical Ecology: Principles and Applications*, 2nd edn, ed. R. M. May, pp. 132–66. Oxford: Blackwell Scientific Publications.

Cavalli-Sforza, L., Piazza, A., Menozzi, P. & Mountain, J. (1989). Genetic and linguistic evolution. *Science*, **244**, 1128–9.

Caws, P. (1969). The structure of discovery. Scientific discovery is no less logical than deduction. *Science*, **166**, 1375–80.

Chamberlain, T. C. (1890). The method of multiple working hypotheses. Reprinted in *Science*, **148**, 754–9 (1965).

Chapman, R. N. (1931). *Animal Ecology, with Especial Reference to the Insects*. New York: McGraw-Hill.

Chapra, S. C. & Reckhow, K. H. (1983). *Engineering Approaches for Lake Management*, vol. 2, *Mechanistic Modeling*. Woburn, Mass.: Butterworth.

Charney, J., Stone, P. H. & Quirk, W. J. (1975). Drought in the Sahara: a biogeophysical feedback mechanism. *Science*, **187**, 434–5.

Cherrett, J. M. (1988). Ecological concepts – a survey of the views of the members of the British Ecological Society. *Biologist*, **35**, 64–6.

(1989). Key concepts, the results of a survey of our members' opinions. *British Ecological Society Symposium*, in press. Oxford: Blackwell.

Chiou, C. T., Freed, V. H., Schmedding, D. W. & Kohnert, R. L. (1977). Partition coefficient and bioaccumulation of selected organic chemicals. *Environmental Science and Technology*, **11**, 475–8.

Chiras, D. D. (1988). *Environmental Science: A framework for decision making*, 2nd edn, Menlo Park, Calif.: Benjamin Cummings.

Choudhury, D. (1984). Aphids and plant fitness: a test of Owen and Wiegert's hypothesis. *Oikos*, **43**, 401–2.

(1985). Aphid honeydew: a re-appraisal of Owen and Wiegert's hypothesis. *Oikos*, **45**, 287–90.

Christiansen, F. B. & Fenchel, T. M. (1977). *Theories of Populations in Biological Communities*. Berlin: Springer-Verlag.

Clark, W. C., Jones, D. D. & Holling, C. S. (1979). Lessons for ecological policy design: a case study of ecosystem management. *Ecological Modelling*, **7**, 1–53.

Clarke, C. W. (1981). Bioeconomics. In *Theoretical Ecology. Principles and Applications*, ed. R. H. May, pp. 387–418. Oxford: Blackwell.

Clements,. F. E. (1905). *Research Methods in Ecology*. Lincoln, Nebraska: University Publishing.

Cody, M. L. (1974). *Competition and the Structure of Bird Communities*. Princeton, N.J.: Princeton University Press.

Cody, M. L & Diamond, J. M., ed. (1975). *Ecology and Evolution of Communities*. Cambridge, Mass.: Belknap Press of Harvard University Press.

Cody, M. L. & Mooney, H. A. (1978). Convergence versus nonconvergence in Mediterranean-climate ecosystems. *Annual Review of Ecology and Systematics*, **9**, 265–321.

Coe, M. J., Cumming, D. H. & Phillipson, J. (1976). Biomass and production of large African herbivores in relation to rainfall and primary production. *Oecologia*, **22**, 341–54.

Cohen, I. B. (1985). *Revolution in Science*. Cambridge, Mass.: Belknap Press of Harvard University Press.

Cohen, J. E. (1976). Irreproducible results and the breeding of pigs, or Nondegenerate limit random variables in biology. *BioScience*, **26**, 391–4.

(1978). *Food Webs and Niche Space*. Princeton: Princeton University Press.

Cold Spring Harbor Symposium in Quantitative Biology. (1957). Population Studies: Animal Ecology and Demography, vol. 22.

Cole, B. J. (1981). Colonizing abilities, island size, and the number of species on archipelagoes. *American Naturalist*, **117**, 629–38.

Cole, H. S. D., Freeman, C., Jahoda, M. & Pavitt, K. L. R. (1973). *Thinking about the Future. A Critique of the Limits to Growth*. London: Chatto & Windus.

Cole, J. R. & Cole S. (1972). The Ortega hypothesis. *Science*, **178**, 368–75.

Colinvaux, P. (1973). *Introduction to Ecology*. New York: John Wiley & Sons.

(1986). *Ecology*. New York: Wiley.

Collingwood, R. G. (1940). *An Essay on Metaphysics*. Oxford: Clarendon Press.

(1946). *The Idea of History*. New York: Oxford University Press.

Collingwood, R. W. (1978). The dissipation of phosphorus in sewage and sewage effluents. In *Phosphorus in the Environment: its chemistry and biochemistry, Ciba Foundation Symposium*, **57**, 229–42. New York: Elsevier.

Colwell, R. K. & Winkler, D. W. (1984). A null model for null models in biogeography. In *Ecological Communities: Conceptual Issues and the Evidence*, ed. D. R. Strong, D. Simberloff, L. G. Abele & A. B. Thistle, pp. 344–59. Princeton, N.J.: Princeton University Press.

Condrey, R. E. (1982). Ingestion-limited growth of aquatic animals: the case for

Blackman kinetics. *Canadian Journal of Fisheries and Aquatic Sciences* **39**, 1585–95.

Condrey, R. E. & Fuller, D. A. (1985). Testing equations of ingestion-limited growth. *Archiv für Hydrobiologie. Beihefte. Ergibnisse der Limnologie*, **21**, 257–67.

Confer, J. (1972). Interrelations among plankton, attached algae and the phosphorus cycle in artificial open systems. *Ecological Monographs*, **42**, 1–23.

Connell, D. W. (1987). Ecotoxicology – a framework for investigations of hazardous chemicals in the environment. *Ambio*, **16**, 47–50.

(1988a). Ecotoxicology – A new approach to understanding hazardous chemicals in the environment. *Search*, **17**, 27–31.

(1988b). Quantitative structure activity relationships and the ecotoxicology of chemicals in aquatic systems. ISI Atlas of Science: *Animal and Plant Sciences*, **1**, 211–5.

Connell, J. H. (1974). Ecology: Field experiments in marine ecology. In *Experimental Marine Biology*, ed. R. Mariscal, pp. 21–54. New York: Academic Press.

(1980). Diversity and the coevolution of competitors, or the ghost of competition past. *Oikos*, **35**, 131–8.

(1983). The prevalence and relative importance of interspecific competition: Evidence from field experiments. *American Naturalist*, **122**, 661–96.

Connell, J. H. & Sousa, W. D. (1983). On the evidence needed to judge ecological stability or persistance. *American Naturalist*, **121**, 709–24.

Connor, E. F. & McCoy, E. D. (1979). The statistics and biology of the species – area relationship. *American Naturalist*, **113**, 791–833.

Connor, E. F. & Simberloff, D. (1979). The assembly of species communities: Chance or competition? *Ecology*, **60**, 1132–40.

(1983). Interspecific competition and species co-occurrence patterns on islands: Null models and the evaluation of the evidence. *Oikos*, **41**, 455–65.

(1984a). Neutral models of species' co-occurrence patterns. In *Ecological Communities: Conceptual Issues and the Evidence*, ed. D. R. Strong, D. Simberloff, L. G. Abele & A. B. Thistle, pp. 316–31. Princeton, N.J.: Princeton University Press.

(1984b). Rejoinders. In *Ecological Communities: Conceptual Issues and the Evidence*, ed. D. R. Strong, D. Simberloff, L. G. Abele & A. B. Thistle, pp. 341–3. Princeton, N.J.: Princeton University Press.

Conover, W. J. (1971). *Practical Non-Parametric Statistics*. New York: Wiley.

Coupland, R. T. (1979). Conclusion. In *Grassland Ecosystems of the World: Analysis of Grasslands and their Uses*, ed. R. T. Coupland, pp. 335–55. Cambridge: Cambridge University Press.

Cousins, S. (1985). Ecologists build pyramids again. *New Scientist*, **106**, 50–4.

(1987). The decline of the trophic level concept. *Trends in Ecology and Evolution*, **2**, 312–16.

Cramer, R. D. (1980). BC(DEF) parameters. I. The intrinsic dimensionality of intermolecular interactions in the liquid state. *Journal of the American Chemical Society*, **102**, 1837–49.

Crane, D. (1972). *Invisible Colleges*. Chicago: University of Chicago Press.

Crecco, V., Savoy, T. & Whitworth, W. (1986). Effect of density-dependent and climatic factors on American shad, *Alosa sapidissima*, recruitment: a predictive approach. *Canadian Journal of Fisheries and Aquatic Sciences*, **43**, 457–63.

Croce, B. (1959). History and chronicle. In *Theories of History*, ed. P. Gardiner, pp.

226–33. New York: The Free Press.

(1960). *History – Its Theory and Practice*. New York: Russell & Russell.

Culver, D. C. (1976). The evolution of aquatic cave communities. *American Naturalist*, **110**, 945–57.

Currie, D. J. & Paquin, V. (1987). Large-scale biogeological patterns of species richness of trees. *Nature*, **329**, 326–7.

Damuth, J. (1981). Population density and body size in mammals. *Nature*, **290**, 699–700.

Darwin, C. (1859). *On the Origin of the Species by Means of Natural Selection, or the preservation of favoured races in the struggle for life*. London: Murray.

Dayton, P. K. (1973). Two cases of resource partitioning in an intertidal community: Making the right prediction for the wrong reason. *American Naturalist*, **107**, 662–70.

(1979). Ecology: A science and a religion. In *Ecological Processes in Coastal and Marine systems*, ed. R., J. Livingston, pp. 3–18. New York: Plenum Press.

Dayton, P. K. & Oliver, J. S. (1980). An evaluation of experimental analyses of population and community patterns in benthic marine environments. In *Marine Benthic Diversity*, ed. K. R. Tenore & B. C. Coull, pp. 93–120. Georgetown, S.C.: University of South Carolina Press.

DeAngelis, D. L. (1988). Strategies and difficulties of applying models to aquatic populations and food webs. *Ecological Modelling*, **43**, 57–73.

DeAngelis, D. L. & Waterhouse, J. C. (1986). Equilibrium and non-equilibrium concepts in ecological models. *Ecological Monographs*, **57**, 1–21.

de Bernardi, R. & Peters, R. H. (1987). Why *Daphnia*? *Memorie dell'Istituto Italiano di Idrobiologia*, **45**, 1–9.

Deevey, E. S. (1987). Of Doomsday and the lower Mississippi. *Science*, **238**, 1215.

de LaFontaine, Y. & Peters, R. H. (1986). Empirical relationships for marine primary production: the effects of environmental variables. *Oceanologica Acta*, **9**, 67–92.

Dennis, R. L., Downton, M. W. & Middleton, P. (1984). Policy making and the role of simplified models: An air quality planning example. *Ecological Modelling*, **24**, 241–63.

Detwiller, R. P. & Hall, C. A. S. (1988). Tropical forests and the global carbon cycle. *Science*, **239**, 42–7.

Dhondt, A. A. (1988). Carrying capacity: a confusing concept. *Acta Oecologica. Oecologia Generalis*, **9**, 337–46.

di Castri, F. & Hadley, M. (1986). Enhancing the credibility of ecology: Is interdisciplinary research for land use planning useful? *GeoJournal*, **13**, 299–325.

Diamond, J. M. (1975). Assembly of species communities. In *Ecology and Evolution of Communities*, ed. M. L. Cody & J. M. Diamond, pp. 342–444. Cambridge Mass.: The Belknap Press of Harvard University Press.

(1978). Niche shifts and the rediscovery of interspecific competition. *American Scientist*, **66**, 322–31.

(1986). Overview: Laboratory experiments, field experiments, and natural experiments. In *Community Ecology*, ed. J. Diamond & T. J. Case, pp. 3–22. New York: Harper & Roe.

Diamond, J. M. & Gilpin, M. E. (1982). Examination of the 'null' model of Connor and Simberloff for species co-occurrences on islands. *Oecologia*, **52**, 64–74.

Diamond, J. M. & May, R. M. (1981). Island biogeography and the design of nature preserves. In *Theoretical Ecology: Principles and applications*, 2nd edn, ed. R. M. May, pp. 228–52. Oxford: Blackwell Scientific Publications.

Dillon, P. J. & Kirchner, W. B. (1975). The effects of geology and land use on the export of phosphorus from watersheds. *Water Research*, **9**, 135–48.

Dillon, P. J. & Rigler, F. H. (1974). The phosphorus–chlorophyll relationship in lakes. *Limnology and Oceanography*, **19**, 767–73.

(1975). A simple method for predicting the capacity of a lake for development based on lake trophic status. *Journal of the Fisheries Research Board of Canada*, **32**, 1519–31.

Donagan, A. (1964). Historical explanation: the Popper-Hempel theory reconsidered. *History and Theory*, **4**, 3–26.

Dorschner, K. W., Fox, S. F., Keener, M. S. & Eikenbary, R. D. (1987). Lotka-Volterra competition revisited: the importance of intrinsic rates of increase to the unstable equilibrium. *Oikos*, **48**, 55–61.

Downing, J. A. (1979). Aggregation, transformation and the design of benthos sampling programs. *Canadian Journal of Fisheries and Aquatic Sciences*, **36**, 1454–63.

(1981). In situ foraging responses of three species of littoral cladocerans. *Ecological Monographs*, **51**, 85–103.

(1985). Review of A Dictionary of Ecology, Evolution, and Systematics. *Experimental Biology*, **43**, 288.

(in press). Comparing apples and oranges. Methodological problems in comparative ecosystem analysis. In *Comparative Analysis of Ecosystems: Patterns, Mechanisms and Theories*, J. J. Cole, G. M. Lovell & S. E. G. Findlay, eds. New York: Springer.

Downing, J. A. & Weber, L. A. (1984). The prediction of forest production from inventory and climatic data. *Ecological Modelling*, **23**, 227–41.

Dray, W. H. (1959). 'Explaining what?' in history. In *Theories of History*, ed. P. Gardiner, pp. 403–8. New York: The Free Press.

(1964). *Philosophy of History*. Englewood Cliffs, N.J.: Prentice Hall.

Dray, W. H., ed. (1966). *Philosophical Analysis and History*. New York: Harper and Row.

Drury, W. H. & Nisbet, I. C. T. (1971). Inter-relations between developmental models in geomorphology, plant ecology, and animal ecology. *General Systems*, **16**, 57–68.

(1973). Succession. *Journal of the Arnold Arboretum of Harvard University*, **54**, 331–68.

Dryzek, J.S. (1983) Ecological rationality. *International Journal of Environmental Studies*, **21**, 5–10.

Ducasse, C. J. (1974). Analysis of the causal relation. In *Philosophical Problems of Causation*, T. L. Beauchamp, ed., pp. 142–53. Encino, California: Dickenson.

Due, D. A. & Polis, G. A. (1986). Trends in scorpion diversity along the Baja California peninsula. *American Naturalist*, **128**, 460–8.

Dunbar, R. I. M. (1982). Adaptation, fitness and the evolutionary tautology. In *Current Problems in Sociobiology*, ed. King's College Sociobiology Group, pp. 9–28. Cambridge: Cambridge University Press.

Dyer, M. I. (1980). Mammalian epidermal growth factor promotes plant growth. *Proceedings of the National Academy of Sciences of the United States of America*, **77**, 4836–7.

Eadie, J. M., Broekhoven, L. & Colgan, P. (1987). Size ratios and artifacts: Hutchinson's rule revisited. *American Naturalist*, **129**, 1–17.

East, R. (1984). Rainfall, soil nutrient status and biomass of large African savanna mammals. *African Journal of Ecology*, **22**, 245–70.

Eberhardt, L. L. (1970). Correlation, regression and density dependence. *Ecology*, **51**, 306–10.

Edmondson, W. T. (1987). *Daphnia* in experimental ecology: notes on historical perspectives. *Memorie dell'Istituto Italiano di Idrobiologia*, **45**, 11–30.

Edmondson, W. T. & Lehman, J. T. (1981). The effect of changes in the nutrient income on the condition of Lake Washington. *Limnology and Oceanography*, **26**, 1–29.

Edmondson, Y. H. (1971). Some components of the Hutchinson legend. *Limnology and Oceanography*, **16**, 157–61.

Edwards, A. W. F. (1986). Are Mendel's results really too close? *Biological Reviews*, **61**, 295–312.

Edwards, R. Y. & Fowle, C. D. (1955). The concept of carrying capacity. *Transactions of the North American Wildlife Conference*, **20**, 589–602.

Egerton, F. N. (1973). Changing concepts in the balance of nature. *Quarterly Review of Biology*, **48**, 322–50.

Egler, F. E. (1947). Arid southeast Oahu vegetation, Hawaii. *Ecological Monographs*, **17**, 383–435.

Ehrlich, P. R. (1989). Discussion: Ecology and management – Is ecological theory any good in practice? In *Perspective in Ecological Theory*, ed. J. Roughgarden, R. M. May & S. A. Levin, pp. 306–18. Princeton, N.J.: Princeton University Press.

Ehrlich, P. R. & Birch, L. C. (1967). The 'balance of nature' and 'population control'. *American Naturalist*, **101**, 97–107.

Ehrlich, P., Harte, J., Harwell, M. A., Raven, P. H., Sagan, C., Woodwell, G. M., Berry, J., Ayensu, E. S., Ehrlich, A. H., Eisner, T., Gould, S. J., Grover, H. D., Herrera, R., May, R. M., Mayr, E., McKay, C. P., Mooney, H. A., Myers, N., Pimentel, D. & Teal, J. M. (1983). Long term biological consequences of nuclear war. *Science*, **222**, 1293–300.

Elliot, E. T., Castañares, L. G., Perlmutter, D. & Porter, K. G. (1983). Trophic level control of production and nutrient dynamics in an experimental planktonic community. *Oikos*, **41**, 7–16.

Elliott, J. M. (1977). *Some Methods for the Statistical Analysis of Samples of Benthic Invertebrates*, 2nd edn. Ambleside, Cumbria: Freshwater Biological Association.

Elster, H. J. (1958). Das limnologische Seetypensystem, Rückblick und Ausblick. *Verhandlungen der Internationalen Vereiningung für theoretische und angewandte Limnologie*, **13**, 101–20.

Elton, C. S. (1927). *Animal Ecology*. London: Sedgewick and Jackson.
 (1958). *The Ecology of Invasions by Animals and Plants*. London: Methuen.

Emlen, J. M. (1973). *Ecology: An Evolutionary Approach*. Reading, Mass.: Addison-Wesley.

(1986). Land-bird densities in matched habitats on six Hawaiian Islands: a test of resource-regulation theory. *American Naturalist*, **127**, 125–39.

Emmel, T. C. (1977). *Global Perspectives on Ecology*. Palo Alto, Calif.: Mayfield.

Endler, J. A. (1982). Problems in distinguishing historical from ecological factors in biogeography. *American Zoologist*, **22**, 441–52.

(1986). *Natural Selection in the Wild*. Princeton, N.J.: Princeton University Press.

Evernden, N. (1984). *The Natural Alien. Humankind and Environment*. Toronto: University of Toronto Press.

Fagerström, T. (1987). On theory, data and mathematics in ecology. *Oikos*, **50**, 258–61.

Farlow, J. O. (1987). Speculations about the diet and digestive physiology of herbivorous dinosaurs. *Paleobiology*, **13**, 60–72.

Ferguson, A. (1976). Can evolutionary theory predict? *American Naturalist*, **110**, 1101–4.

Feyerabend, P. K. (1975). *Against Method: Outline of an Anarchistic Theory of Knowledge*. London: NLB.

Fleming, T. H. (1973). Numbers of mammal species in north and central American forest communities. *Ecology*, **54**, 555–63.

Flew, A. (1975). *Thinking about Thinking or Do I Sincerely Want to be Right?* London; Fontana.

Forbes, S. A. (1887). The lake as a microcosm. *Bulletin Science Association of Peoria, Illinois*, 1887, 77–87.

Forsberg, C. (1987). Evolution of lake restoration in Sweden. *Schweizerische Zeitschrift für Hydrologie*, **49**, 260–74.

Foster, W. A. (1984). The distribution of the sea-lavender aphid *Staticobium staticis* on a marine saltmarsh and its effect on host plant fitness. *Oikos*, **42**, 97–104.

Fowler, S. V. & Lawton, J. H. (1985). Rapidly induced defenses and talking trees: the devil's advocate position. *American Naturalist*, **126**, 181–95.

Frank, P. W. (1952). A laboratory study of intraspecies and interspecies competition in *Daphnia pulicaria* (Forbes) and *Simocephalus vetulus* O. F. Muller. *Physiological Zoology*, **25**, 178–204.

Franklin, A. (1986). *The Neglect of Experiment*. Cambridge: Cambridge University Press.

Fretwell, S. D. (1972). *Populations in a Seasonal Environment*. Princeton, N.J.: Princeton University Press.

(1975). The impact of Robert MacArthur on ecology. *Perspectives in Biology and Medicine*, **6**. 1–13.

(1977). The regulation of plant communities by the food chains exploiting them. *Perspectives in Biology and Medicine*, **20**, 169–85.

(1987). Food chain dynamics: the central theory of ecology? *Oikos*, **50**, 291–301.

Friant, S. L. & Henry, L. (1985). Relationship between toxicity of certain organic compounds and their concentrations in tissues of aquatic organisms: a perspective. *Chemosphere*, **14**, 1897–907.

Frost, B. W. (1985). Food limitations of the planktonic marine copepods *Calanus pacificus* and *Pseudocalanus* sp. in a temperate fjord. *Archiv für Hydrobiologie. Beihefte. Ergebnisse der Limnologie*, **21**, 1–13.

Futuyma, D. (1979). *Evolutionary Biology*. Sunderland, Mass.: Sinauer.

Gallie, W. B. (1959). Explanations in history and the genetic sciences. In *Theories of*

History, ed. P. Gardiner, pp. 386–402. New York: Free Press.

Ganong, W. F. (1904). The cardinal principles of ecology. *Science*, **19**, 493–8.

Gardiner, P., ed. (1959). *Theories of History*. New York: Free Press.

Garfield, E. (1972). Citation analysis as a tool in journal evaluation, *Science*, **178**, 471–8.

(1985). Uses and misuses of citation frequency. *Current Contents*, **16**, 3–9.

(1988). Update on the most cited papers in the SCI, 1965–1986. Part 2. Sixty years of research, from insecticides to aids. *Current Contents*, **19**, 3–10.

Gates, D. (1980). *Biophysical Ecology*. Berlin: Springer.

Gauch, H. G., Jr. (1982). *Multivariate Analysis in Community Ecology*. Cambridge: Cambridge University Press.

Gaudet, C. L. & Keddy, P. A. (1988). Predicting competitive ability from plant traits: a comparative approach. *Nature*, 334, 242–3.

Gause, G. F. (1935). Vérifications expérimentale de la théorie mathématique de la lutte pour la vie. *Actualités des Sciences*, **277**, 1–61.

(1970). Criticism of invalidation of the principle of competitive exclusion. *Nature*, **227**, 89.

Geiger, R. (1965). *The Climate near the Ground*, 2nd edn. Cambridge, Mass.: Harvard University Press.

Geist, V. (1987). Bergmann's rule is invalid. *Canadian Journal of Zoology*, **65**, 1035–8.

Geller, W. (1985). Production, food utilization and losses of two coexisting, ecologically different *Daphnia* species. *Archiv für Hydrobiologie. Beihefte. Ergebnisse der Limnologie*, **21**, 67–79.

Ghiselin, M. T. (1969). *The Triumph of the Darwinian Method*. Berkeley: University of California Press.

Gilbert, F. S. (1980). The equilibrium theory of island biogeography: fact or fiction? *Journal of Biogeography*, **7**, 209–35.

Gilpin, M. E. & Diamond, J. M. (1982). Factors contributing to non-randomness in species co-occurrences on island. *Oecologia*, **52**, 75–84.

(1984a). Are species co-occurrences on islands non-random, and are null hypotheses useful in community ecology? In *Ecological Communities: Conceptual Issues and the Evidence*, ed. D. R. Strong, D. Simberloff, L. G. Abele & A. B. Thistle, pp. 297–315. Princeton, N.J.: Princeton University Press.

(1984b). Rejoinders. In *Ecological Communities: Conceptual Issues and the Evidence*, ed. D. R. Strong, D. Simberloff, L. G. Abele & A. B. Thistle, pp. 332–41. Princeton, N.J.: Princeton University Press.

Gimingham, C. H. (1987). Harnessing the winds of change: heathland ecology in retrospect and prospect. *Journal of Ecology*, **75**, 895–914.

Glaser, B. G. (1964). Comparative failure in science. *Science*, **143**, 1012–14.

Gleick, J. (1987). *Chaos: Making a New Science*. New York: Viking.

Goda, T. & Matsuoka, Y. (1986). Synthesis and analysis of a comprehensive lake model – with the evaluation of diversity of ecosystems. *Ecological Modelling*, **31**, 11–32.

Godbout, L. (1987). Empirical models predicting catch of brook trout (*Salvelinus fontinalis*) in Quebec sport fishery lakes. Ph.D. Thesis. McGill University.

Godbout, L. & Peters, R. H. (1988). Potential determinants of stable catch in the brook trout (*Salvelinus fontinalis*) sport fishery in Quebec. *Canadian Journal of Fisheries and Aquatic Sciences*, **45**, 1771–8.

Goldman, C. R. (1965). Micronutrient limiting factors and their detection in natural phytoplankton populations. *Memorie dell'Istituto Italiano di Idrobiologia*, **18** (supplement), 121–35.

Goodman, D. (1975). The theory of diversity-stability relationships in ecology. *Quarterly Review of Biology*, **50**, 237–66.

Gorham, E. (1979). Shoot height, weight and standing crop in relation to density in monospecific plant stands. *Nature*, **279**, 148–50.

Goudge, T. A. (1961). *The Ascent of Life*. Toronto: University of Toronto Press.

Gould, S. J. (1974). The evolutionary significance of 'bizarre' structures: antler size and skull size in the 'Irish Elk', *Megaloceros giganteus*. *Evolution*, **28**, 191–220.

(1976). Darwin's untimely burial. *Natural History*, **85**, 24–30.

(1978). The panda's peculiar thumb. *Natural History*, **87**, 20–30.

(1981). *The Mismeasure of Man*. New York: Norton.

(1982). Darwinism and the expansion of evolutionary theory. *Science*, **216**, 380–7.

(1983). The hardening of the modern synthesis. In *Dimensions of Darwinism*, ed. M. Grene, pp. 71–93. Cambridge: Cambridge University Press.

(1986). Context of justification vs context of discovery. *Yearbook of Science and the Future*, pp. 236–48. Chicago: Encyclopedia Britannica.

Gould, S. J. & Lewontin, R. C. (1979). The spandrels of San Marco and the Panglossian paradigm: a critique of the adaptionist programme. *Proceedings of the Royal Society of London B, Biological Sciences*, **205**, 581–98.

Grant, P. R. (1972a). Convergent and divergent character displacement. *Biological Journal of the Linnean Society*, **4**, 39–68.

(1972b). Interspecific competition among rodents. *Annual Review of Ecology and Systematics*, **3**, 79–106.

(1975). The classical case of character displacement. In *Evolutionary Biology*, vol. 3, ed. T. Dobzhansky, M. K. Hecht & W. C. Steere, pp. 237–337. New York: Plenum Press.

Grant, P. R. & Schluter, D. (1984). Interspecific competition inferred from patterns of guild structure. In *Ecological Communities: Conceptual Issues and the Evidence*, ed. D. R. Strong, D. Simberloff, L. Abele & A. B. Thistle, pp. 201–33. Princeton: Princeton University Press.

Gray, R. D. (1987). Faith and foraging. In *Foraging Behavior*, ed. A. C. Kamil, J. R. Krebs & H. R. Pulliam, pp. 69–140. New York: Plenum Press.

Green, R. F. (1979). *Sampling Design and Statistical Methods for Environmental Biologists*. New York: John Wiley & Sons.

Greenslade, P. J. M. (1983). Adversity selection and the habitat templet. *American Naturalist*, **122**, 352–65.

Grene, M. (1983). Introduction. In *Dimensions of Darwinism*, ed. M. Grene, pp. 1–15. Cambridge: Cambridge University Press.

Griesbach, S., Peters, R. H. & Youakim, S. (1982). An allometric model for pesticide bioaccumulation. *Canadian Journal of Fisheries and Aquatic Sciences*, **39**, 727–35.

Grime, J. P. (1974). Vegetation classification by reference to strategies. *Nature*, **250**, 26–31.

(1977). Evidence for the existence of three primary strategies in plants and its relevance to ecological and evolutionary theory. *American Naturalist*, **111**, 1169–94.

Gruber, H. E. & Barrett, P. H. (1974). *Darwin on Man: A Psychological Study of Scientific Creativity, Together with Darwin's Early and Unpublished Notebooks.* New York: Dutton.

Guhl, W. (1987). Aquatic ecosystem characterization by biotic indices. *Internationale Revue gesamten Hydrobiologie,* **72**, 431–55.

Gulland, J. A. (1974). *The Management of Marine Fisheries.* Bristol: Scientechnica.

Hailman, J. P. (1982). Evolution and behaviour: an iconoclastic view. In *Learning, Development and Culture: Essay in Evolutionary Epistemology,* ed. H. C. Plotkin, pp. 205–54. New York: John Wiley & Sons.

Hairston, N. G. (1959). Species abundance and community organization. *Ecology,* **40**, 404–16.

(1969). On the relative abundance of species. *Ecology,* **50**, 1091–4.

Hairston, N. G., Smith, F. E. & Slobodkin, L. B. (1960). Community structure, population control, and competition. *American Naturalist,* **94**, 421–5.

Hairston, N. G., Jr. (1979). Fitness, survival, and reproduction. *Systematic Zoology,* **28**, 392–95.

Hall, C. A. S. (1988). An assessment of several of the historically most influential theoretical models used in ecology and of the data provided in their support. *Ecological Modelling,* **43**, 5–31.

Hall, C. A. S. & DeAngelis,. D. L. (1985). Models in ecology. Paradigms found or paradigms lost? *Bulletin of the Ecological Society of America,* **66**, 339–46.

Hall, D. J., Threlkeld, S. T., Burns, C. W. & Crowley, P. H. (1976). The size-efficiency hypothesis and the size structure of zooplankton communities. *Annual Review of Ecology and Systematics,* **7**, 177–208.

Hamilton, W. D. (1972). Altruism and related phenomena, mainly in social insects. *Annual Review of Ecology and Systematics,* **3**, 193–232.

Hanlon, R. T. & Budelmann, B.-U. (1987). Why cephalopods are probably not deaf. *American Naturalist,* **129**, 312–17.

Hanson, J. M. & Leggett, W. C. (1982). Empirical prediction of fish biomass and yield. *Canadian Journal of Fisheries and Aquatic Sciences,* **39**, 257–63.

Hardin, G. (1957). The threat of clarity. *American Journal of Psychiatry,* **114**, 392–6.

(1961). The competitive exclusion principle. *Science,* **131**, 1292–7.

Hardy, A. C. (1924). The herring in relation to its animate environment, part 1. Ministry of Agriculture and Fisheries. *Fishery Investigations Series 2,* **7**, 1–53.

Harper, J. L. (1977). Review of theoretical ecology. *Journal of Theoretical Ecology,* **65**, 1009–12.

(1982). After description. In *The Plant Community as a Working Mechanism,* ed. E. I. Newman, pp. 11–26. Oxford: Blackwell Scientific Publications.

Harris, C. L. (1975). An axiomatic interpretation of the neo-Darwinian theory of evolution. *Perspectives in Biology and Medicine,* **18**, 179–84.

Harrison, A. F. (1978). Phosphorus cycles of forest and upland grass land ecosystems and some effects of land management practices. In *Phosphorus in the Environment: its Chemistry and Biochemistry, Ciba Foundation Symposium,* **57**, 175–99. New York: Elsevier.

Hart, R. (1977). Why are biennials so few? *American Naturalist,* **111**, 792–9.

Harvey, H. H. (1975). Fish populations in a large group of acid-stressed lakes. *Verhandlungen der Internationale Vereinigen für theoretische und angewandte Limno-*

logie, **19**, 2406–17.

Harvey, P. H., Colwell, R. K., Silvertown, J. W & May, R. M. (1983). Null models in ecology. *Annual Review of Ecology and Systematics*, **14**, 189–211.

Haskell, F. E. (1940). Mathematical systematization of 'environment', 'organism', and 'habitat'. *Ecology*, **21**, 1–16.

Hassell, M. P. (1978). *The Dynamics of Arthropod Predator Prey Systems*. Princeton, N.J.: Princeton University Press.

Hassell, M. P., Southwood, T. R. E. & Reader, P. M. (1987). The dynamics of the viburnum white fly (*Aleurotrachelus jelinekii*): a case study of population regulation. *Journal of Animal Ecology*, **56**, 283–300.

Hawkins, C. P. & MacMahon J. A. (1989). Guilds: The multiple meanings of a concept. *Annual Review of Ecology and Systematics*, **34**, 423–51.

Healy, M. C. (1984). Multiattribute analysis and the concept of optimum yield. *Canadian Journal of Fisheries and Aquatic Sciences*, **41**, 1393–406.

Heath, R. (1980). Are microcosms useful for ecosystem analysis? In *Microcosms in Ecological Research*, ed. J. P. Giesy, pp. 333–47. Washington: U.S. Department of Energy, Symposium Series 52.

Hecky, R. E., Newbury, R. W., Bodaly, R., Patalas, K. & Rosenburg, D. M. (1984). Environmental impact prediction and assessment: the Southern Indian lake experience. *Canadian Journal of Fisheries and Aquatic Sciences*, **41**, 720–32.

Hedgpeth, J. W. (1977). Models and muddles. Some philosophical observations. *Helgoländer wissenschaftliche Meeresuntersuchungen*, **30**, 92–104.

Hemmingsen, A. M. (1960). Energy metabolism as related to body size and respiratory surfaces, and its evolution. *Reports of the Steno Memorial Hospital and Nordinsk Insulin Laboratorium*, **9**, 6–110.

Hempel, C. G. (1942). The function of general laws in history. *Journal of Philosophy*, **39**, 35–48.

   (1962). Explanation in science and history. In *Frontiers of Science and Philosophy*, ed. R. G. Colodny, pp. 9–33. Pittsburgh: University of Pittsburgh Press.

   (1966). *Philosophy of Natural Science*. Englewood Cliffs, N.J.: Prentice-Hall.

Hempel, C. G. & Oppenheim, P. (1948). Studies in the logic of explanation. *Philosophy of Science*, **15**, 135–75.

Henderson, L. J. (1913). *The Fitness of the Environment: An Inquiry into the Biological Significance of the Properties of Matter*. New York: MacMillan.

Hilborn, R. (1987). Living with uncertainty in resource management. *North American Journal of Fisheries Management*, **7**, 1–5.

Hilborn, R. & Stearns, S. C. (1983). On inference in ecology and evolutionary biology: the problem of multiple causes. *Acta Biotheoretica*, **31**, 145–64.

Hjort, J. (1914). Fluctuations in the great fisheries of northern Europe. *Rapports et Process-Verbaux des Réunions, Conseils international pour l'Exploration de la Mer*, **20**, 1–228.

Holling, C. S. (1959). Some characteristics of simple types of predation and parasitism. *Canadian Entomologist*, **91**, 385–98.

   (1973). Resilience and stability of ecological systems. *Annual Reviews of Ecology and Systematics*, **4**, 1–23.

Holton, G. J. (1978). *The Scientific Imagination: Case Studies*. Cambridge, New York: Cambridge University Press.

Horn, H. S. (1974). The ecology of secondary succession. *Annual Review of Ecology and Systematics*, **5**, 25–37.

(1981a). Succession. In *Theoretical Ecology: Principles and Applications*, ed. R. M. May, pp. 253–71. Oxford: Blackwell Scientific Publications.

(1981b). Sociobiology. In *Theoretical Ecology: Principles and Applications*, ed. R. M. May, pp. 272–94. Oxford: Blackwell Scientific Publications.

Horne, A. J. (1985). Applied limnology, *Ecology*, **66**, 318–19.

Howe, H. F. (1985). Gomphothere fruits: a critique, *American Naturalist*, **125**, 853–65.

Hrbàček, J. (1987). Systematics and biogeography of *Daphnia* species. *Memorie dell'Istituto Italiano di Idrobiologia*, **45**, 31–5.

Hughes, A. J. & Lambert, D. M. (1984). Functionalism, structuralism, and 'ways of seeing'. *Journal of Theoretical Biology*, **111**, 787–800.

Hull, D. L. (1968). The operational imperative: Sense and nonsense in operationism. *Systematic Zoology*, **17**, 438–57.

(1981). Historical narratives and integrating explanations. In *Pragmatism and Purpose: Essays Presented to Thomas A. Goudge*, ed. L. W. Sumner, J. G. Slater & F. Wilson, pp. 172–88. Toronto: University of Toronto Press.

Hume, D. (1739, 1740). *A Treatise of Human Nature*, 2 volumes. London: Dent.

(1748). An enquiry concerning human understanding. In *Theory of Knowledge*, ed. D. C. Yalden-Thomas. Edinburgh: Nelson, 1951.

Humphreys, W. F. (1979). Production and respiration in animal populations. *Journal of Animal Ecology*, **48**, 427–53.

Hurlbert, S. H. (1971). The nonconcept of species diversity: A critique and alternative parameters. *Ecology*, **52**, 577–86.

(1981). A gentle depilation of the niche. Dicean resource sets in resource hyperspace. *Evolutionary Theory*, **5**, 177–84.

(1984). Pseudoreplication and the design of ecological field experiments. *Ecological Monographs*, **54**, 187–211.

Hutchinson, G. E. (1953). The concept of pattern in ecology. *Proceedings of the Academy of Natural Sciences of Philadelphia*, **105**, 1–12.

(1957). Concluding remarks. *Cold Spring Harbor Symposium on Quantitative Biology*, **22**, 415–27.

(1959a). Il concetto moderno di nicchia ecologica. *Memorie Dell'Istituto Italiano di Idrobiologia*, **11**, 9–22.

(1959b). Homage to Santa Rosalia, or Why are there so many kinds of animals? *American Naturalist*, **93**, 145–59.

(1963). The naturalist as an art critic. *Proceedings of the Academy of Natural Sciences of Philadelphia*, **115**, 99–111.

(1969). Eutrophication past and present. In *Eutrophication Causes, Consequences Correctives*, pp. 17–28. Washington: National Academy of Science.

(1978). *An Introduction to Population Ecology*. New Haven: Yale University Press.

Hutchinson, G. E. & MacArthur, R. H. (1959). A theoretical model of size distributions among species of animals. *American Naturalist*, **93**, 145–9.

Hynes, H. B. N. (1970). *The Ecology of Running Waters*. Toronto: University of Toronto Press.

Ivlev, V. S. (1943). The biological productivity of waters. *Uspekhi Sovremannai*

Biologii, **19**, 98–120. (in Russian) reprinted in English in *Journal of the Fisheries Research Board of Canada*, **23**, 1727–59. (1966).

Jackson, J. B. C. (1981). Interspecific competition and species' distributions: The ghosts of theories and data past. *American Zoologist*, **21**, 889–901.

Jacob, F. (1977). Evolution and tinkering. *Science*, **196**, 1161–6.

Jacobs, J. (1975). Diversity, stability and maturity in ecosystems influenced by human activities. In *Unifying Concepts in Ecology*, ed. W. H. Van Dobben & R. H. Lowe-McConnell, pp. 187–207. The Hague: Dr W. Junk.

Jacobsen, T. (1975). A quantitative method for the separation of chlorophylls a and b from phytoplankton pigments by high pressure liquid chromatography. *Marine Science Communications*, **4**, 33–47.

James, F. C. & Boecklen, W. J. (1984). Interspecific morphological relationships and the densities of birds. In *Ecological Communities: Conceptual Issues and the Evidence*, ed. D. R. Strong, D. Simberloff, L. Abele & A. B. Thistle, pp. 458–77. Princeton: Princeton University Press.

Janzen, D. H. (1983). No park is an island: increase in interference from outside as park size decreases. *Oikos*, **41**, 402–10.

Janzen, D. H. & Martin, P. S. (1982). Neotropical anachronisms: the fruits the gomphotheras ate. *Science*, **215**, 19–27.

Joergensen, S. E. (1984a). Modelling the eutrophication of shallow lakes. *Water Quality Bulletin*, **9**, 438–59, 471.

(1984b). *Modelling the Fate and Effect of Toxic Substances in the Environment*. Amsterdam: Elsevier.

(1986). Developments in ecological modelling. In *Agricultural Non-point Source Pollution. Model Selection and Application*, ed. A. Giorgini & F. Zingales, pp. 37–53. Amsterdam: Elsevier.

Jones, J. R. & Hoyer, M. V. (1982). Sportfish harvest predicted by summer chlorophyll-a concentration in midwestern lakes and reservoirs. *Transactions of the American Fisheries Society*, **111**, 176–9.

Joynt, C. B. & Rescher, N. (1961). The problem of uniqueness in history. *History and Theory*, **1**, 150–62.

Juanes, F. (1986). Population density and body size in birds. *American Naturalist*, **128**, 921–9.

Jumars, P. A. (1987). Editorial comment: The evolving natural history of a manuscript. *Limnology and Oceanography*, **32**, 1011–14.

Kaiser, K. L. E., ed. (1984). *QSAR in Environmental Toxicology: Proceedings of a Workshop on Quantitative Structure-Activity Relationships in Environmental Toxicology*. Dordrecht, Holland: D. Reidel.

Kajak, A. (1980). Invertebrate predator subsystem. In *Grasslands Systems Analysis and Man*, ed. A. F. Breymeyer & G. M. Van Dyne, pp. 539–89. Cambridge: Cambridge University Press.

Kalff, J. (1989–MS). The limits of the possible in aquatic plant research.

Keddy, P. A. (1987). Beyond reductionism and scholasticism in plant community ecology. *Vegetatio*, **69**, 209–11.

(1989a). *Competition*. London: Chapman and Hall.

(1989b). Competitive hierarchies in herbaceous plant communities. *Oikos*, **54**, 234–41.

Keddy, P. A. & Shipley, B. (1989). Competitive heirarchies in herbaceous plant communities. *Oikos*, **54**, 234–41.

Kempthorne, O. (1983). Logical, epistomological and statistical aspects of nature-nurture data interpretation. *Biometrics*, **34**, 1–23.

Kenney, B. C. (1982). Beware of spurious self-correlations! *Water Resource Research*, **18**, 1041–8.

Kerr, R. A. (1987). Has stratospheric ozone started to disappear? *Science*, **237**, 131–2.

Kerr, S. R. (1980). Niche theory in fisheries ecology. *Transactions of the American Fisheries Society*, **109**, 254–7.

Kettlewell, H. B. D. (1973). *The Evolution of Melanism*. Oxford: Clarendon.

Kier, L. B. & Hall, L. H. (1976). *Molecular Connectivity in Chemistry and Drug Design*. New York: Academic.

Kimball, K. D. & Levin, S. A. (1985). Limitations of laboratory bioassays: The need for ecosystems-level testing. *BioScience*, **35**, 165–71.

King, J. L. & Jukes, T. H. (1969). Non-Darwinian evolution. *Science*, **164**, 788–98.

Kingsland, S. (1982). The refractory model: the logistic curve and the history of population ecology. *The Quarterly Review of Biology*, **57**, 29–52.

Kingsland, S. E. (1985). *Modeling Nature. Episodes in the History of Population Ecology*. Chicago: University of Chicago Press.

Kirby, A. J. (1978). The organic chemistry of phosphate transfer. In *Phosphorus in the Environment: its chemistry and biochemistry, Ciba Foundation Symposium*, **57**, 117–34. New York: Elsevier.

Kirkendall, L. R. & Stenseth, N. C. (1985). On defining 'breeding once'. *American Naturalist*, **125**, 189–204.

Kitts, D. B. (1983). The complexity of living bodies and the structure of biological theories. *Acta Biotheoretica*, **32**, 195–205.

Knowlton, M. F., Hoyer, M. V. & Jones, J. R. (1984). Sources of variabilty in phosphorus and chlorophyll and their effects on use of lake survey data. *Water Resources Research*, **20**, 397–407.

Koch, A. L. (1966). The logarithm in biology. I. Mechanisms generating the log-normal distribution exactly. *Journal of Theoretical Biology*, **12**, 276–90.

(1969). The logarithm in biology. II. Distributions simulating the log-normal. *Journal of Theoretical Biology*, **23**, 251–68.

Kochanski, Z. (1973). Conditions and limitations of prediction making in biology. *Philosophy of Science*, **40**, 29–51.

Koestler, A. (1969). *The Act of Creation*. London, N.Y.: Macmillan.

Kohn, A. J. (1971). Phylogeny and biogeography of *Hutchinsonia*: G. E. Hutchinson's influence through his doctoral students. *Limnology and Oceanography*, **16**, 173–6.

Kozlowsky, D. G. (1968). A critical evaluation of the trophic level concept. *Ecology*, **49**, 48–60.

Krebs, C. J. (1985). *Ecology: The Experimental Analysis of Distribution and Abundance*, 3rd edn. New York: Harper and Row.

Krebs, J. R. (1980). Ornithologists as unconscious theorists. *Auk*, **97**, 409–12.

Krebs, J. R. & Avery, M. I. (1985). Central place foraging in the European bee-eater, *Merops apiaster. Journal of Animal Ecology*, **54**, 459–72.

Krebs, J. R., Kacelnik, A. & Taylor, P. (1978). Test of optimal sampling by foraging great tits. *Nature*, **275**, 27–31.

Kuhn, T. S. (1970). *The Structure of Scientific Revolutions*, 2nd edn. Chicago: University of Chicago Press.

(1977). *The Essential Tension: Selected Studies in Scientific Tradition and Change*. Chicago: University of Chicago Press.

Kuttner, R. E. (1975). On infant convenience and female breast form. *American Naturalist*, **109**, 596.

Lahti, T. & Ranta, E (1986). Island biogeography and conservation: a reply to Murphy and Wilcox. *Oikos*, **47**, 388–9.

Lakatos, I. (1978). Falsification and the methodology of scientific research programmes. In *The Methodology of Scientific Research Programmes*, ed. J. Worrall & G. Currie, pp. 8–101. Cambridge: Cambridge University Press.

Lampert, W. (1987). Feeding and nutrition in *Daphnia*. *Memorie dell'Istituto Italiano di Idrobiologia*, **45**, 143–92.

(1988). The relation between zooplankton biomass and grazing: a review. *Limnologica*, **19**, 11–20.

Larkin, P. (1977). An epitaph for the concept of maximum sustainable yield concept. *Transactions of the American Fisheries Society*, **106**, 1–11.

(1978). Fisheries management – An essay for ecologists. *Annual Reviews of Ecology and Systematics*, **9**, 57–74.

Larkin, P. A. (1984). A commentary on environmental impact assessment for large projects affecting lakes and streams. *Canadian Journal of Fisheries and Aquatic Sciences*, **41**, 1121–7.

Leblanc, S. A. & Barnes, E. (1974). On the adaptive significance of the female breast. *American Naturalist*, **108**, 577–8.

Lechowicz, M. J. (1982). The sampling characteristics of electivity indices. *Oecologia*, **52**, 22–30.

(1984). Why do temperate deciduous trees leaf out at different times? Adaptation and ecology of forest communities. *American Naturalist*, **124**, 821–42.

Lee, K. K. (1969). Popper's falsifiabilty and Darwin's natural selection. *Philosophy*, **44**, 291–302.

Leggett, W. C., Frank, K. T. & Carscadden, J. E. (1984). Meteorological and hydrographic regulation of year-class strength in capelin (*Mallotus villosus*). *Canadian Journal of Fisheries and Aquatic Sciences*, **41**, 1193–201.

Lehman, J. T. (1986). The goal of understanding in limnology. *Limnology and Oceanography*, **31**, 1143–59.

Leopold, A. (1943). A deer eruption. *Wisconsin Conservation Publications*, **321**, 3–11.

Levin, S. A. (1981a). The role of theoretical ecology in the description and understanding of populations in heterogeneous environments. *American Zoologist*, **21**, 865–75.

(1981b). Mathematics, ecology, ornithology. *Auk*, **97**, 422–5.

Levins, R. (1968). *Evolution in Changing Environments: Some Theoretical Explorations*. Princeton, N.J.: Princeton University Press.

Levins, R. & Lewontin, R. (1980). Dialectics and reductionism in ecology. In *Conceptual Issues in Ecology*, ed. E. Saarinen, pp. 107–38. Dordrecht: D. Reidel.

Lewin, R. (1983a). Santa Rosalia was a goat. *Science*, **221**, 636–9.

(1983b). Predators and hurricanes change ecology. *Science*, **221**, 737–40.

(1989). Biologists disagree over bold signature of nature. *Science*, **244**, 27–8.

Lewontin, R. C. (1968). Introduction. In *Population Biology and Evolution*, ed. R. C.

Lewontin, pp. 1–4. Syracuse, N.Y.: University of Syracuse Press.

Lewontin, R. C. (1969). The bases of conflict in biological explanation. *Journal of the History of Biology*, **2**, 35–45.

(1972). Testing the theory of natural selection. *Nature*, **236**, 181–2.

(1978). Adaptation. *Scientific American*, **239:3**, 212–30.

(1979). Sociobiology as an adaptationist program. *Behavioral Science*, **24**, 5–14.

Lieth, H. & Box, E. O. (1972). Evapotranspiration and primary productivity. C. W. Thornthwaite Memorial Model. *Publications in Climatology*, **25**, 37–46.

Lieth, H. & Whittaker, R. H. (1975). *Primary Productivity of the Biosphere*. Berlin: Springer-Verlag.

Likens, G. E. (1985). An experimental approach for the study of ecosystems. *Journal of Ecology*, **73**, 381–96.

(1989). *Long-Term Studies in Ecology*. New York: Springer-Verlag.

Lincoln, R. J., Boxshall, G. A. & Clark, P. F. (1982). *A Dictionary of Ecology, Evolution and Systematics*. Cambridge, N.Y.: Cambridge University Press.

Lindeman, R. L. (1942). The trophic-dynamic aspect of ecology. *Ecology*, **23**, 399–418.

Lindsey, C. C. (1966). Body sizes of poikilotherm vertebrates at different latitudes. *Evolution*, **20**, 456–65.

Livingstone, R. J. 1988. Inadequacy of species-level designations for ecological studies of coastal migratory fishes. *Environmental Biology of Fishes*, **22**, 225–34.

Loehle, C. (1983). Evaluation of theories and calculation tools in ecology. *Ecological Modelling*, **19**, 239–47.

(1987a). Errors of construction, evaluation, and inference: a classification of sources of error in ecological models. *Ecological Modelling*, **36**, 297–314.

(1987b). Hypothesis testing in ecology: psychological aspects and the importance of theory maturation. *Quarterly Review of Biology*, **62**, 397–409.

Lotze, J.-H. & Anderson, S. (1979). Procyon lotor. Mammalian Species. *American Society of Mammalogists*, **119**, 1–8.

Lovelock, J. E. (1979). *Gaia, a New Look at Life on Earth*. Oxford: Oxford University Press.

Lynch, M. & Shapiro, J. (1981). Predation, enrichment and phytoplankton community structure. *Limnology and Oceanography*, **26**, 86–102.

MacArthur, R. H. (1957). On the relative abundance of bird species. *Proceedings of the National Academy of Sciences of the United States*, **43**, 293–5.

(1958). Population ecology of some warblers of north-eastern coniferous forests. *Ecology*, **39**, 599–619.

(1966). Notes on Mrs. Pielou's Comments. *Ecology*, **47**, 1074.

(1972a) *Geographical Ecology: Patterns in the Distribution of Species*. Princeton, N.J.: Princeton University Press.

(1972b). Coexistence of species. In *Challenging Biological Problems – Directions towards their Solution*, ed. J. A. Behnke, pp. 253–259. New York: Oxford University Press.

MacArthur, R. H. & Wilson, E. O. (1967). *The Theory of Island Biogeography*. Princeton, N.J.: Princeton University Press.

MacBeth, N. (1971). *Darwin Retried: an Appeal to Reason*. Boston, Mass.: Harvard Common Press.

Macevicz, S. & Oster, G. (1976). Modeling social insect populations II: Optimal reproductive strategies in annual eusocial insect colonies. *Behavioural Ecology and Sociobiology*, **1**, 265–82.

MacFadyen, A. (1957). *Animal Ecology*. London: Pitman.

Mackay, D. (1982). Correlation of bioconcentration factors. *Environmental Science and Technology*, **16**, 274–8.

Mackay, D. & Patterson, S. (1982). Fugacity revisited. *Environmental Science and Technology*, **16**, 654A–60A.

Maelzer, D. A. (1965a). A discussion of components of environment in ecology. *Journal of Theoretical Biology*, **8**, 141–62.

(1965b). Environment, semantics, and system theory in ecology. *Journal of Theoretical Biology*, **8**, 395–402.

(1970). The regression of Log N_{n+1} on Log N_n as a test of density dependence: an exercise with computer-constructed density independent populations. *Ecology*, **51**, 810–22.

Mailhot, H. (1987). Prediction of algal bioaccumulation and uptake rate of nine organic compounds by ten physiochemical properties. *Environmental Science and Technology*, **21**, 1009–13.

Mailhot, H. & Peters, R. H. (1988). Empirical relationships between the 1-octanol/ water partician coefficient and nine physiochemical properties. *Environmental Science and Technology*, **22**, 1479–88.

Mandelbrot, B. B. (1982). *The Fractal Geometry of Nature*. San Francisco: Freeman.

Manser, A. R. (1965). The concept of evolution. *Philosophy*, **40**, 18–34.

Margalef, R. (1968). *Perspectives in Ecological Theory*. Chicago: University of Chicago Press.

Marshall, T. M., Morin, A. & Peters, R. H. (1988). Precision and accuracy in estimates of mean chlorophyll-a concentration: implications for sampling design. *Water Resources Research*, **24**, 1027–34.

Marshall, C. T. & Peters, R. H. (1989). General pattern in the seasonal development of chlorophyll-a for temperate lakes. *Limnology and Oceanography*, **34**, 856–67.

Marshall, E. (1987). Armageddon revisited. *Science*, **236**, 1421–2.

Martin, A. C., Zim, H. S. & Nelson, A. L. (1951). *American Wildlife and Plants. A Guide to Wildlife Food Habits*. New York: Dover.

Mason, H. L. (1947). Evolution in certain floristic associations in western North America. *Ecological Monographs*, **17**, 201–10.

Mason, H. L. & Langenheim, J. H. (1958). Language analysis and the concept environment. *Ecology*, **38**, 325–40.

Mather, J. R. (1978). *The Climatic Water Budget in Environmental Analysis*. Lexington Mass: Lexington Books.

May, R. M. (1974). *Stabilty and Complexity in Model Ecosystems*, 2nd edn. Princeton: Princeton University Press.

(1975). Stability in ecosystems: Some concepts. In *Unifying Concepts in Ecology*, ed. W. H. Van Dobben & R. H. Lowe-McConnell, pp. 161–8, The Hague: Dr W. Junk.

(1976a). *Theoretical Ecology: Principles and Applications*, ed. R. M. May. Oxford: Blackwell Scientific Publications.

May, R. M. (1976b). Simple mathematical models with very complicated dynamics. *Nature*, **261**, 459–67.

(1977). Mathematical models and ecology: Past and future. In *The Changing Scenes in the Natural Sciences 1776–1976*, ed. C. E. Goulden, pp. 189–201. Philadelphia: Academy of Natural Sciences. Special Publication 12.

(1981a). Patterns in multi-species communities. In *Theoretical Ecology: Principles and Applications*, ed. R. M. May, pp. 197–227. Oxford: Blackwell.

(1981b). The role of theory in ecology. *American Zoologist*, **21**, 903–10.

(1981c). Models for two interacting populations. In *Theoretical Ecology. Principles and Applications*, ed. R. M. May, pp. 78–104. Oxford: Blackwell Scientific Publications.

(1984). An overview: Real and apparent patterns in community structure. In *Ecological Communities: Conceptual Issues and the Evidence*, ed. D. R. Strong, D. Simberloff, L. Abele & A. B. Thistle, pp. 3–16. Princeton: Princeton University Press.

(1986). The search for patterns to the balance of nature: Advances and retreats. *Ecology*, **67**, 1115–26.

(1988). How many species are there on earth? *Science*, **241**, 1441–9.

May, R. M. & Seger, J. (1986). Ideas in ecology. *American Scientists*, **74**, 256–67.

Mayer, W. V. (1984). The arrogance of ignorance – Ignoring the ubiquitous. *American Zoologist*, **24**, 423–31.

Maynard Smith, J. (1972). *On Evolution*. Edinburgh: Edinburgh University Press.

Mayr, E. (1960). Cause and effect in biology. *Science* , **134**, 1501–6.

(1983). How to carry out the adaptationist program? *American Naturalist*, **121**, 324–34.

McCarty, L. S. (1986). The relationship between aquatic toxicity, QSARs and bioconcentration for some organic-chemicals. *Environmental Toxicology and Chemistry*, **5**, 1071–80.

McCauley, E. & Kalff, J. (1981). Empirical relationships between phytoplankton and zooplankton biomass in lakes. *Canadian Journal of Fisheries and Aquatic Sciences*, **38**, 458–63.

McLellan, G. H. & Hignett, T. P. (1978). Some economic and technical factors affecting use of phosphate raw materials. in *Phosphorus in the Environment: its chemistry and biochemistry, Ciba Foundation Symposium*, **57**, 49–73. New York: Elsevier.

McIntosh, R. P. (1970). Community, competition, and adaptation. *Quarterly Review of Biology*, **45**, 259–80.

(1976). Ecology since 1990. In *Issues and Ideas in America*, ed. B. J. Taylor & T. J. White, pp. 353–72. Norman; University of Oklahoma Press.

(1982). The background and some current problems of theoretical ecology. In *Conceptual Issues in Ecology*, ed. E. Saarinen, pp. 1–62. Dordrecht, Holland: D. Reidl.

(1985). *The Background of Ecology: Concept and Theory*. Cambridge, N.Y.: Cambridge University Press.

(1987). Pluralism in ecology. *Annual Reviews of Ecology and Systematics*, **18**, 321–41.

McNab, B. K. (1971). On the ecological significance of Bergmann's rule. *Ecology*, **52**, 845–54.

McNaughton, S. J. (1977). Diversity and stability of ecological communities: a comment on the role of empiricism in ecology. *American Naturalist*, **111**, 515–25.

—— (1984). Grazing lawns: animals in herds, plant form, and coevolution. *American Naturalist*, **124**, 863–86.

—— (1985). The ecology of a grazing ecosystem: The Serengeti. *Ecological Monographs*, **55**, 259–94.

—— (1986a). On plants and herbivores. *American Naturalist*, **128**, 765–70.

—— (1986b) Grazing lawns: On domesticated and wild grazers. *American Naturalist*, **128**, 937–9.

McNaughton, S. J. & Wolf, L. L. (1979). *General Ecology*, 2nd edn. New York: Holt, Rinehart and Winston.

McQueen, D. J., Post, J. R. & Mills, E. L. (1986). Trophic relationships in freshwater pelagic ecosystems. *Canadian Journal of Fisheries and Aquatic Sciences*, **43**, 1571–81.

Meadows, D. H., Meadows, D. L., Randers, J. & Behrens, W. W. (1972). *The Limits to Growth*. New York: Universe.

Medawar, P. (1967). *The Art of the Soluble*. London: Methuen.

Meentemeyer, V. (1978). Macroclimate and lignin control of litter decomposition rates. *Ecology*, **59**, 465–72.

Mellanby, K. (1987). Review of P. R. Ehrlich (1986), The machinery of life: the living world around us and how it works. *Journal of Animal Ecology*, **56**, 369–70.

Mentis, M. T. (1988). Hypothetico-deductive and inductive approaches in ecology. *Functional Ecology*, **2**, 5–14.

Merton, R. K. (1973). *The Sociology of Science: Theoretical and Empirical Investigations*. Chicago: University of Chicago Press.

Mertz, D. B. & McCauley, D. E. (1980). The domain of laboratory ecology. In *Conceptual Issues in Ecology*, ed. E. Saarinen, pp. 229–44. Dordrecht: D. Reidl.

Metcalf, R. L. (1971). Model ecosystems for the evaluation of pesticide biodegradability and ecological magnification. *Environmental Science and Technology*, **5**, 709–15.

Metcalf, R. L., Sanborn, J. R., Lu, P. Y. & Nye, D. (1975). Laboratory model ecosystem studies of the degradation and fate of radiolabeled tri-, tetra-, and pentachlorobiphenyl compared with DDE. *Archives of Environmental Contamination and Toxicology*, **3**, 151–65.

Michod, R. E. (1981). Positive heuristics in evolutionary biology. *British Journal of the Philosophy of Science*, **32**, 1–36.

Miles, D. B., Ricklefs, R. E. & Travis, J. (1987). Concordance of ecomorphological relationships in three assemblages of passerine birds. *American Naturalist*, **129**, 347–64.

Millar, J. S. (1977). Adaptive features of mammalian reproduction. *Evolution*, **31**, 370–86.

Milne, A. (1961). Definition of competition among animals. *Society of Experimental Biology, Symposium*, **15**, 40–61.

Minshall, G. W., Cummins, K. W., Petersen, R. C., Cushing, C. E., Bruns, D. A., Sedell, J. R. & Vannote, R. L. (1985). Developments in stream ocosystem theory. *Canadian Journal of Fisheries and Aquatic Sciences*, **42**, 1045–55.

Mitchell, R. (1974). Scaling in ecology. *Science*, **184**, 1131.

Moen, A. N. (1973). *Wildlife Ecology: An Analytical Approach*. San Francisco: Freeman.

Mohr, C. O. (1940). Comparative populations of game, fur, and other mammals. *The American Midland Naturalist*, **24**, 581–4.

Monteith, J. L. (1973). *Principles of Environmental Physics*. London: Arnold.

Moriarty, F. (1983). *Ecotoxicology. The Study of Pollutants in Ecosystems*. London: Academic Press.

Morin, A. (1985). Variabilty of density estimates and the optimization of sampling programs for stream benthos. *Canadian Journal of Fisheries and Aquatic Sciences*, **42**, 1530–4.

Morin, A. & Peters, R. H. (1988). Effect of microhabitat features, seston quality and periphyton on abundance of overwintering blackfly larvae in Southern Quebec. *Limnology and Oceanography*, **33**, 431–46.

Morris, D. (1967). *The Naked Ape*. London: Cape.

Moynihan, M. (1985). Why are cephalopods deaf? *American Naturalist*, **125**, 465–9.

Murdoch, W. W. (1966). 'Community structure, population control and competition' – a critique. *American Naturalist*, **100**, 219–26.

(1970). Population regulation and population inertia. *Ecology*, **51**, 497–502.

(1979). Predation and the dynamics of prey populations. *Fortischritte der Zoologie*, **25**, 295–310.

Murdoch, W. W., Chesson, J. & Chesson, P. L. (1985). Biological control in theory and practice. *American Naturalist*, **125**, 344–66.

Murphy, D. D. & Wilcox, B. A. (1986). On island biogeography and conservation. *Oikos*, **47**, 385–7.

Murray, B. G., Jr. (1986). The structure of theory and the role of competition in community dynamics. *Oikos*, **46**, 145–58.

Nagel, E. & Newman, J. R. (1958). *Gödel's Proof*. New York: New York University Press.

Naumann, E. (1930). Die Hauptypen der Gewässer in produktionsbiologischer Hinsicht. *Verhandlungen der Internationalen Vereinigung für theoretische und angewandte Limnologie*, **5**, 72–4.

Naylor, B. G. & Handford, P. (1985). In defence of Darwin's theory. *BioScience*, **35**, 478–84.

Neely, W. B. (1979). Estimating rate constants for the uptake and clearance of chemicals by fish. *Environmental Science and Technology*, **13**, 1506–10.

Neely, W. B., Branson, D. R. & Blau, G. E. (1974). Partition coefficient to measure bioconcentration potential of organic chemicals in fish. *Environmental Science and Technology*, **8**, 1113–15.

Nelson, P. (1985) Naturalness in theoretical physics. *American Scientist*, **73**, 60–7.

Nirmalakhandan, N. & Speece, R. E. (1988). Structure–activity relationships. *Environmental Science and Technology*, **22**, 606–15.

Nisbett, R. & Ross, L. (1980). *Human Inference: Strategies and Shortcomings of Social Judgement*. Engelwood Cliffs, N.J.: Prentice-Hall.

Niven, B. S. (1980). The formal definition of the environment of an animal. *Australian Journal of Ecology*, **5**, 37–46.

(1982). Formalization of the basic concepts of animal ecology. *Erkenntnis*, **17**, 307–20.

Nürnberg, G. K. (1984). The prediction of internal P loads in lakes with anoxic hypolimnia. *Limnology and Oceanography*, **29**, 111–24.

Oakeshotte, M. (1966). Historical continuity and causal analysis. In *Philosophical Analysis and History*, ed. W. H. Dray, pp. 193–212. New York: Harper & Row.

Odum, E. (1953). *Fundamentals of Ecology*. Philadelphia: Saunders.

Odum, E. P. (1969). The strategy of ecosystem development. *Science*, **164**, 262–70.

(1971). *Fundamentals of Ecology*, 3rd edn. Philadelphia: W. B. Saunders.

(1985). Trends expected in stressed ecosystems. *BioScience*, **35**, 419–22.

Odum, H. T. (1957). Trophic structure and productivity of Silver Springs, Florida. *Ecological Monographs*, **57**, 111–28.

OECD (1982). *Monitoring of Inland Waters (Eutrophication Control). Synthesis Report*. Paris: OECD.

Oglesby, R. T. (1977). Relationship of fish yield to lake phytoplankton standing crop, production and morphometric factors. *Journal of the Fisheries Research Board of Canada*, **34**, 2271–9.

O'Grady, R. T. (1984). Evolutionary theory and teleology. *Journal of Theoretical Biology*, **107**, 563–78.

O'Hara, R. J. (1988). Homage to Clio, or, Toward an historical philosophy for evolutionary biology. *Systematic Zoology*, **37**, 142–55.

Oliver, B. G. & Niimi, A. J. (1983). Bioconcentration of chlorobenzene from water by rainbow trout: Correlations with partition coefficients and environmental residues. *Environmental Science and Technology*, **17**, 287–91.

Ollason, J. G. (1987). Artificial design in natural history: why it's so easy to understand animal behaviour. In *Perspectives in Ecology*, vol. 7, *Alternatives*, ed. P. P. G. Bateson & P. H. Klopfer, pp. 233–57. New York: Plenum Press.

Olsen, Y. & Østgaard, K. (1985). Estimating release rates of phosphorus from zooplankton: model and experimental verification. *Limnology and Oceanography*, **30**, 844–52.

O'Neill. R. V., DeAngelis, D. L., Waide, J. B. & Allen, T. F. H. (1986). *A Hierarchical Concept of Ecosystems*. Princeton: Princeton University Press.

Orians, G. (1975). Diversity, stability and maturity in natural ecosystems. In *Unifying Concepts in Ecology*, ed. W. H. Van Dobben & R. H. Lowe-McConnell, pp. 139–58. The Hague: Dr W. Junk.

Orians, G. H. (1980). Micro and macro in ecological theory. *BioScience*, **30**, 79.

Oster, G. F. (1981). Predicting populations. *American Zoologist*, **21**, 831–44.

Oster, G. F. & Wilson, E. O. (1978). *Caste and Ecology in the Social Insects*. Princeton, N.J.: Princeton University Press.

Ostrofsky, M. L. (1978). Modification of phosphorus retention models for use with lakes with low areal water loading. *Journal of the Fisheries Research Board of Canada*, **35**, 1532–6.

Owen, D. F. & Weigert, R. G. (1976). Do consumers maximize plant fitness? *Oikos*, **27**, 488–92.

(1984). Aphids and plant fitness: 1984. *Oikos*, **43**, 403.

Owen-Smith, N. (1987). Pleistocene extinctions: the pivotal role of megaherbivores. *Paleobiology*, **13**, 351–62.

Pace, M. L. (1984). Zooplankton community structure but not biomass influences the phosphorus–chlorophyll *a* relationship. *Canadian Journal of Fisheries and Aquatic Sciences*, **41**, 1089–96.

Paine, R. T. (1977). Controlled manipulation in the marine intertidal zone and their contributions to ecological theory. In *The Changing Scenes in Natural Sciences*, ed. C. E. Goulden, pp. 245–70. Philadelphia: Academy of Natural Sciences.

(1988). Food webs: road maps of interactions or grist for theoretical development. *Ecology*, **69**, 1648–54.

Park, T. (1962). Beetles, competition, and populations. *Science*, **138**, 1369–75.

Parker, G. A. (1978). Searching for mates. In *Behavioral Ecology: an Evolutionary Approach*, ed. J. R. Krebs & N. B. Davies, pp. 214–44. Oxford: Blackwell Scientific Publications.

Parry, G. D. (1981). The meanings of r- and K-selection. *Oecologia*, **48**, 260–4.

Partridge, D., Lopez, P. D. & Johnstone. (1984). Computer programs as theories in biology. *Journal of Theoretical Biology*, **108**, 539–64.

Patrick, R. (1963). The structures of diatom communities under varying ecological conditions. *Annals of the New York Academy of Sciences*, **108**, 353–8.

Patten, B. C. (1968). Mathematical models of plankton production. *International Revue gesamten Hydrobiologie*, **53**, 357–408.

(1975). *Systems Analysis and Simulation in Ecology, vol. 3*. New York: Academic Press.

(1985). Energy cycling in the ecosystem. *Ecological Modelling*, **28**, 1–72.

Paul, E. A. & Robertson, G. P. (1989). Ecology and the agricultural sciences: a false dichotomy? *Ecology*, **70**, 1594–7.

Peet, R. (1974). The measurement of species diversity. *Annual Review of Ecology and Systematics*, **5**, 285–302.

Penman, H. L. (1963). *Vegetation and Hydrology*. Commonwealth Bureau of Soils Technical Communication, 53: Harpenden.

Penry, D. L. & Jumars, P. A. (1987). Modeling animal guts as chemical reactors. *American Naturalist*, **129**, 69–96.

Petelle, M. (1980). Aphids and melezitose: A test of Owen's 1978 hypothesis. *Oikos*, **35**, 127–8.

Peterman, R. M. (1990). Statistical power analysis can improve fisheries research and management. *Canadian Journal of Fisheries and Aquatic Sciences*, **47**, 2–15.

Peters, R. H. (1971). Ecology and the world view. *Limnology and Oceanography*, **16**, 143–8.

(1972). Phosphorus Regeneration by Zooplankton. Ph. D. Thesis, University of Toronto.

(1976). Tautology in evolution and ecology. *American Naturalist*, **110**, 1–12.

(1977). Unpredictable problems with tropho-dynamics, *Environmental Biology of Fishes*, **2**, 97–102.

(1978). Predictable problems with tautology in evolution and ecology. *American Naturalist*, **112**, 759–62.

(1980a). Useful concepts for predictive ecology. In *Conceptual Issues in Ecology*, ed. E. Saarinen, pp. 215–27. Dordrecht: D. Reidel.

(1980b). From natural history to ecology. *Perspectives in Biology and Medicine*, **23**, 191–203.

(1983). *The Ecological Implications of Body Size*. Cambridge: Cambridge University Press.

(1984). Methods for the study of feeding, grazing and assimilation by zooplankton. In *A Manual on Methods for the Assessment of Secondary Productivity in Fresh Waters, IBP Handbook*, 2nd edn, ed. J. A. Downing & F. H. Rigler, pp. 336–412. Oxford: Blackwell Scientific Publications.

(1986). The role of prediction of limnology. *Limnology and Oceanography*, **31**, 1143–59.

(1986/7). Los objectivos de la investigacion y la naturaleza de la ciencia. *Alquibla*, **10/11**, 23–9.

(1987). Metabolism in *Daphnia*. *Memorie dell'Istituto Italiano di Idrobiologia*, **45**, 193–243.

(1988a). The relevance of allometric comparisons to growth, reproduction, and nutrition in primates and man. In *Symposium on Comparative Nutrition*, ed. K. Blaxter & I. Macdonald, pp. 1–19. London: Libbey.

(1988b). Some general problems for ecology illustrated by food web theory. *Ecology*, **69**, 1673–6.

(1989a). Some pathologies in limnology. *Memorie dell'Istituto Italiano di Idrobiologia*, **47**, 175–212.

(1989b). L'energetica ecologica: passato, presente, e futuro. *Atti della Società Italiana di Ecologia*, **37**, 259–66.

Peters, R. H. & Downing, J. A. (1983). Empirical analysis of zooplankton filtering and feeding rates. *Limnology and Oceanography*, **29**, 763–84.

Peters, R. H. & Raelson, J. V. (1984). Relations between individual size and mammalian population density. *The American Naturalist*, **124**, 498–517.

Peters, R. H. & Rigler, F. H. (1973). Phosphorus excretion by *Daphnia*. *Limnology and Oceanography*, **18**, 821–39.

Peters, R. H. & Wassenberg, K. (1983). The effect of body size on animal abundance. *Oecologia*, **60**, 89–96.

Peterson, R. (1975). The paradox of the plankton: An equilibrium hypothesis. *American Naturalist*, **109**, 35–49.

Pianka, E. R. (1966). Latitudinal gradients in species diversity: a review of concepts. *American Naturalist*, **100**, 33–46.

Pianka, E. R. (1970). On r- and K-selection. *American Naturalist*, **104**, 592–7.

(1981). Competition and niche theory. In *Theoretical Ecology: Principles and Applications*, ed. R. M. May, pp. 167–96. Oxford: Blackwell.

(1988). *Evolutionary Ecology*, 4th edn. New York: Harper and Row.

Pielou, E. C. (1974). *Population and Community Ecology*. New York; Garden and Beach Science.

(1981a). The usefulness of ecological models: A stock-taking. *The Quarterly Review of Biology*, **56**, 17–31.

(1981b). The broken-stick model: a common misunderstanding. *American Naturalist*, **117**, 609–10.

Pierce, G. J. & Ollason, J. G. (1987). Eight reasons why optimal foraging theory is a complete waste of time. *Oikos*, **49**, 111–17.

Pilson, M. E. Q., Oviatt, C. A. & Nixon, S. W. (1980). Annual nutrient cycles in a marine microcosm. In *Microcosms in Ecological Research*, ed. J. P. Giesy, pp. 753–78. Augusta, Georgia: U.S. Department of Energy, Symposium Series 52.

Pimm, S. L. (1980). Properties of food webs. *Ecology*, **61**, 219–25.

   (1982). *Food Webs*. London: Chapman and Hall.

Pimm, S. L. & Lawton, J. H. (1978). On feeding on more than one trophic level. *Nature*, **275**, 542–4.

Platt, J. R. (1964). Strong inference. *Science*, **146**, 347–53.

Poole, R. W. (1978). The statistical prediction of population fluctuations. *Annual Review of Ecology and Systematics*, **9**, 427–48.

Popper, K. R. (1957). The aim of science. Reprinted in *Popper Selections*, ed. D. Miller (1985), pp. 162–70. Princeton, N.J.: Princeton University Press.

   (1960). *The Poverty of Historicism*, 2nd edn. London: Routledge and Kegan Paul.

   (1968). *Conjectures and Refutations: The Growth of Scientific Knowledge*. New York: Harper & Row.

   (1974a). Darwinism as a metaphysical research programme. In *The Philosophy of Karl Popper*, ed. P. A. Schilpp, pp. 133–43. LaSalle Illinois: Open Court.

   (1974b). Scientific reductionism and the essential incompleteness of all science. In *Studies in the Philosophy of Biology*, ed. F. J. Ayala & T. Dobzhansky, pp. 259–84. Berkeley: University of California Press.

   (1978). Natural selection and the emergence of mind. *Dialectica*, **32**, 339–55.

   (1979). *Objective Knowledge: an Evolutionary Approach*. Oxford: Clarendon Press.

   (1983). *Realism and the Aim of Science*. Totowa, N.J.: Rowman & Littlefield.

   (1985). *Selections*, ed. D. Miller. Princeton, N.J.: Princeton University Press.

Popper, K. R. & Miller, D. (1983). A proof of the impossibility of inductive probability. *Nature*, **302**, 687–8.

Porter, K. G., Gerritsen, J. & Orcutt, J. D., Jr. (1982). The effect of food concentration on swimming patterns, feeding behaviour, ingestion, assimilation and respiration by *Daphnia*. *Limnology and Oceanography*, **27**, 935–49.

Porter, K. G., Orcutt, J. D., Jr. & Gerritsen, J. (1983). Functional response and fitness in a generalist filter feeder. *Daphnia magna* (Cladocera: Crustacea). *Ecology*, **64**, 735–42.

Prairie, Y. T. (1987). A test of the sedimentation assumptions of phosphorus input-output models. *Archiv für Hydrobiologie*, **111**, 321–7.

Prairie, Y. T. & Bird, D. (1989). The pseudoproblem of spurious correlation: examples from ecology. *Oecologia*, **81**, 285–8.

Pramer, D. (1985), Terminal science. *BioScience*, **35**, 141.

Price, D. J. de Solla (1986). *Little Science, Big Science . . . and Beyond*. New York: Columbia University Press.

Price, M. V. (1986). Structure of desert rodent communities: A critical review of questions and approaches. *American Zoologist*, **26**, 39–40.

Price, P. W. (1984). Communities of specialists: Vacant niches in ecological and evolutionary time. In *Ecological Communities: Conceptual Issues and the Evidence*, ed. D. R. Strong, D. Simberloff, L. Abele & A. B. Thistle, pp. 510–23. Princeton: Princeton University Press.

Price, P. W. Slobodchikoff, C. N. & Gaud, W. S. (1984). *A New Ecology: Novel Approaches to Interactive Systems*. New York: John Wiley & Sons.

Prothero, J. (1986). Methodological aspects of scaling in biology. *Journal of Theoretical Biology*, **118**, 259–86.

Quin, J. F. & Dunham, A. E. (1983). On hypothesis testing in ecology and evolution. *American Naturalist*, **122**, 602–17.

Reckhow, K. H. & Chapra, S. C. (1983). *Engineering Approaches to Lake Management, vol. 1, Data Analysis and Empirical Modelling*. Woburn, Mass.: Butterworth.

Reckhow, K. H. & Simpson, J. T. (1980). A procedure using modeling and error analysis for the prediction of lake phosphorus concentration from land use information. *Canadian Journal of Fisheries and Aquatic Sciences*, **37**, 1439–48.

Redfield, A. C. (1960). The inadequacy of experiment in marine biology. In *Perspectives in Marine Biology*, ed. A. A. Buzzati-Traverso, pp. 17–26. Berkeley: University of California Press.

Reich, C. A. (1970). *The Greening of America*. New York: Random House.

Reichenbach, H. (1951). *The Rise of Scientific Philosophy*. Berkeley: University of California Press.

Reimers, W. A. (1986). Complementary models for ecosystems. *American Naturalist*, **127**, 59–73.

Rensch, B. (1960). *Evolution above the Species Level*. New York: Columbia University Press.

Rey, J. R. (1984). Experimental tests of island biogeographic theory. In *Ecological Communities: Conceptual Issues and the Evidence*, ed. D. R. Strong, D. Simberloff, L. G. Abele & A. B. Thistle, pp. 101–12. Princeton, N.J.: Princeton University Press.

Reynolds, C. S. (1984). *The Ecology of the Freshwater Phytoplankton*. Cambridge: Cambridge University Press.

Reynolds, J. F. & Acock, B. (1985). Predicting the resoponse of plants to increasing carbon dioxide: A critique of plant growth models. *Ecological Modelling*, **29**, 107–29.

Reynoldson, T. B. & Bellamy, L. S. (1971). The establishment of interspecific competition in field populations, with an example of competition in action between *Polycelis nigra* (Mull.) and *P. tenuis* (Ijima) (Turbellaria, Tricladia). In *Proceedings of the Advanced Study Institute on Dynamics and Numbers in Populations*, ed. P. J. den Boer & G. R. Gradwell, pp. 282–97. Wageningen: Pudoc.

Ricker, W. E. (1954). Stock and recruitment. *Journal of the Fisheries Research Board of Canada*, **11**, 559–623.

(1958). Production, reproduction, and yield. *Verhandlungen der Internationalen Vereiningung für theoretische und angewandte Limnologie*, **13**, 84–100.

Ricklefs, R. E. (1979). *Ecology*, 2nd edn. New York: Chiron Press.

Riggs, D. S. (1970). *The Mathematical Approach to Physiological Problems: a Critical Primer*, 2nd edn. Cambridge, Mass.: MIT Press.

Rigler, F. H. (1971). Feeding rates. In *Secondary Productivity of Fresh Waters*, ed. W. T. Edmondson & G. G. Winberg, pp. 228–55. Oxford: Blackwell Scientific Publications.

(1973). A dynamic view of the phosphorus cycle in lakes. In *Environmental Phosphorus Handbook*, ed. E. J. Griffith, A. Beeton, J. M. Spencer & D. T. Mitchell, pp. 539–72. New York: John Wiley & Sons.

(1975a). The concept of energy flow and nutrient flow between trophic levels. In *Unifying Concepts in Ecology*, ed. W. H. Van Dobben & R. H. Lowe-McConnell, pp. 15–26. The Hague: Dr W. Junk.

(1975b). Nutrient kinetics and the new typology. *Verhandlungen der Internationale Vereinigung für theoretische und angewandte Limnologie*, **19**, 197–210.

(1976). Review of Systems Analysis and Simulation in Ecology, Vol. 3. *Limnology and Oceanography*, **21**, 481–3.

(1978). Limnology in the high Arctic: a case study of Char Lake. *Verhandlungen der Internationale Vereinigung der theoretische und angewandte Limnologie*, **20**, 127–40.

(1982a). Recognition of the possible: an advantage of empiricism in ecology. *Canadian Journal of Fisheries and Aquatic Sciences*, **39**, 1323–31.

(1982b). The relation between fisheries management and limnology. *Transactions of the American Fisheries Society*, **111**, 121–32.

Riley, G. A. (1946). Factors controlling phytoplankton populations on George's Bank. *Journal of Marine Research*, **6**, 54–73.

(1965). A mathematical model of regional variations in plankton. *Limnology and Oceanography*, **10** (supplement), R202–15.

Roff, D. A. (1983). Analysis of catch/effort data: A comparison of three methods. *Canadian Journal of Fisheries and Aquatic Sciences*, **40**, 1496–1506.

Romesburg, H. C. (1981). Wildlife science: Gaining reliable knowledge. *Journal of Wildlife Management*, **45**, 293–313.

Rose, H. & Rose, S. (1969). *Science and Society*. Harmondsworth: Pelican.

Rotenbury, J. T. & Wiens, J. A. (1985). Statistical power analysis and community-wide patterns. *American Naturalist*, **125**, 164–8.

Rothschild, B. J. (1986). *Dynamics of Marine Fish Populations*. Cambridge, Mass.: Harvard University Press.

Roughgarden, J. (1983). Competition and theory in community ecology. *American Naturalist*, **122**, 583–601.

Roughgarden, J., May, R. M. & Levin, S. A. (1989). *Perspectives in Ecological Theory*. Princeton, N.J.: Princeton University Press.

Ruse, M. (1977). Is biology different from physics? In *Logic, Laws and Life*, ed. Colmodny, R. G., pp. 89–127. Pittsburgh: University of Pittsburgh.

(1982). *Darwinism Defended: A Guide to the Evolution Controversies*. Reading, Mass.: Addison-Wesley.

Russell, B. (1917). *Mysticism and Logic*. London: Allen & Unwin.

Ryan, P. M. (1986). Prediction of angler success in an Atlantic salmon, *Salmo salar*, fishery two fishing seasons in advance. *Canadian Journal of Fisheries and Aquatic Sciences*, **43**, 2531–4.

Ryder, R. A. (1965). A method for estimating the potential fish production of north temperate lakes. *Transactions of the American Fisheries Society*, **94**, 214–18.

(1982). The morphoedophic index – use, abuse and fundamental concepts. *Transactions of the American Fisheries Society*, **111**, 154–64.

Ryther, J. H. & Dunstan, W. M. (1971). Nitrogen, phosphorus, and eutrophication in the coastal marine environment. *Science*, **171**, 1008–13.

Saarinen, E., ed. (1982). *Conceptual Issues in Ecology*. Dordrecht: D. Reidel.

Saether, B.-E. (1987). The influence of body weight on the covariation between reproductive traits in European birds. *Oikos*, **48**, 79–88.

Sala, O. E., Parton, W. J., Joyce, L. A. & Lauenroth, W. K. (1988). Primary production of the central grasslands region of the United States. *Ecology*, **69**, 40–5.

Salt, G. W. (1979). A comment on the use of the term emergent properties. *American Naturalist*, **113**, 145–61.

Salt, G. W., ed. (1984). *Ecology and Evolutionary Biology*. Chicago: University of Chicago Press.

Sanders, H. L. (1969). Benthic marine diversity and the stability time hypothesis. *Brookhaven Symposium in Biology*, **22**, 71–81.

Sartory, D. P. (1985). The determination of algal chlorophyllous pigments by high performance liquid chromatography and spectrophotometry. *Water Research*, **19**, 605–10.

Sattler, R. (1986) *Biophilosophy: Analytic and Holistic Perspectives*. Berlin: Springer–Verlag.

Scavia, D. & Fahnenstiel, G. L. (1987). Dynamics of Lake Michigan pytoplankton: Mechanisms controlling epilimnetic communities. *Journal of Great Lakes Research*, **13**, 103–20.

Schaefer, M. B. (1954). Some aspects of the dynamics of populations important to the management of the commercial marine fisheries. *Bulletin of the Inter-American Tropical Tuna Commission*, **1**, 27–56.

Schindler, D. W. (1971). Carbon, nitrogen and phosphorus and the eutrophication of freshwater lakes. *Journal of Phycology*, **7**, 321–2.

(1976). The impact statement boondoggle. *Science*, **192**, 509.

(1978). Evolution of phosphorus limitation in lakes. *Science*, **196**, 260–2.

(1987). Detecting ecosystem responses to anthropogenic stress. *Canadian Journal of Fisheries and Aquatic sciences*, **44**, 6–25.

(1988). Experimental studies of chemical stressors on whole lake ecosystems. *Verhandlungen der Internationale Vereinigen für theoretische und angewandte Limnologie*, **23**, 11–41.

(1989). The role of ecologists in the face of global change. *Ecology*, **70**, 1.

Schlesinger, W. H. (1989). The role of ecologists in the face of global change. *Ecology*, **70**, 1.

Schoener, T. W. (1972). Mathematical ecology and its place among the sciences. I. The biological domain. *Science*, **178**, 389–91.

(1982). The controversy over interspecific competition. *American Scientist*, **70**, 586–95.

(1983). Field experiments on interspecific competition. *American Naturalist*, **122**, 240–85.

(1984). Size differences among sympatric, bird-eating hawks: a worldwide survey. In *Ecological Communities: Conceptual Issues and the Evidence*, ed. D. R. Strong, D. Simberloff, L. G. Abele & A. B. Thistle, pp. 254–81. Princeton, N.J.: Princeton University Press.

Schoenly, K. & Reid, W. (1987). Dynamics of heterotrophic succession in carrion arthropod assemblages: discrete seres or a continuum of change? *Oecologia*, **73**, 192–202.

Scriven, M. (1959a). Truisms as the grounds for historical explanation. I *Theories of*

History, ed. P. Gardiner, pp. 443–75, New York: The Free Press.

(1959b). Explanation and prediction in evolutionary theory. *Science*, **130**, 477–82.

Seip, K. L. & Ibrekk, H. (1988). Regression equations for lake management. *Verhandlungen des Internationale Vereinigen für theoretische und angewandte Limnologie*, **23**, 778–85.

Shapiro, J. & Wright, D. I. (1984). Lake restoration by biomanipulation: Round Lake, Minnesota, the first two years. *Freshwater Biology*, **14**, 371–83.

Shaw, E. M. (1983). *Hydrology in Practice*. Wokingham: Van Nostrand Reinhold.

Shipley, B. & Keddy, P. A. (1987). The individualistic and community-unit concepts as falsifiable hypotheses. *Vegetatio*, **69**, 47–55.

Shugart, H. H. & West, D. C. (1980). Forest succession models. *BioScience*, **30**, 308–13.

Silvertown, J. W. (1983). Why are biennials sometimes not so few? *American Naturalist*, **121**, 448–53.

Simberloff, D. S. (1974). Equilibrium theory of island biogeography and ecology. *Annual Review of Ecology and Systematics*, **5**, 161–82.

(1976a). Species turnover and equilibrium island biogeography. *Science*. **194**, 572–8.

(1976b). Experimental zoogeography of islands: effects of island size. *Ecology*, **57**, 629–48.

(1978). Using island biogeography distributions to determine if colonization is stochastic. *American Naturalist*, **112**, 713–26.

(1980a). A succession of paradigms in ecology: Essentialism to materialism and probalism. In *Conceptual Issues in Ecology*, ed. E. Saarinen, pp. 63–99. Dordrecht: Holland: D. Reidel.

(1980b). Reply. In *Conceptual Issues in Ecology*, ed. E. Saarinen, pp. 139–53. Dordrecht: D. Reidel.

(1982). The status of competition theory in ecology. *Annales Zoologici Fennici*, **19**, 241–53.

(1983). Competition theory, hypothesis testing and other community ecology buzzwords. *American Naturalist*, **122**, 626–35.

(1987). The spotted owl fracas: Mixing academic, applied and political ecology. *Ecology*, **68**, 766–72.

Simberloff, D. S. & Abele, L. G. (1976). Island biogeography theory and conservation practice. *Science*, **191**, 285–6.

(1982). Refuge design and island biogeography theory: Effects of fragmentation. *American Naturalist*, **120**, 41–50.

(1984). Conservation and obfuscation: subdivision of reserves. *Oikos*, **42**, 399–401.

Simberloff, D. S. & Boecklen, W. (1981). Santa Rosalia reconsidered. *Evolution*, **35**, 1206–28.

Simpson, G. G. (1963). Historical science. In *Fabric of Geology*, ed. C. C. Albritton, pp. 24–48, London: Addison-Wesley.

Sims, P. L. & Coupland, R. T. (1979). Producers. In *Grassland Ecosystems of the World: Analysis of Grasslands and their Uses*, ed. R. T. Coupland, pp. 49–72. Cambridge: Cambridge University Press.

Sinclair, A. R. E. (1975). The resource limitation of trophic levels in tropical grassland ecosystems. *Journal of Ecology*, **44**, 497–520.

Singh, J. S., Trlica, M. J., Risser, P. G., Redmann, R. E. & Marshall, J. K. (1980). Autotrophic subsystem. In *Grassland Systems Analysis and Man*, ed. A. F. Breymeyer & G. M. Van Dyne, pp. 59–200. Cambridge: Cambridge University Press.

Sissenwine, M. P. (1978). Is MSY an adequate foundation for optimum yield? *Fisheries*, **3(6)**, 22–42.

Slobodkin, L. B. (1961). *Growth and Regulation of Animal Populations*. New York: Holt, Reinhart, and Winston.

(1972). On the inconstancy of ecological efficiency and the form of ecological theories. *Transactions of the Connecticut Academy of Arts and Science*, **44**, 292–305.

(1975). Comments from a biologist to a mathematician. In *Ecosystem analysis and prediction*, ed. S. A. Levin, pp. 318–29. Proceedings SIAM-SIMS Conference on Ecosystems, Alta, Utah, July 1–5, 1974, SIAM Institute for Mathematics and Society.

(1986a). Natural philosophy rampant. *Paleobiology*, **12**, 111–18.

(1986b). The role of minimalism in art and science. *American Naturalist*, **127**, 257–65.

(1988). Intellectual problems of applied ecology. *BioScience*, **38**, 337–42.

Slobodkin, L. B., Smith, F. E. & Hairston, N. G. (1967). Regulation in terrestrial ecosystems, and the implied balance of nature. *American Naturalist*, **101**, 109–24.

Smith, F. E. (1952). Experimental methods in population dynamics: A critique. *Ecology*, **33**, 441–50.

(1970). Analysis of ecosystems. In *Ecological Studies. I. Analysis of Temperate Forest Ecosystems*, ed. D. Reichle, pp. 7–18. Berlin: Springer-Verlag.

Smith, V. H. (1982). The nitrogen and phosphorus dependence of algal biomass in lakes: An empirical and theoretical analysis. *Limnology and Oceanography*, **27**, 1101–11.

(1983). Low nitrogen to phosphorus ratios favor dominance by bluegreen algae in lake phytoplankton. *Science*, **221**, 669–71.

(1985). Predictive models for the biomas of bluegreen algae in lakes. *Water Resources Research*, **21**, 433–9.

Sober, E. (1984). *The Nature of Selection: Evolutionary Theory in Philosophical Focus*. Cambridge, Mass.: MIT Press.

Sommer, U., Gliwicz, Z. M., Lampert, W. & Duncan, A. (1986). The PEG-model of seasonal succession of planktonic events in fresh waters. *Archiv für Hydrobiologie*, **106**, 433–71.

Southern, H. N. (1970). Ecology at the crossroads. *Journal of Ecology*, **58**, 1–11.

Southwood, T. R. E. (1966). *Ecological Methods*. London: Methuen.

(1977). Habitat, the templet for ecological strategies? *Journal of Animal Ecology*, **46**, 337–65.

(1980). Ecology – A mixture of pattern and probabilism. In *Conceptual Issues in Ecology*, ed. E. Saarinen, pp. 203–14. Dordrecht: D. Reidel.

(1981). Bionomic strategies and population parameters. In *Theoretical Ecology: Principles and Applications*, 2nd. edn, ed. R. M. May, pp. 30–52. Oxford: Blackwell Scientific Publications.

(1988). Tactics, strategies and templets. *Oikos*, **52**, 3–17.

Spomer, G. G. (1973). The concept of interaction and operational environment in environmental analysis. *Ecology*, **54**, 200–4.

Sprules, W. G. (1975). Factors affecting the structure of limnetic crustacean zooplankton communities in central Ontario lakes. *Verhandlungen der Internationale Vereinigen für theoretische und angewandte Limnologie*, **19**, 635–43.

St Amant, J. L. S. (1970). The detection of regulation in animal populations. *Ecology*, **51**, 823–8.

Stanley, S. M. (1973). An explanation for Cope's rule. *Evolution*, **27**, 1–26.

Statzner, B. & Higler, B. (1985). Questions and comments on the river continuum concept. *Canadian Journal of Fisheries and Aquatic Sciences*, **42**, 1038–44.

Stearns, S. C. (1976). Life-history tactics: a review of the ideas. *Quarterly Review of Biology*, **51**, 3–47.

(1982). The emergence of evolutionary and community ecology as experimental sciences. *Perspectives in Biology and Medicine*, **25**, 621–48.

(1983). The influence of size and phylogeny on patterns of covariation among life-history traits in the mammals. *Oikos*, **41**, 173–87.

Stearns, S. C. & Koella, J. C. (1986). The evolution of phenotypic plasticity in life-history traits: Predictions of reaction norms for age and size at maturity. *Evolution*, **40**, 893–913.

Stearns, S. C. & Schmid-Hempel, P. (1987). Evolutionary insights should not be wasted. *Oikos*, **49**, 118–25.

Stebbins, G. L. (1977). In defense of evolution: Tautology or theory. *American Naturalist*, **111**, 386–90.

Stemberger, R. S. & Gilbert, J. J. (1985). Assessment of threshold food levels and population growth in planktonic rotifers. *Archiv für Hydrobiologie. Beihefte. Ergebnisse der Limnologie*, **21**, 269–75.

Stephens, D. W. & Krebs, J. R. (1986). *Foraging Theory*. Princeton, N.J.: Princeton University Press.

Stockner, J. (1988). Phototrophic picoplankton: An overview from marine and freshwater ecosystems. *Limnology and Oceanography*, **33**, 765–75.

Strong, D. R. Jr. (1980). Null hypotheses in ecology. In *Conceptual Issues in Ecology*, ed. E. Saarinen, pp. 245–59. Dordrecht, Holland: D. Reidel.

(1983). Natural variability and the manifold mechanisms of ecological communities. *American Naturalist*, **122**, 636–60.

(1984). Density-vague ecology and liberal population regulation in insects. In *A New Ecology: Novel Approaches to Interactive Systems*, ed. P. W. Price, C. N. Slobodchikoff & W. S. Gaud, pp. 313–27. New York: Wiley.

(1986a). Population theory and understanding pest outbreaks. In *Ecological Theory and Integrated Pest Management*, ed. M. Kogon, pp. 37–58. New York: Wiley.

(1986b). Density vague population change. *Trends in Ecology and Evolution*, **1**, 39–42.

Strong, D. R., McCoy, E. D. & Rey, J. R. (1977). Time and number of herbivore species: The pests of sugarcane. *Ecology*, **58**, 167–75.

Strong, D. R., Simberloff, D., Abele, L. G. & Thistle, A. B. (1984). *Ecological Communities: Conceptual Issues and the Evidence*. Princeton, N.J.: Princeton University Press.

Stumm, W., Schwarzenbach, R. & Sigg, L. (1983). From environmental analytical chemistry to ecotoxicology – a plea for more concepts and less monitoring and testing. *Angewandte Chemie. International Edition in English*, **22**, 380–9.

Sutcliffe, W. H. Jr. (1972). Some relations of land drainage, nutrients, particulate material, and fish catch in two Eastern Canadian Bays. *Journal of the Fisheries Research Board of Canada*, **29**, 357–62.

Sverdrup, H. (1983). Lake liming. *Chemica Scripta*, **22**, 12–18.

Sverdrup, H. & Warfvinge, P. (1985). A reacidification model for acidified lakes neutralized with calcite. *Water Resources Research*, **21**, 1374–80.

Sverdrup, H. U., Warfvinge, P. G. & Bjerle, I. (1986). A simple method to predict the time required to reacidify a limed lake. *Vatten*, **42**, 10–15.

Symanski, R. (1976). The manipulation of ordinary language. *Annals of the Association of American Geographers*, **66**, 605–14.

Taylor, L. R. (1961). Aggregation, variance and the mean. *Nature*, **189**, 732–5.

Taylor, P. (1988). Revising models and generating theory. *Oikos*, **54**, 121–6.

Taylor, St. C. S. (1968). Time taken to mature in relation to mature weight for sexes, strains, and species of domesticated mammals and birds. *Animal Production*, **10**, 157–69.

Teal, J. M. (1957). Community metabolism in a temperate cold spring. *Ecological Monographs*, **27**, 283–302.

Thom, R. (1975). *An Outline of a General Theory of Models*. Reading, Mass.: Benjamin.

Thompson, K. (1984). Why Biennials are not so few as they ought to be. *American Naturalist*, **125**, 854–61.

Thompson, N. S. (1981). Towards a falsifiable theory of evolution. In *Perspectives in Ethology*, vol. 4, ed. P. P. G. Bateson & P. H. Klopfer, pp. 51–73. New York: Plenum Press.

    (1987). The misappropriation of teleonomy. In *Perspectives in Ethnology*, vol. 7, ed. P. P. G. Bateson & P. H. Klopfer, pp. 259–74. New York: Plenum.

Thompson, P. (1983). Historical laws in modern biology. *Acta Biotheoretica*, **32**, 167–77.

Thornhill, R. (1984). Scientific methodology in entomology. *Florida Entomologist*, **67**, 74–96.

Thornthwaite, C. W. & Mather, J. R. (1957). Instructions and tables for computing potential evapotranspiration and the water balance. *Publication in Climatology*, **10**, 185–311.

Tillman, D. A. (1978). *Wood as an Energy Source*. New York: Academic Press.

Tilman, D. (1982). *Resource Competition and Community Structure*. Princeton, N.J.: Princeton University Press.

    (1987). The importance of mechanisms of interspecific competition. *American Naturalist*, **129**, 769–74.

    (1988). *Dynamics and Structure of Plant Communities*. Princeton, N.J.: Princeton University Press.

Toffler, A. (1970). *Future Shock*. New York: Bantam.

Toft, C. A. & Shea, P. J. (1983). Detecting community-wide patterns: estimating power strengthens statistical inference. *American Naturalist*, **122**, 618–25.

Tribbia, J. J. & Antheis, R. A. (1987). Scientific basis of modern weather prediction. *Science*, **237**, 493–9.

Trivers, R. L. & Hare, H. (1976). Haplodiploidy and the evolution of the social insects. *Science*, **191**, 249–63.

Tuomi, J. (1981). Structure and dynamics of Darwinian evolutionary theory. *Systematic Zoology*, **30**, 22–31.

Tuomi, J. & Haukioja, E. (1979). Predictability of the theory of natural selection: An analysis of the structure of the Darwinian theory. *Savonia*, **3**, 1–8.

Turco, R. P., Toon, O. B., Ackerman, T. P., Pollack, J. B. & Sagan, C. (1983). Nuclear winter: Global consequences of multiple nuclear explosions. *Science*, **222**, 1283–92.

Turner, N. C. (1988). Effects of water stress on plant growth and yield. *ISI Atlas of Science: Animal and Plant Sciences*, **1**, 241–4.

Twitchell, A. R. & Dill, H. H. (1949). One hundred raccoons from one hundred and two acres. *Journal of Mammalogy*, **30**, 130–3.

Ulanowicz, R. E. (1986). *Growth and Development. Ecosystems Phenology*. New York: Springer-Verlag.

Umpleby, S. A. (1987). World population: still ahead of schedule. *Science*, **237**, 1555–6.

Underwood, A. J. & Denley, E. J. (1984). Paradigms, explanations and generalizations in models for the structure of intertidal communities on rocky shores. In *Ecological Communities: Conceptual Issues and the Evidence*, ed. D. R. Strong, D. Simberloff, L. G. Abele & A. B. Thistle, pp. 151–80. Princeton, N.J.: Princeton University Press.

Vallentyne, J. (1974). Who's who in ecology 1973. *Limnology and Oceanography*, **9**, 875.

van der Beken, A. & Hermann, A. (1985). *New Approaches in Water Balance Computations, IAHS Publication, 148*. Wallingford, U. K.: IAHS Press.

van der Steen, W. J. (1983a). Methodological problems in evolutionary biology I. Testability and tautologies. *Acta Biotheoretica*, **32**, 207–15.

(1983b). Methodological problems in evolutionary biology II. Appraisal of arguments against adaptationism. *Acta Biotheoretica*, **32**, 219–22.

van Keulen, H. (1974). Evaluation of models. *Proceedings of the International Congress of Ecology*, **1**, 250–2.

Vannote, R. L., Minshall, G. W., Cummins, K. W., Sedell, J. R. & Cushing, C. E. (1980). The river continuum concept. *Canadian Journal of Fisheries and Aqauatic Sciences*, **37**, 130–7.

Van Soest, P. J. (1988). Fibre in the diet. In *Comparative Nutrition*, ed. K. Blaxter & I. Macdonald, pp. 215–25. London: Libby.

Van Valen, L. (1973). Pattern and the balance of nature. *Evolutionary Theory*, **1**, 31–49.

Van Valen, L. & Pitelka, F. (1974). Commentary: intellectual censorship in ecology. *Ecology*, **55**, 925–6.

Veith, G. D., Call, D. J. & Brooke, L. T. (1983). Structure-toxicity relationships for the fathead minnow, *Pimephales promelas*: Narcotic industrial chemicals. *Canadian Journal of Fisheries and Aqauatic Sciences*, **40**, 743–8.

Velikofsy, I. (1950). *Worlds in Collision*. New York: Doubleday.

Verhulst, P.-F. (1838). Notice sur la loi que la population suit dans son accroissement. *Correspondence mathématique et physique*, **10**, 113–21.

(1845). Recherches mathématiques sur la loi d'accroissement de la population. *Memoirs de l'Académie Royale de la Belge*, **18**, 1–38.

Verschueren, K. (1983). *Handbook of Environmental Data on Organic Chemicals*, 2nd edn. New York: Van Nostrand.

Vogt, K. A., Grier, C. C. & Vogt, D. J. (1986). Production, turnover and nutrient dynamics of above- and below-ground detritus of world forests. *Advances in Ecological Research*, **15**, 303–77.

Vollenweider, R. A. (1968). *Scientific Fundamentals of Eutrophication of Lakes and Flowing Waters with Special Reference to Phosphorus and Nitrogen*. OECD Paris. OECD/DAS/CSI/68.27.

(1971). *A Manual on Methods for Measuring Primary Production in Aquatic Environments*. (*IBP Handbook 12*). Oxford: Blackwell Scientific Publications.

(1987). Scientific concepts and methodologies pertinant to lake research and lake restoration. *Schweizerische Zeitschrift für Hydrologie*, **49**, 129–47.

Von Foerster, H., Mora, P. M. & Amiot, L. W. (1960). Doomsday: Friday 13 November, A.D. 2026. *Science*, **132**, 1291–5.

Walter, G. G. (1986). A robust approach to equilibrium yield curves. *Canadian Journal of Fisheries and Aquatic Sciences*, **43**, 1332–9.

Walter, H. (1973). *Vegetation of the Earth in Relation to Climate and Eco-physiological Conditions*. New York: Springer-Verlag.

Walters, C. J. (1971). Systems ecology: The systems approach and mathematical models in ecology. In *Fundamentals of Ecology*, E. P. Odum, 3rd edn, pp. 267–92. Philadelphia: W. B. Saunders.

(1986). *Adaptive Management of Renewable Resources*. New York: Macmillan.

Walters, C. J. & Hilborn, R. (1978). Ecological optimization and adaptive management. *Annual Review of Ecology and Systematics*, **9**, 157–88.

Walters, C. J., Park, R. A. & Koonce, J. F. (1980). Dynamic models of lake ecosystems. In *The Functioning of Freshwater Ecosystems*, ed. E. D. LeCren & R. H. Lowe-McConnell, pp. 455–79. Cambridge: Cambridge University Press.

Wangersky, P. J. (1978). Lotka-Volterra population models. *Annual Review of Ecology and Systematics*, **9**, 189–218.

Ware, D. M. (1978). Bioenergetics of pelagic fish: Theoretical change in swimming speed and ration with body size. *Journal of the Fisheries Research Board of Canada*, **35**, 220–8.

Waring, R. H. (1983). Estimating forest growth and efficiency in relation to canopy leaf area. *Advances in Ecological Research*, **13**, 327–54.

Wartenberg, D., Ferson, S. & Rohlf, F. J. (1987). Putting things in order: a critique of detrended correspondence analysis. *American Naturalist*, **129**, 434–48.

Wassermann, G. D. (1981). On the nature of the theory of evolution. *Philosophy of Science*, **48**, 16–37.

Watson, J. D. (1968). *The Double Helix: a personal account of the discovery of the structure of DNA*. New York: Atheneum.

Watson, S. & Kalff, J. (1981). Relationships between nannoplankton and lake trophic status. *Canadian Journal of Fisheries and Aquatic Sciences*, **38**, 960–7.

Watt, A. S. (1947). Pattern and process in the plant community. *Journal of Ecology*, **35**, 1–22.

Watt, K. E. F. (1975). Critique and comparison of biome ecosystem modelling. In

Systems Analysis and Simulation in Ecology, ed. B. C. Patten, pp. 139–52. New York: Academic.

Weatherhead, P. J. (1986). How unusual are unusual events? *American Naturalist*, **128**, 150–4.

Weaver, W. (1964). Scientific explanation. *Science*, **143**, 1297–300.

Weisskopf, V. F. (1984). The frontiers and limits of science. *Daedalus, Journal of the American Academy of Arts and Sciences*, **113**, 177–95.

Welch, D. W. (1986). Identifying the stock-recruitment relationships for age structured populations using time invariant matched linear filters. *Canadian Journal of Fisheries and Aquatic Sciences*, **43**, 108–23.

Weller, D. E. (1987). A reevaluation of the − 3/2 power rule of plant self-thinning. *Ecological Monographs*, **57**, 23–43.

Werner, E. E. (1980). Niche theory in fisheries ecology. *Transactions of the American Fisheries Society*, **109**, 257–60.

Westman, W. E. (1978). Measuring the inertia and resilience of ecosystems. *BioScience*, **28**, 705–10.

Westoby, M. (1984). The self-thinning rule. *Advances in Ecological Research*, **14**, 1677–225.

(1985). Does heavy grazing usually improve the food resource for grazers? *American Naturalist*, **126**, 870–1.

White, E. G. (1984). A multispecies simulation model of grassland producers and consumers. II Producers. *Ecological Modelling*, **24**, 241–62.

White J. (1981). The allometric interpretation of the self-thinning rule. *Journal of Theoretical Biology*, **89**, 475–500.

White, T. C. R. (1978). The importance of a relative shortage of food in animal ecology. *Oecologia*, **33**, 71–86.

Whitney, G. G. (1987). An ecological history of the Great Lakes forest of Michigan. *Journal of Ecology*, **75**, 667–84.

Whittaker, R. H. (1953). A consideration of climax theory: The climax as a population and pattern. *Ecological Monographs*, **23**, 41–78.

(1957). Recent evolution of ecological concepts in relation to the eastern forests of North America. *American Journal of Botany*, **44**, 197–296.

(1961). Experiments with radio-phosphorus tracer in aquarium microcosms. *Ecological Monographs*, **31**, 157–88.

(1975a). *Communities and Ecosystems*, 2nd edn. New York: MacMillan.

(1975b). The design and stability of plant communities. In *Unifying Concepts in Ecology*, ed. W. H. Van Dobben & R. H. Lowe-McConnell, pp. 169–81. The Hague: Dr W. Junk.

Whittaker, R. H., Levins, S. A. & Root, R. B. (1973). Niche, habitat and ecotope. *American Naturalist*, **107**, 321–38.

Wiegert, R. G. (1975). Simulation modeling of the algal-fly components of a thermal ecosystem: Effects of spatial heterogeneity, time delays and model condensation. In *Systems Analysis and Simulation in Ecology*, vol. 3, ed. B. C. Patten, pp. 157–81. New York: Academic Press.

Wiegert, R. G. & Owen, D. F. (1971). Trophic structure, available resources, and population density in terrestrial vs aquatic ecosystems. *Journal of Theoretical Biology*, **30**, 69–81.

Wiens, J. A. (1983). Avian community ecology: An iconoclastic view. In *Perspectives in Ornithology*, ed. A. H. Brush & G. A. Clark, pp. 355–403. Cambridge: Cambridge University Press.

(1984). On understanding a non-equilibrium world: Myth and reality in community patterns and processes. In *Ecological Communities: Conceptual Issues and the Evidence*, ed. D. R. Strong, D. Simberloff, L. Abele & A. B. Thistle, pp. 439–57. Princeton: Princeton University Press.

Wilcox, B. A. & Murphy, D. D. (1985). Conservation strategy: The effects of fragmentation on extinction. *American Naturalist*, **125**, 879–87.

Wildavsky. A. (1973). [Book review of] Politicians, bureaucrats and the consultant. *Science*, **182**, 1335–8.

Williams, C. B. (1964). *Patterns in the Balance of Nature and Related Problems in Quantitative Ecology*. London: Academic.

Williams, M. B. (1973). Falsifiable predictions of evolutionary theory. *Philosophy of Science*, **40**, 518–37.

(1985). The scientific status of evolutionary theory. *The American Biology Teacher*, **47**, 205–10.

Williams, R. J. P. (1978). Phosphorus biochemistry. In *Phosphorus in the Environment: its chemistry and biochemistry, Ciba Foundation Symposium*, **57**, 95–116. New York: Elsevier.

Willis, E. O. (1984). Conservation, subdivision of reserves and the anti-dismemberment hypothesis. *Oikos*, **42**, 396–8.

Wilmott, C. J. (1977). Watbug: A FORTRAN IV algorithm for calculating the climatic water budget. *Publications in Climatology*, **30**, 1–55.

Wilson, D. S. (1983). The group selection controversy: History and Current Status. *Annual Review of Ecology and Systematics*, **14**, 159–87.

Wilson, E. O. (1971). *The Insect Societies*. Cambridge, Mass.: Belknap Press of Harvard University Press.

(1975). *Sociobiology: The New Synthesis*. Cambridge, Mass.: Belknap Press of Harvard University Press.

(1987). Causes of ecological success: the case of the ants. *Journal of Animal Ecology*, **56**, 1–9.

Wilson, E. O. & Willis, E. O. (1975). Applied biogeography. In *Ecology and Evolution of Communities*, ed. M. L. Cody & J. M. Diamond, pp. 522–34. Cambridge, Mass.: The Belknap Press of Harvard University Press.

Wilson, S. D. & Keddy, P. A. (1986a). Measuring diffuse competition along an environmental gradient: Results from a shoreline plant community. *American Naturalist*, **127**, 862–9.

(1986b). Species competitive ability and position along a natural stress/disturbance gradient. *Ecology*, **67**, 1236–42.

Wimsatt, W. C. (1980). Reductionistic research strategies and their biases in the units of selection controversy. In *Conceptual Issues in Ecology*, ed. E. Saarinen, pp. 155–202. Dodrecht: D. Reidel.

Winterhalder, B. P. (1980). Canadian fur bearer cycles and Cree-Ojibwa hunting and trapping practices. *American Naturalist*, **115**, 870–9.

Wise, D. H. (1984). The role of competition in spider communities: insights from field experiments with a model organism. In *Ecological Communities: Conceptual*

Issues and the Evidence, ed. D. R. Strong, D. Simberloff, L. G. Abele & A. B. Thistle, pp. 42–53. Princeton, N.J.: Princeton University Press.

Woodford, P. F., ed. (1968). *Scientific Writing for Graduate Students*. Bethesda, Md: Council of Biology Editors.

Woodwell, G. F. (1989). On causes of biological impoverishment. *Ecology*, **70**, 14–15.

Woolfenden, G. E. & Fitzpatrick, J. V. (1984). *The Florida Scrub-Jay: Demography of a Cooperatively Breeding Bird*. Princeton: Princeton University Press.

Wroblewski, J. S. (1983). The role of modeling in biological oceanography. *Ocean Science and Engineering*, **8**, 245–85.

Yagil, R. (1988). Comparative salt and water metabolism. In *Symposium on Comparative Nutrition*. ed. K. Blaxter & I. Macdonald, pp. 117–32. London: John Libbey.

Yan, N. D. (1986). Empirical prediction of crustacean zooplankton biomass in nutrient-poor Canadian shield lakes. *Canadian Journal of Fisheries and Aquatic Sciences*, **43**, 788–96.

Yodzis, P. (1980). The connectance of real ecosystems. *Nature*, **284**, 544–5.

Zuckerman, H. (1977). *Scientific Elite: Nobel Laureates in the United States*. New York: Free Press.

Index of names and first authors

Abbott, I., 189
Abrahamson, W.G., 226
Abrams, P., 97, 264
Alexander the Great, 163
Alexander, R. D., 72, 158
Alexander, R. McN., 150
Alexander, V., 150, 223
Allee, W. C., 88
Allen, T. F. H., 114, 161, 162
Altman, P. L., 296
Andersson, G., 143
Andrewartha, H. G., 87, 89, 103, 142, 300
Andrzejewska, L., 151
Antonovics, J. 161, 186
Arthur, W., 97
Auerbach, M. J., 46, 47, 96

Baird, D., 85
Ball, I. R., 168
Barker, A. D., 64, 65
Barr, T. C., 41
Barton, D. R., 92
Baudo, R., 215
Beatty, J. 160, 169
Beauchamp, T. L., 131, 172
Beck, M. B., 116, 122, 217
Begon, M., 3, 15, 57, 170
Belovsky, G. E., 110
Belsky, A. J., 182, 246
Bender, E. A., 139, 268
Benndorf, J., 143, 226, 291
Berger, T. R., 223
Berkowitz, A. R., 173
Berlinski, D., 116
Bethell, T., 63, 65
Beverton, R. J. H., 103, 191, 192, 250
Billings, W. D., 51
Birks, H. J. B., 100, 210
Bishop, J. A., 65
Biswas, A. K., 116, 118, 186, 188
Bliss, L. C., 171
Bloesch, J., 214
Blueweiss, L., 210
Bohr, N., 293

Bonacina, C., 130
Bonner, J. T., 253
Botkin, D. B., 116
Box, E. O., 183, 274, 282–4, 302
Box, G. E. P., 92, 186–8, 231, 238
Bradie, M., 61, 73, 82
Brady, R. H., 65
Briand, F., 47, 86–7
Brocksen, R. W., 143
Brown, J. H., 114, 157, 161, 190, 272, 289, 303–4
Brown, S., 10
Brown, W. L., 157
Burkholder, P. R., 49, 50
Burt, C., 242, 251
Busack, S. D., 198
Bysshe, S. E., 183, 285

Caesar, J., 163
Cain, A., J., 67
Cain, S. A., 80
Cairns, J., Jr., 215
Calder, W. A., 88, 94, 128, 203, 210, 274
Calvo, J. C., 282
Canale, R. P., 116
Canfield, D. E. Jr., 279, 294
Cantlon, J. E., 2, 11, 13
Caplan, A., 65, 68
Caraco, N., 277
Carlson, R. E., 82–3, 235
Carpenter, S. R., 83, 115, 143–4, 174, 186, 226, 231, 291
Cartwright, N., 3, 109, 235
Case, T. J., 152
Castrodeza, C., 65
Caswell, H., 5, 95, 108–9, 126–7, 175, 253
Cates, R. G., 224
Caughley, G., 130, 207, 209
Cavalli-Sforza, L., 72
Caws, P., 16, 22–3
Chamberlain, T. C., 22, 56, 81, 294
Chapman, R. N., 24
Chapra, S. C., 174, 211
Charney, J., 153

Cherrett, J. M., 75, 111
Chiou, C. T., 285
Chiras, D. D., 11
Choudhury, D., 180–1
Christiansen, F. B., 57
Clark, C. W., 192, 195
Clark, W. C., 116, 118, 217, 270, 303–4
Clements, F. E., 40
Cody, M. L., 5, 14, 75, 184, 215
Coe, M. J., 151, 274, 276
Cohen, I. B., 23
Cohen, J. E., 116
Cole, B. J., 188
Cole, H. S. D., 116
Cole, J. R., 113
Colinvaux, P., 3, 15, 264
Collingwood, R. G., 132, 166–7, 175
Collingwood, R. W., 135
Colwell, R. K., 198–9
Condrey, R. E., 254, 281
Confer, J., 214
Connell, D. W., 183, 284
Connell, J. H., 95, 138, 140, 143, 157, 162, 239, 256, 258, 267
Connor, E. F., 5, 188, 197–8, 206, 217, 251, 294
Conover, W. J., 238
Copernicus, N., 109
Coupland, R. T., 276
Cousins, S., 83–5, 199
Cramer, R. D., 285
Crane, D., 185
Crecco, V., 195
Croce, B., 166
Culver, D. C., 41
Currie, D. J., 183, 274, 289

Damuth, J., 88, 142
Darwin, C., 161, 220
Dayton, P. K., 4, 9, 96, 108, 139, 212, 219, 224–5, 246, 293, 302
De Bernardi, R., 214
De LaFontaine, Y., 235
DeAngelis, D. L., 95, 116, 128
Deevey, E. S., 175, 254
Dennis, R. L., 118, 298
Detwiller, R. P., 115, 125, 173
Dhondt, A. A., 82
Di Castri, F., 1, 10, 104
Diamond, J. M., 16, 21, 140, 161, 165, 188, 197, 199, 206, 210, 258, 268
Dillon, P. J., 133, 146, 182, 210, 274, 278–9

Donagan, A., 149
Dorschner, K. W., 55
Downing, J. A., 41, 102, 194, 198, 205, 234, 237
Dray, W. H., 148–9, 151, 155, 160
Drury, W. H., 24, 87, 152–4, 201
Dryzek, J. S., 12
Ducasse, C. J., 74
Due, D. A., 188
Dunbar, R. I. M., 69, 73
Dyer, M. I., 180–1

Eadie, J. M., 217, 238
East, R., 151, 274, 276
Eberhardt, L. L., 251
Edmondson, W. T., 133, 168, 180
Edmondson, Y. H., 52
Edwards, A. W. F., 31, 179
Edwards, R. Y., 77, 82
Egerton, F. N., 92, 95
Egler, F. E., 80
Ehrenhaft, F., 31
Ehrlich, P. R., 60, 95, 116, 173
Einstein, A., 109, 179, 223
Elizabeth I., 164–5
Elliot, E. T., 143, 214
Elliott, J. M., 233, 234
Elster, H. J., 88
Elton, C. S., 40, 91, 96, 175, 220
Emlen, J. M., 55
Emlen, J. T., 233
Emmel, T. C., 98
Endler, J. A., 63, 66–7, 73, 162, 168, 258
Ethelred the Unready, 164
Evernden, N., 98

Fagerström, T., 17, 22, 31, 55, 218
Farlow, J. O., 151
Ferguson, A., 65–7
Feyerabend, P. K., 25
Fleming, T. H., 190
Flew, A., 182, 189, 240–6, 268
Forbes, S. A., 247
Forsberg, C., 12, 211
Foster, W. A., 181
Fowler, S. V., 5, 234, 249
Frank, P. W., 214
Franklin, A., 22, 31, 158
Fretwell, S. D., 14, 20–1, 48, 75, 83, 86–7, 143, 215, 222, 225, 252
Friant, S. L., 286
Frost, B. W., 254

Futuyma, D., 101

Galileo, 22, 223
Gallie, W. B., 157–8
Ganong, W. F., 5
Gardiner, P., 154–5, 196–7
Garfield, E., 6, 179, 230
Gates, D., 92, 110, 150, 224
Gauch, H. G., Jr., 184
Gaudet, C. L., 262
Gause, G. F., 57, 97
Geiger, R., 153
Geist, V., 190–1
Geller, W., 254
Ghiselin, M. T., 63, 65–6, 68, 70, 73, 161
Gilbert, F. S., 256
Gilpin, M. E., 5, 198–9, 251
Gimingham, C. H., 186
Glaser, B. G., 223
Gleick, J., 230
Goda, T., 164
Godbout, L., 103, 146
Goldman, C. R., 138
Goodman, D., 24, 96, 288
Gorham, E., 45
Goudge, T. A., 160, 163
Gould, S. J., 5, 63–6, 69, 152, 162, 249, 251, 256
Grant, P. R., 157, 161, 168, 198, 236, 258
Gray, R. D., 24, 36, 113, 159, 167, 219, 222, 231, 237, 251, 256
Green, R. F., 232, 233
Greenslade, P. J. M., 202
Grene, M., 60
Griesbach, S., 115
Grime, J. P., 202
Gruber, H. E., 22
Guhl, W., 185
Gulland, J. A., 103

Haeckel, E., 5, 88
Hailman, J. P., 5
Hairston, N. G., 191
Hairston, N. G., Jr., 65–6, 83, 86, 143
Hall, C. A. S., 5, 14, 16, 24, 54, 55, 186, 192, 226, 231, 292
Hamilton, W. D., 66
Hanlon, R. T., 157
Hanson, J. M., 195
Hardin, G., 90, 97, 253
Hardy, A. C., 83, 200
Harper, J. L., 52, 63

Harris, C. L., 65
Harrison, A. F., 135
Hart, R., 142
Harvey, H. H., 289
Harvey, P. H., 199
Haskell, F. E., 21, 89
Hassell, M. P., 14, 234, 239, 254
Hawkins, C. P., 80, 82, 95
Healy, M. C., 103
Heath, R., 143, 216
Hecky, R. E., 223
Hedgpeth, J. W., 118
Hemmingsen, A. M., 85
Hempel, C. G., 148–50
Henderson, L. J., 150
Henson, V., 5
Hilborn, R., 131, 137, 141, 175
Hjort, J., 195
Holling, C. S., 93, 204–5, 209
Holton, G. J., 31
Horn, H. S., 94, 154, 156, 201
Horne, A. J., 88
Howe, H. F., 5, 247–9
Hrbàček, J., 180
Hughes, A. J., 88
Hull, D. L., 30, 77, 149, 152, 160
Hume, D., 30, 131, 214
Humphreys, W. F., 274, 281
Hurlbert, S. H., 24, 77, 81–2, 91, 95, 185, 233, 256, 288
Hutchinson, G. E., 21, 35–6, 48, 52, 56–7, 76, 82, 88, 91, 96–7, 141–2, 175, 179, 217, 220, 224–5, 230, 257, 272, 290
Hynes, H. B. N., 96

Ivlev, V. S., 85

Jackson, J. B. C., 257–8
Jacob, F., 297
Jacobs, J., 175
Jacobsen, T., 235
James, F. C., 97
Janzen, D. H., 189, 224, 248–9
Joergensen, S. E., 115–18, 135, 270
Jones, J. R., 235
Joynt, C. B., 150
Juanes, F., 190, 213
Jumars, P. A., 230

Kaiser, K. L. E., 274, 286
Kajak, A., 145, 151, 276, 281
Kalff, J., 181, 246

Keddy, P. A., 226, 256–60, 262, 264, 267, 272
Kekulé, F. A., 22
Kelvin, Lord, 33
Kempthorne, O., 114
Kenney, B. C., 46
Kepler, J., 109
Kerr, R. A., 173
Kerr, S. R., 102
Kettlewell, H. B. D., 65
Kier, L. B., 284
Kimball, K. D., 215
King, J. L., 63
Kingsland, S. E., 5, 21, 24, 48, 54–5, 155, 225, 230
Kirby, A. J., 135
Kirkendall, L. R., 87
Kitts, D. B., 118
Knowlton, M. F., 231
Koch, A. L., 238
Kochanski, Z., 154
Koestler, A., 23
Kohn, A. J., 52
Kozlowsky, D. G., 85
Krebs, C. J., 15, 170
Krebs, J. R., 3, 170, 196, 289
Kuhn, T. S., 16–17, 20, 23, 33, 35, 78, 114, 133, 158, 218
Kuttner, R. E., 157

Lack, D., 224
Lahti, T., 189
Lakatos, I., 60, 70, 190
Lampert, W., 110, 205, 213
Larkin, P. A., 83, 102–3, 200, 223
Leblanc, S. A., 157
Lechowicz, M. J., 152, 185
Lee, K. K., 65
Leggett, W. C., 195
Lehman, J. T., 5–6, 17, 21, 164, 224, 242, 301
Leopold, A., 130
Levin, S. A., 52, 231
Levins, R., 32, 117, 225
Lewin, R., 140, 258, 289, 301
Lewontin, R. C., 4, 52, 65–8, 81, 95, 111, 256
Lieth, H., 88, 103, 183, 277, 282
Likens, G. E., 173, 179, 231
Lincoln, R. J., 40
Lindeman, R. L., 83–5
Lindsey, C. C., 190

Livingstone, R. J., 83
Loehle, C., 20, 105, 108, 116, 128, 246
Lotze, J.-H., 19–20
Lovelock, J. E., 5
Lynch, M., 151

MacArthur, R. H., 22, 31, 48, 52, 57, 79, 95, 154, 170, 191, 198, 205–6, 214–15, 220, 224–5, 252, 264, 266, 289, 303
MacBeth, N., 65
Macevicz, S., 224
MacFadyen, A., 82, 91
Mackay, D., 286
Maelzer, D. A., 89, 251
Mailhot, H., 285–6
Mandelbrot, B. B., 230
Manser, A. R., 64–5
Margalef, R., 24
Marshall, C. T., 152, 231
Marshall, E., 173
Martin, A. C., 83–4
Marx, K., 154
Mason, H. L., 80, 89
Mather, J. R., 281–2
May, R. M., 5, 21, 24, 46, 52, 57, 80, 93–4, 96, 116, 142, 170, 220, 224, 257–8, 264, 290, 303
Mayer, W. V., 65
Maynard Smith, J., 24
Mayr, E., 5, 63, 129, 131
McCarty, L. S., 286
McCauley, E., 151, 235
McClellan, G. H., 135
McIntosh, R. P., 5–6, 11, 16, 40, 52, 60, 79–80, 89, 111, 116, 186, 258, 289–90
McNab, B. K., 190
McNaughton, S. J., 3, 15, 96, 101, 153, 180–1, 276
McQueen, D. J., 143, 226
Meadows, D. H., 116
Medawar, P., 13, 23, 146, 297
Meentemeyer, V., 183, 282
Mellanby, K., 12
Mendel, G., 31, 179
Mentis, M. T., 21, 253
Merton, R. K., 9–10, 23
Mertz, D. B., 138–9
Metcalf, R. L., 214, 285
Michod, R. E., 55, 73
Miles, D. B., 184
Millar, J. S., 84
Millikan, R. A, 31, 179

Milne, A. 77, 81–2, 259, 261
Minshall, G. W., 226
Mitchell, R., 59
Moen, A. N., 116
Mohr, C. O., 194
Monteith, J. L., 282
Moriarty, F., 285
Morin, A., 146, 234, 274, 287–8, 292
Morris, D., 22, 243
Mother Teresa, 215
Moynihan, M., 157
Murdoch, W. W., 86, 95, 193, 224, 251
Murphy, D. D., 189
Murray, B. G., Jr., 97, 228, 262

Nagel, E., 38
Napoleon, 163
Naumann, E., 82
Naylor, B. G., 65, 70
Neely, W. B., 281, 285
Newton, I., 22, 35, 223
Nelson, P., 109
Nicholson, A. J., 224
Nirmalakhandan, N., 284
Nisbett, R., 129, 178
Niven, B. S, 89
Nürnberg, G. K., 279

O'Grady, R. T., 72
O'Hara, R. J., 147
O'Neill, R. V., 91, 111, 114, 291
Oakeshott, M., 163
Odum, E. P., 3, 83, 87, 199, 201–2, 209, 224, 257–8
Odum, H. T., 83, 224
OECD, 274, 280
Oglesby, R. T., 146
Oliver, B. G., 285
Ollason, J. G., 69, 156, 159, 167, 178, 214, 249, 256
Olsen, Y., 122, 213
Oppenheimer, R., 109
Orians, G. H., 93, 300, 303
Oster, G. F., 20, 113, 116, 125
Ostrofsky, M. L., 279
Owen, D. F., 180–1
Owen-Smith, N., 153

Pace, M. L., 144
Paine, R. T., 46, 86–7, 138, 143, 199–200, 210, 256
Park, T., 214

Parker, G. A., 161, 219
Parry, G. D., 87, 210
Partridge, D., 108
Patrick, R., 289
Patten, B. C., 84, 115, 230
Paul, E. A., 3
Peet, R., 24, 82, 95, 288
Penman, H. L., 183, 282
Penry, D. L., 161
Petelle, M., 180
Peterman, R. M., 233, 239
Peters, R. H., 5, 7, 9, 32, 46, 65, 72, 82–6, 88, 95, 97, 114, 117, 121–3, 127–8, 142, 151, 162, 175–6, 185–6, 190, 201, 203, 205, 213–14, 226, 230, 258, 274–6, 291, 294–6
Peterson, R., 97
Pianka, E. R., 3, 15, 20, 87, 97, 101, 202, 209, 215, 242, 263–4, 289
Pielou, E. C., 114, 126, 128, 145, 191
Pierce, G. J., 5, 24, 36, 113, 251
Pilson, M. E. Q., 214
Pimm, S. L., 83, 86, 199–200, 210, 242
Platt, J. R., 22, 294
Poole, R. W., 108, 114, 128, 301
Popper, K. R., 16, 21–2, 25, 30, 65, 67, 72, 76, 93, 105–6, 109, 111, 114, 131, 148, 154, 171, 194, 212, 217
Porter, K. G., 204–5, 254
Prairie, Y. T., 43, 280
Pramer, D., 144, 186
Price, D. J. de Solla, 6, 142, 180, 185
Price, P. W., 16
Prothero, J., 237
Proxmire, Senator, 178

Quinn, J. F., 133, 139, 143

Reckhow, K. H., 211, 279
Redfield, A. C., 138
Reed, W., 178
Reich, C. A., 160
Reichenbach, H., 39
Reimers, W. A., 289, 293
Rensch, B., 151, 153
Rey, J. R., 206
Reynolds, C. S., 43, 226
Reynolds, J. F., 103
Reynoldson, T. B., 260
Ricker, W. E., 103, 191–2
Ricklefs, R. E., 3, 15
Rigler, F. H., 4, 13, 78, 82–4, 86, 100, 102,

Rigler, F. H. (*cont.*)
  108, 113–14, 116, 126, 138, 142, 200, 205, 210, 213–14, 226–7, 230, 248, 303
Riley, G. A., 83, 207, 209
Roff, D. A., 145, 195
Romesburg, H. C., 21, 33
Rose, H., 179
Rotenbury, J. T., 239
Rothschild, B. J., 251
Roughgarden, J., 5, 21–2, 25, 52, 135, 140, 143, 217–18, 303
Ruse, M., 64–5, 67, 70, 118, 242
Russell, B., 133
Rutherford, E., 179
Ryan, P. M., 195
Ryder, R. A., 103, 195
Ryther, J. H., 277

Saarinen, E., 5
Saether, E. -E., 203
Sala, O. E., 276
Salt, G. W., 5, 95, 110
Sanders, H. L., 289
Sartory, D. P., 235
Sattler, R., 20–1, 72, 77, 107, 110, 112
Scavia, D., 168
Schaefer, M. B., 103
Schindler, D. W., 94, 133, 138, 186, 211, 216, 223, 231, 233, 281, 288
Schlesinger, W. H., 11, 171
Schoener, T. W., 48, 86, 140, 162, 198, 203 207, 217, 225, 248, 256, 258, 262–4, 266, 289, 300
Schoenly, K., 80, 100
Schopenhauer, A., 240
Scriven, M., 149, 155, 158
Seip, K. L., 276, 296
Shapiro, J., 143, 226
Shaw, E. M., 24
Shipley, B., 80, 100
Shugart, H. H., 125
Silvertown, J. W., 142
Simberloff, D. S., 4, 80, 104, 128–9, 140, 162, 186, 188–9, 206, 212, 225, 246, 249, 256, 267
Simpson, G. G., 190
Sims, P. L., 276
Sinclair, A. R. E., 276
Singh, J. S., 276
Sissenwine, M. P., 103, 192
Slobodkin, L. B., 6, 21, 53, 68, 79, 86, 109, 171, 231, 264, 297–8
Smith, F. E., 5, 56, 86, 111

Smith, V. H., 92
Sober, E., 39, 55, 58, 69
Sommer, U., 152, 291
Southern, H. N., 11, 103, 171
Southwood, T. R. E., 21, 56, 87, 156, 202–3, 234
Spencer, H., 161
Spengler, O., 154
Spomer, G. G., 89
Sprules, W. G., 289
St. Amant, J. L. S., 251
Stanley, S. M., 153
Statzner, B., 226
Stearns, S. C., 5, 24, 42, 66, 69, 70, 161–2, 184, 186, 203, 218, 251
Stebbins, G. L., 65
Stemberger, R. S., 254
Stephens, D. W., 159, 161–2, 169–71, 186, 195, 209, 214, 251
Stockner, J., 226
Strong, D. R., Jr., 5, 16, 53, 54, 129, 140, 198, 212, 224, 256, 264–5, 274, 288
Stumm, W., 284
Sutcliffe, W. H. Jr., 195
Sverdrup, H., 211, 221
Symanski, R., 218

Taylor, L. R., 49
Taylor, P., 108
Taylor, St. C. S., 152
Teal, J. M., 83
Thom, R., 230
Thompson, K., 142
Thompson, N. S., 25, 65–6, 69, 153
Thompson, P., 108, 149, 154
Thompson, W. R., 5
Thornhill, R., 219
Thornthwaite, C. W., 182–3, 281
Tillman, D. A., 194
Tilman, D., 91, 161, 206, 209, 224, 262, 266, 268
Toffler, A., 161
Toft, C. A., 233, 239
Toynbee, A., 154
Tribbia, J. J., 216
Trivers, R. L, 66
Tuomi, J., 70
Turco, R. P., 173
Turner, N. C., 274, 281
Twichell, A. R., 31

Ulanowicz, R. E., 85, 304
Umpleby, S. A., 253

Underwood, A. J., 97, 140, 260, 267

Vallentyne, J., 12
Van der Beken, A., 282
Van der Steen, W. J., 63, 65, 70
Van Keulen, H., 116
Van Soest, P. J., 84
Van Valen, L., 9, 83, 293
Vannote, R. L., 226
Veith, G. D., 286
Velikofsy, I., 162
Verhulst, P. -F., 54
Verschueren, K., 285
Vogt, K. A., 103
Vollenweider, R. A., 24, 82, 116, 122, 187, 226, 230, 274, 302
Von Foerster, H., 253

Wallace, A., 181
Walter, G. G., 193
Walter, H., 276
Walters, C. J., 116, 127, 141, 175, 210, 299, 303
Wangersky, P. J., 52, 54
Ware, D. M., 116, 209
Waring, R. H., 282
Wartenberg, D., 184
Wassermann, G. D., 70
Watson, J. D., 22
Watson, S., 295
Watt, A. S., 80
Watt, J., 109
Watt, K., 5, 16, 224, 116, 118, 217, 270
Weatherhead, P. J., 158, 247
Weaver, W., 64

Weisskopf, V. F., 16, 247
Welch, D. W., 193
Weller, D. E., 45
Werner, E. E., 102
Westman, W. E., 93
Westoby, M., 45, 181
White, E. G., 164
White, J., 45
White, T. C. R., 151
Whitney, G. G., 168
Whittaker, R. H., 15, 24, 77, 80, 91, 94–5, 153, 214
Wiegert, R. G., 83, 85, 143, 214
Wiens, J. A., 4–5, 80, 95, 162, 207, 212
Wilcox, B. A., 189
Wildavsky, A., 118
Williams, C. B., 303
Williams, M. B., 64
Williams, R. J. P., 135
Willis, E. O., 189
Wilmott, C. J., 282
Wilson, D. S., 4–5, 66, 72, 161, 262
Wimsatt, W. C., 111, 118
Winterhalder, B. P., 168
Wise, D. H., 260
Woodford, P. F., 28, 221, 228, 252
Woodwell, G. F., 11
Woolfenden, G. E., 219
Wroblewski, J. S., 106, 116

Yagil, R., 110
Yan, N. D., 144
Yodzis, P., 86

Zuckerman, H., 23

Subject index

abstraction, 101–2, 118, 25–7, 298
abundance, *see also* density, omnivore,
    population, 28, 108, 142, 224, 274
  allometry, 19–20, 32, 88, 275, 296
  birds, 190, 213, 233
  instrumentalism, 142
  relative, 101, 141
academic colleges, *see* invisible colleges
acid rain, 12, 30–1, 130, 183, 211, 221–2
accuracy, 17, 30–1, 178, 189–96, 275
  limiting similarity, 265
adaptation, 61, 63–4, 75
  hard core, 70–1
  non adaptive traits, 67
  normic explanations and, 157
  over-interpretation, 249
  ubiquity, 66–7
adaptive management, 141, 175, 274, 299
ad hoc hypotheses, 194–5, 252
  community assembly rules, 199
  food web analysis, 199–201
  genetics, 70–1
  historical explanation, 168
  logical fallacy and, 244, 247
  other factors, 88, 152
  soft core, 71
  special pleading, 247
  system complexity, 88
  theoretical reformulations, 195
aesthetics, *see* appeal
affirming the consequent, 182, 240–1
aggregation
  food web analysis, 199
  resource response, 300
agriculture, 2, 102–4, 270, 279, 281, 299
algae
  bluegreen, 92
  sinking, 207
  surface: volume ratio, 42–3
algebraic models, 196, 207–10
  competition, 263–6
  limiting similarity, 264–5
Allen's rule, 190–1
allolimy, 49–50

allometry, 210, 247–75
  competition, 258
  constraint, 63
  general theory, 213–14
  gestation time, 247
  predictive matrix for, 297
  qualitative analyses, 210
  r-K-selection, 202–3
  succession, 201–2
allopatry, 198, 241
American Naturalist, 3, 9–10
amplitude, and stability, 93–5
analogy and metaphor, 18, 24, 225
  causal connection and, 129
  induction, 214
  model systems, 214–16
  nature as a machine, 112–3
  pathetic fallacy, 245
  prediction, 215–16
  science as a cathedral, 113–14
analysis, 21, 25–7, 107, 128, 237–9
anarchism, 24, 490
anecdote, 143
Annual Reviews in Ecology and Systematics,
  10, 296
anthropomorphism, 153, 165
aphorism, *see* truism
appeal, 33–6, 178, 218–19
  elements of, 218
  evolution, 71–3
  fisheries models, 192, 250–1
  food web analysis, 200
  predictive ecology, 290–304
applicability and application, 53, 104,
  108–9, 186–7
  competition, 265
  mechanism, 224
  predictive ecology, 297
arithmetic, 38, 53, 72
art and historical explanation, 147, 169
Art of the soluble, 13, 146, 297
artefact, 22, 140
  Hutchinsonian ratio, 217
  special pleading, 247

association, 80
atheory, 97–100
autecology, 150
authority, 245, 298–9
    hypothesis selection, 223–5
autodefinition
    by invisible colleges, 185
available water, 276
average, as a theory, 18
axioms, *see also* premises
    deduction, 26
    limiting similarity, 265
axes, unscaled, 59, 196
    graphical analyses, 203–7
    island biogeography, 205

Bacillus thuringinensis, 287
balance of nature, 92–5
beauty, 17, 265
beaver, 84–5
begging the question, 244,301
belief, 23, 166, 301
Bergmann's rule, 190 1
bias
    cultural, 258
    data interpretation, 246, 254
    disabling, 251
    peak analysis, 237–8
    sampling, 22, 234
    special pleading, 247
    tunnel vision, 251
bioaccumulation, 164, 214, 284–6
biocoenosis, 80–1
    conflation, 91
BioCo-TIE, 41, 49–50
biogeographic rules, 190–1
biogeography, 41, 210, 217
    competition theory, 161
    historical explanation, 168–70
    typology, 100
biological control, 193
biomanipulation, 143, 226
biomass relations, 275–7
biomes, 75, 80–1
birds
    model organisms, 215, 308
Biston, see peppermoth
black box, 268, 303
black flies, 286–8
Bonferroni correction, 239
bottle effects, 139
boundary conditions, 27–8, 31–2, 62, 216,

298
    predictive matrices, 294
breasts, 157
brittleness and stability, 93
broken-stick, 79, 191
busy-work, 219, 273
but-they-never-will-agree diversion, 245
but-you-can-understand-why evasion, 244

calculating tools, 20, 105, 230
calculus, 35
carrying capacity, 54, 56, 75
    conflation of, 82
Cartesian model, 27–8, 95, 99, 275
Castor, *see* beaver
casual processes, 163, 198, 217
catastrophe theory, 24, 339
catch-effort, 103, 192
causal analysis
    alternative to action, 136
    factors, 163
causality, 128–36
causation, causal connection, cause, 14,
        105–46, 244, 269
    commonsense notions, 129–30, 141
    definition, 107, 132–3
    -effect, 134–72
    explanation, 133
    Humean, 131–4
    instrumentalism, 141–6
    manipulability, 132, 143
    multiple, 64, 269
    physics, 133, 158, 187
    regularity, 132, 143
    scientific notions, 130–6
    ultimate and proximal, 131, 133–4
    unobservability, 64, 128
cave communities, 61
cephalopods, 157
ceteris paribus, 90, 158, 189
chaos, 230
Char Lake, 214
chipmunk, 157
chlorophyll, *see also* phosphorus response,
        226, 235, 287–8
circularity, *see* tautology
citation, 2, 6, 179
    frequency, 6–8
    immediacy and impact, 8–10
    methods papers, 230
    selective, 225
Citation Classic, 8

classification, 48–51, 74, 80–92, 153
  biogeography, 210
  competition, 262
  dichotomous, 87–9
  lake trophy, 82–3
  logically-black-is-white slide, 242–3
  qualitative theories, 210
  trophic level, 82–6
climate and plant form, 283–4
climax, 24, 153
clines, see gradients
coactions 49–50
coevolution, see herbivore-plants,
  megafaunal dispersal
coexistence, see competitive exclusion
coherence, 292–4
  conceptual, 91–2
colligative explanation, 155, 160–3, 167
  competition, 267
collinearity, 138, 145, 302
colonization, 165, 205–6
common knowledge, 193
community, 13, 75
  conflation, 80–2, 91
community assembly rules, 197–9, 210, 218
comparison
  among laboratories, 77
  across literature, 248–50
compartments, 117, 119
competition, 75, 115, 140, 244, 256–73
  coefficient, 56, 264, 268
  concept, 261
  conflation, 81
  definitions, 259
  diffuse, 97, 226
  evolutionary force, 263
  fad, 225
  ghost of competition past, 157, 263, 267
  interference-exploitative, 262
  intractability, 79–80
  microcosms, 214
  over-interpretation, 241, 249, 265
  prevalence, 257–9
  propositions, 262–3, 271
  resource-space, 262
  scramble-interference, 262
  theoretical status, 270–2
  theory, 161, 257, 259, 264
competitive efficiency, 101
competitive exclusion, 74–5, 95, 157, 226,
  251, 263–4
complementarity, 274, 290

predictive ecology, 292–4
completeness in narrative explanation,
  164–5
complexity, 11, 74, 119–28, 146, 272–3, 298
  ad hoc explanation, 88
  modelling, 118–19
  temporal trends in modelling, 123
concept, see also operationalization, 23, 150,
  74–104
  basis for communication, 292
  competition, 261
  definition, 78
  empty, 98
  environment, 75, 91–2
  environmentalism, 98–100
  explanatory, 105
  limits in discussion, 100–4
  most popular, 75
  scale, 81
  stability, 95
  temporal, 150
concept cluster, 74, 82, 87, 91–2, 131
confidence limits, 19, 27–8, 32, 122, 190,
  196
  qualitative analysis, 210
confirmation, 23, 59, 86, 197
  competition theory, 266
  partial, 173
  prejudices, 228
  qualitative, 206
conflation, 74, 81
conjecture, 16, 22
connectance, 46–7
conservation, 13, 75, 92–3, 210
  of soil, 299
  of species, 96
consistency, 34–6
constancy and stability, 93–5
constant conjunction and cause, 131–2
constraint
  logistic compromises, 63, 113
  prey choice model, 159
consumers and trophic dynamics, 84–5
contaminant uptake, 115, 281
contexts of justification and discovery, 16
contiguity and cause, 131–2
continuum, 48–9, 81, 87
controls, 138, 231–3
convergence, see ecological convergence
Cope's law, 152–3
coprophages and trophic dynamics, 85
correlation, 143

cause, 131–2, 143–4, 244, 246–7, 267
  coefficient, 46
  field data, 140
corroboration, *see* confirmation
covering law explanation, 148–50, 152,
    155, 156, 163
creationism, 64–5, 72
creativity, 23–4, 291
credibility, 104
criteria (*see also* individual listings), 17–37
  accuracy, 17, 30–1
  appeal, 33–4
  beauty, 17
  cohesion, 292–4
  consistency, 17, 33–5
  economy, 33
  explained variation, 136
  generality, 31–2
  goal definition, 28–9
  heuristic, 17, 27, 35–6
  hierarchy, 36–7
  immediacy, 29–30
  logical, 27
  operationalization, 30
  plausibility, 131
  practicability, 34
  precision, 32
  predictive power, 36
  quantification, 32–3
  relevance, 29
  simplcity, 34–5
  testability, 17
  understanding, 17
  validation, 290
criticism
  advantages, 1–3
  categorical, 256
  definition, 17
  need for, 15–16
  role of, 3–4, 24–5, 220, 226, 272
  topical, 256
crop harvest, 142
crossover, genetic, 71
currency, prey choice model, 159
Current Contents, 8
cycling, 75
  biochemical, 134
  biogeochemical, 115
  microcosms, 214
  phosphorus, 115–17, 134–5

Daphnia, 213, 254

Darcy's law, 24
Darwin's finches, 267–8
data
  availability, 236–7
  banks, 237
  hypothesis, 246
  requirements for simulation, 217
  as theory, 31
debate, 4–6
decomposers, *see* detritivores
decomposition and evapotranspiration, 183
deduction, *see also* tautology, 23, 25–6
  prey choice model, 159
definition, used as a theory, 77
degrees of freedom, 138, 163
density, dependence and independence, 48
  biotic vs abiotic regulation, 145–6
  dichotomous classification, 87
  population, 19
  raccoons, 19, 26
  stability 75, 94–5
  tunnel vision, 250–1
denying the antecedent, 240–1
dependent variable, *see* response variable
description, epistomological 25
design in nature, 69–70, 113
determinism,
  definition, 107
  food web analysis, 200
detritivores and trophodynamics, 83–6
development of organism, 152
developmental law, 150, 152–4
  historical processes, 154
dichotomy, 87–92
  organism-environment, 74, 87–92
diet choice, 170
  North American birds and mammals, 84
digestion and optimality, 161
digestive capacity, 110
dilution and pollution, 98
disc equation, 205, 209–10
discussion, 100–1, 236, 239–54
disease control, 2, 102
discovery, *see* context of
disequilibrium, 97
distributions and aggregations, 49, 266
  competition, 267
disturbance regime, 204
diversity, *see also* species number, 24, 75,
    141–2, 289–90, 304
  crop pests, 289
  evapotranspiration, 289–90

diversity (*cont.*)
  indices, 185
  natural historians, 289
  stability, 74, 92–5, 204, 289
Dollo's law, 151
domain of theory, *see* boundary conditions
Doomsday, 253
Drosophila, model system, 214
dung flies, 219
doubt, *see* uncertainty

ecological efficiencies, 79
ecological convergence, divergence and
    equivalents, 157, 184, 215, 226
ecology and ecologists
  academic, 1, 13–14
  applied, 2, 12, 101, 224
  debates, 4–6
  definitions, 2–3, 17, 40, 88, 175, 183
  evidence of weakness, 1–2, 6–18
  goal, 14, 183
  shortcomings, 12
  specialists-generalists, 101–2, 220
  statistical-empirical holists, 111
  symposia, 5–6
  terms, 40–1
Ecology, 3, 7, 9–10
economy, 33–4, 178, 216–18
  competition, 260, 272
ecosystem
  conflation, 80–1, 91–2
ecotoxicology, 274, 284–6
  costs of tests, 284
edge effects, 139
elasticity and stability, 93–5
electronic particle analysis, 230
elegance, 265, 267
emergent properties, 112
empathetic explanation, *see* rational
    reconstruction
empirical rules, 150
empiricism, 60, 108–11, 114, 274–5, 300–3
  definition, 107
  predictive power, 301
emulation and hypothesis selection, 225
energetics, 71
energy flow, 75, 165
engineering in nature, 69–70
environment, 79–80
  definitions, 89–92
  operationalization, 87–92, 300
  theory of, 434, 89

environmental impact assessment, 164, 223
environmental determinism
  ad hoc explanation, 99
  IQ, 145–5
environmental problems, *see also* global
    problems, 2–3, 10–13, 102
environmentalism, 74
equilibrium, 94–5, 188, 208, 246
error
  propagation, 122–6, 146
  pure, 136
  Type I, 63, 239
  Type II, 233, 238–9
essentialism, definition, 107
ethics, 176, 254
eusociality, 65–6, 224
eutrophication, *see also* phosphorus, 82–3,
    165, 297
  abatement, 2, 24, 138, 211
  causes, 133
  models, 164
  top down control, 143
evapotranspiration, 8, 182–3, 281–3, 290
evolution, 24, 60–73, 158
  defenses of, 65–71
  non-Darwinian, 63–4
  regressive, 41
  synthesis and dys-synthesis, 161
evolutionary steady state, 61–2
  social insects, 224
ex cathedra statement, 224–5
existential statement, 97
experiment
  field, 138–40, 143, 268
  instrumentalism, 137–40
  laboratory, 16, 137–8, 140, 249
  marine biology, 138
  natural, *see* correlation
  test, 269
expert systems, 214, 296
explanation, *see also* coligative, genetic,
    historical, individual, narrative, and
    normic explanations and rational
    reconstruction, 14, 29, 105–77, 290,
    301
  competition, 260
  deep, 270
  definition, 107, 156
  dynamical, 106, 108
  evolutionary, 158, 265
  falsification, 90
  goal, 106, 108–10

integrating, 160
non-adaptationist, 67
non-empirical, 301
scientific, 148
sketches, 149, 151, 166
explanatory models, 20, 105–6
empirical support, 302
extinction, 61–2
island biogeography, 205–6
preserve design, 188
extraneous material, 252
extrapolation, 253–4

facts, 22
fads, 225–7
faith and science, 194
fallacies, 240–6
falsification, 18, 23, 28, 59, 104, 190, 197
explanation, 90
naive, 245
non-operational theories, 96–8
verbal models, 202
verification, 228
familiarity, 194, 267
fanaticism, 71
feedback, 94–5, 125
feeding rate, 114
finaglers, 251
first maxim of Balliol men, 244
fish yield, 102–4
chlorophyll, 235
fisheries
collapse, 164, 195
models, 191–3, 195, 204, 250–1
fitness, 63
definition, 68–70
environment, 150
herbivore and plant, 181
Florida scrub-jay, 219
flushing of lakes, 222, 280
fly–algae model system, 214
folk wisdom, 130
food chains, 285
food web, 75
analysis, 46–8
biomanipulation, 143, 226
classification, 120–1
tool, 201, 210–11
verbal model, 199–201
foraging theory, 20, 170, 214
forcing functions, 115, 119
forecasting, 216, 301

forest production, 193–4
forestry, 2, 102–3, 299
founder effects, 63
fractals, possible fad, 230
fragility, 75
fraud, 251
fugacity, 286
functional definition, *see* operationalization
functional response
graphical analysis, 204–5
algebraic analysis, 204–10
fundamentalism, Christian, 64, 242
funding and research, 11, 111, 179, 187,
237, 303

Gaia, 5
game theory, 24
Gause's law, *see* competitive exclusion
generalist species, 48, 87
slugs, 224
generality, 17, 31–2, 34, 36, 178, 211–16,
275
generation time, 42
generic theory, *see* hypertheory
genetic drift, 63
genetic explanation, 155, 157–60
competition, 265
genetic fallacy, 242–3
giant axon, model system, 214
global problems, 11, 102, 173–4, 211
future historical scenarios, 175
goals
causal explanation, 270
confusion, 146
introduction, 221–2
in science, 1, 27–30, 106–10
nebulous, 74
realism, 118–19
Golden Fleece Award, 178
golden mean, 243
gomphotheres, *see* megafaunal dispersal
gradients
analysis, 140
competition, 267
peak analysis, 238
typology, 100
graphical analyses, 59, 196, 203–7
competition, 263
limiting, 264–6
gravity, 22, 156, 219
grain, environmental, 48
Green's statistical rules, 231–3

ground water flow, 24
group selection, 5
growth rate, 42, 44
    ingestion rate, 281
    of science and technology, 179
guild
    conflation, 80–2
    discussion, 101
    economy, 210, 217

habitat, 75, 92
habitat selection, 161
half-life
    biological, 42
    citation, 9
handling time, 209
haplodiploidy, 66
Hardy-Weinberg model, 55
heaper, the, 242–3
heat exchange, 111
Henry's law, 287
herbivores, 84, 276
    algebraic models, 207–8
    exudates, 180–2
    feeding preferences, 110, 224
    and plants, 75, 151, 233
    relevance of data, 180–2
heredity, 61–2
herring, 192, 195
heterogeneity and diversity, 75, 94–5
heuristics, 17, 22, 27, 35–7, 105, 171, 225
    causation, 135–6
    competition, 267
    historical explanation, 171
    microcosms, 216
    natural selection, 73
    positive, 55–6
    predictive ecology, 291
    r-, K-selection, 203, 210
    succession, 203, 210
hierarchy, 114, 161, 274, 291–2, 299–300, 303
historians of science, 23
historical explanation, 147–77
    competition, 266–8
    ecology, 154–5, 167–72
    non-historical explanation, 154
    null models, 218
    role of, 167, 170–3
    six modes, 155–67
historical laws, 149–54
historicism, definition, 107

history, 147
holism, 110–11, 114, 274, 303–4
    defined, 107
homeostasis, 75, 94–5
Hubbard Brook, 173
Hutchinsonian niche, 91
Hutchinsonian ratios, 217, 226, 238
hydrology, 183, 277–81
hymenopteran sociality, 66
hypertheory, 707
hypothesis, 13–15, 20, 22–3, 101, 246, 290, 294
    competing and working, 56, 81, 274, 290, 294
    evaluation, 24–5
    frequency in ecology, 322
    selection, 223–8
    inferential, 181
hypothetico–deductive science, 18, 21–6, 52–3, 225
hysteresis and stability, 94

iconoclasm, 5, 225
impact factor, 8–10
    assessment, see environmental impact assessment
immediacy, 181–4
    citation, 8–10
    predictive ecology, 304
    theory, 29–30, 178
immigration, 205–6
importance, see relevance
inbreeding depression, 156
independent variable, see predictor variable
indicator organism, 75
indices, 184–5
indirect effects, 269
individual explanation, 155, 164–6
    tautology, 241
individualistic view, see continuum view
induction, 131, 214
industrial melanism, see peppermoth
inferential test, see superficial and partial tests
information theory, 24
ingestion, 281
insect damage, 102
insight, 23, 225
instructions to authors, 220
instrumentalism, 105–6, 201, 274
    and causation, 129, 133, 136–46
    definition, 107

and experiment, 137–40
  IQ, 144–5
  predictive ecology, 296–304
interactions
  environmental, 50–1
  trophic, 86
International Biological Program, 16, 83, 116, 127
intrinsic rate of increase, 54, 56
Introduction, 221–9
  discussion, 222, 240
  model, 226–7
  proposal, 222
introductory texts, 15
invisible colleges, 136, 185–6, 297
island biogeography, 188–9, 199, 205–6, 246, 289

jargon, 40–1, 50
  conceptual conflation, 104
jerryrigging
  evolutionary, 69
Journal of Ecology, 3
just-so stories, 161

Kaibab plateau, 130
keystone species, 75
kinetic relationships, 154
knowledge, 21, 29–31
  hypothetical nature of, 30
  scientific, *see* theory
K, *see* carrying capacity
K-selection, 48
  energetic efficiency, 156
  unAmerican fallacy, 242

Lago d'Orta, 130
Lake Constance, 110
Lakes Erie and Ontario, 174
lake loading, *see* phosphorus
Lamarckism, 64
land management, 2
land use and phosphorus loading, 135
law, 20, 106, 149–54
  developmental, 152–4
  empirical, 150–2
  historical and scientific, 148–50
  limited generality, 151
  physics, 109
  universal, 150
leafing out, 152
life history, 75, 186, 210

generality, 214
  optimality, 161
liming of lakes, 12, 211, 221–2, 299
limited generalizations, 155, 163, 167
limiting factors, 75, 276
  and experiment, 138
limiting similarity, *k*, 97, 226, 297
  theory of, 263–4, 271
limits of science, 1
Limnology and Oceanography, 7, 9–10
linkage, 63, 71
lobbies, 136, 179
Loch Ness monster, 97
logarithmic transformation, 238
logic, 18, 21–7, 221, 240–5, 255, 263
  definition, 18
  lapses, 228
  models, 39, 192
  nature, 72
  theory, 38, 53
logical
  comparators, 55
  fallacies and flaws, 182, 240–50
logically-black-is-white slide, 242–3
logico-deductive constructs, 52, 60
logistic equation, 24, 38, 54–6, 72, 179
  possible fad, 230
long term ecosystem study, 173
Lotka-Volterra, 24, 56–8, 160, 193, 262, 264–6
Louis IX, Saint, 164
lumber yield, 102–4

macrophytes, 114
maleability and stability, 94–5
mammal, definition, 77
management, 12, 187
  black flies, 288
  fisheries, 102–3, 191, 187
  forest, 2
  resources, 92
manipulability, manipulation, 132, 143
mate competition and optimality, 161
materialism, 107, 112
mathematical theory, 16
  as a deduction, 26
  and simplicity, 35
mathematistry, 230–1
maximum sustainable yield, 75, 103, 192, 204
  graphical analysis, 204
mean-variance ratios, 49

measurement,22
  costs, *see* economy
  error, 136
  required replication, 120–1
mechanism, 14, 111–28, 140, 146, 290
  alternatives, 246
  applied ecology, 224
  begging the question, 301
  competition, 257, 260, 268–70
  definition, 107
  irrelevance, 181–4
mechanistic explanation, 206
medicine, 174–5
Mediterranean-type ecosystems, 184
megafaunal dispersal, 224, 247, 248–9
megalomania, 252
megaparameters, 299–300
Mendelian genetics, 31, 71
mercury, 164
mesocosms, 214
metabolic rate, 281
  contaminant uptake, 281
  homeotherms, 294
  man, 126
  poikilotherm, 213
  production, 244, 281
metaphor, *see* analogy
metaphysics
  defined, 107
  Popper, 106, 107–10
  universality, 150
metatheory, *see* hypertheory
Methods, 229–35
microcosms, 143, 214, 285
milk
  dairy farmers, 130
  trophic dynamics, 84–5
minim, 253
mistraining, 188
model, *see also* simulation
  calculating, 20, 205
  calibration, 116
  complicated, 298
  computer, 246
  continuously stirred mixed reactor, 281, 290
  explanatory 20, 106, 108, 142, 301
  semi-empirical, 282
  support, 437
  statistical and empirical, 108, 114, 301
  systems models, 115–17, 128, 146
  theory, 20, 105

model systems, 214–16
morphology
  competition, 161
  passerine foraging, 184
mosaics and typology, 100
multiple causes, 131, 134, 241, 269
multiple working hypothesis, *see* hypotheses, competing
multivariate statistics, 184–5
mutualism, 244, 257–8, 270

narrative explanation, 155, 163–4, 167, 267–8
natural history, 72, 147, 149, 175–6, 219
  art of, 147, 175
  folk art, 176
natural selection, 38, 60–73
  appeal, 71–3
  application to other fields, 72
  conflation, 82
  design, 69–70
  evidence for, 67–8
  explanatory power, 65–6
  heuristics, 73
  logical model, 72
  prediction, 70
  tautology, 60–5
  ubiquity, 66–7
Nature, 7, 227
nature-nurture, 144–5
necessity, 131–2, 181–4
negative evidence, 148, 238–9
neutralization, *see* liming
neutral selection, 67
new synthesis, 186
niche, 75
  bluegreen algae, 92
  breadth, 265, 293, 300
  conflation, 80–1, 91–2
  differences, 97
  n-dimensional, 91
  overlap zones, 101
  theory, 20
  thermal, 92
nitrogen, 180, 277
nominalism, 107
non-concept, 74, 81
  cause, 128
  dichotomies, 87
  diversity, 95
  lake trophic classes, 83
non-operational relations

competitive exclusion, 97
diversity-stability, 96–7
non-predictive constructs, 20–6, 36
non-sequitur, see logical lapses
notation, 35
normality and log-normality, 238
normal science, 20, 56, 220
normic explanation, 155–7
competition, 267
no-true-Scotsman move, 241–2
novelty, see originality
null models, 55, 198–9
economy, 217
nutrient, see also phosphorus
free levels, 88
possible fads, 226

observation, 23
Ockham's razor, 35, 132, 157, 217, 267, 298
octanol–water partition coefficient, see
partition coefficient
Oecologia, 3
Ohm's law, 24
oil drop experiment, 31, 158
Oikos, 3
oligotrophy, 82–3
omnibus terms, 74, 81
competition, 261
dichotomization, 87
omnivore
density, 19, 240
classification of, 83
food web analysis, 200
North American birds and mammals, 84
no-true-Scotsman move, 242
Ontario lake management, 281
open-endedness, 74
competition, 259–63
concept of environment, 90
operationalization, 30, 32, 34, 74–104, 275
competition, 259–63
definition, 74–7
importance, 76
maintenance, 77–8
and theory, 76–7
opinion, see authority
optimal foraging, 75, 161, 186
purpose of, 112

optimality, 159, 161
diet theory, 222
organicism, 107, 110

organism and environment, see dichotomy
organic chemicals, 284–5
originality, 226–8, 244
outliers
historical explanation, 168
logical fallacy, 254
peak analysis, 238
over-interpretation, 165, 248–9
overlap, see resources
ozone, 11, 73, 102

palynology and economy, 216
panchestron, 74
competition, 267
environment, 90–1
niche, 91
panselectionism, 64, 67
paradigm, 186, 188
parameters, 115, 119
parametric analysis, 238
parasitism, 270
parsimony, see Ockham's razor and cause,
132
partition coefficient, 183, 285–6
pathetic fallacy, 245
pattern, 303
peak analysis, 236–8
penetrance, genetic, 71
peninsular effect, 108
peppermoth, 61, 65, 150
persistance and stability, 93
perturbation
correlation and manipulation, 143
and stability, 93–5
pest control, 102, 288
phases in research, 21–3, 25
personal, 79
public, 26
phenology, 281
phenomenology, 23, 107
phosphorus, 92
abatement and loading, 12, 24, 88, 117,
122, 138, 174, 182, 211, 274–5, 277–81,
302, 304
concentration, 279
cycling, 115–17
export, 278
response, 133, 143, 150, 182, 210, 275–7,
294
phylogeny, 153, 158
historical explanation, 168
physical processes, 183

physics, 3, 109, 133, 152
physiological relations, 246, 281–3
phytometer, 262–3
pilot study, 233
piscivore and planktivore, 143
plankton size, possible fad, 226
plantations, 193
plant life form, 183, 274, 283–4, 297, 302
plausibility and rationalization, 39, 157,
    194, 247, 362–3
  competition, 260, 286–8
  false test for cause, 136
  natural selection, 72
  top-down control, 144
pleiotropism, 63, 71
pluralism, 25, 270, 289–91
policy and politics, 12–13
  qualitative theories, 210
pollution, 98
population
  as a class, 80–1
  cycles, 75
  distributions, 49
  fish, 103
  growth, 54–8
  human, 253
  stability, 95
  regulation, 5, 15, 145
post-hoc-ergo-propter-hoc, see whatever-
    follows-must-be-the-consequence
    fallacy
power analysis, 233
practicality, 34, 37, 219, 274
  competition, 265, 272
  predictive ecology, 296–304
precipitation, 275–6, 278–80, 283
precision, 32, 34, 36, 136, 178, 189, 196–
    211, 271, 275
  historical explanation, 169
  predictive matrices, 294
preconditions, see boundary conditions
predation, 257, 270
predator-prey relation, 52, 75, 276–7
prediction, 23
  Darwinian, 158
  ecology, 1
  explanation, 106–10
  future scenarios, 175
  global, 12
  inability, 5
  vs logic, 53
  operationalization, 78

weak, 178–219
predictive ecology, 274–304
predictive information value, 196
predictive matrices, 290, 294–6
predictive power, 34, 106, 147, 149, 178,
    301
  limiting similarity, 265
predictor variable, 27–8, 31–2, 216, 300
premises, 39, 51
  natural selection, 61, 68
prescription, epistomological, 25
prestige, 189, 194
prey choice, 159, 169–70
primary production, 103, 226, 230, 235
  evapotranspiration, 88, 183, 230, 282
primates, 214
prim science, 53–4, 265
priority, temporal, 131, 244
Procyon, see raccoon
productivity, 103, 115
  respiration, 274, 281–2
  scientific, 187
progress, scientific, 16
proof, 38
proposal and introduction, 222
pseudocognate, 74
  stability, 954
pseudo-explanation, 108
pseudo-refuting description, 245
pseudoreplication, 233–4
publication, necessity, 218, 222
public health, 2
pure research, 179, 187
purpose see goals, relevance

Q_{10}, 241
QSARs, 284–6
quackery, 31–3, 174
qualitative theories, 196–211
  competition, 271
  costs, 210–11
  predictive ecology, 283–4
qualitative trends, 126
quantification, 32, 37, 178, 196, 275
  allometry, 275
  comparisons, 248
questions
  for discussion, 101
  fallacy of many, 245
  intractable, 13–15, 258–9, 297
  tractable, 224, 297
  unit, 114, 128

rabbits, 26, 53
raccoon abundance, 18–21, 28–33
  maximum, 31
random
  communities, 96
  distribution, 49
  models, 217
ranking, 49
rate constant of contaminant clearance, 42
rate maximizers, 169
rationalization, *see* plausibility
rational reconstruction, 155, 166–7, 175
  competition, 267–8
realism, *see also* operationalization, 105–10
  and cause, 128
  goal in modelling, 118–119
reality, 53, 105
reassembly, mechanistic models, 269
reconceptualization, 102–4
reduction and reductionism, 105–46
  competition, 269
  dead issue, 111
  definition, 107
  holism and, 111
  methodological rule, 111
redundancy, 48, 189
reference system, 231
refutation, 16, 22, 26–7, 197
regress
  cause, 133, 142
  instrumentalism, 142
  mechanism, 269
  specificity, 212
regression
  all subset regression, 239, 302
  imprecision, 196, 210
  peak analysis, 238
  predictive ecology, 290, 301–2
  qualitative theory, 210
  replication, 139
  representativity, 139
  and theory, 18–20, 275
regularity, 128, 132–42
regulation of population, 4, 145–6
  intractability, 79–80, 94–5
regulations, hunting and fishing, 299
rejection, *see* refutation
rejection rates, 8–10
relativism, 74
  competition, 259
  environment concept, 90–1
relevance, 29, 34, 36, 137, 178–89, 254

data and theory, 180–1
  field, 252
  introduction, 221–3
  predictive ecology, 304
  societal, 186–9
  vacuous contrast, 249
reliability, 234
renaissance, 163
replication, 121, 139, 231–3
representativity, 139, 232, 234–5, 249
reproductive investment, 219
research topic choice, 187
resemblance and causal connection, 129
resilience and stability, 93–5
resources, 102, 274–7, 304
  conflation, 91
  graphical analyses, 207–9, 264–5
  limitation, 101, 260, 276
  megaparameter, 299–300
  overlap, 264–6, 271
  partitioning, 97
  utilization function, 264–5
respiration, *see* metabolism
response
  linear vs interactive, 119–22
response variable, 27–9
Results, 235–9
retention 279–80, 294
reviews
  qualitative nature of, 296
revolution in science, 16
Rhine River, 11
Ricker models, 191–3
rigor in ecology, 1
river blindness, 287
river continuum, 226
r, K-selection, 48, 75
  continuum, 209
  dichotomous classification, 48, 87
  qualitative analyses, 202–3, 210
  trend analysis, 202–3
rock nuthatch, 229
rooting depth, 282
rules, empirical, 150
runoff, 278–80

salience and causal attribution, 129
salivary factors, 180–2
sampling bias, 22, 232
satiation, 204
scale, 74
  algology, 161

scale (*cont.*)
  black fly models, 287–8
  competition, 260–1
  confusion, 182, 246
  effects, 139
  extrapolation, 253
  environmental definition, 90
  logic, 241
  phosphorus loading, 135
  theory,114
scatter diagram, trend analysis, 238
scenarios, 175
  competition, 266–8
scholasticism, 60
  abstraction, 101
  stability, 93
science
  components, 21
  definition, 18, 25
  demarcation, 148
  hard and soft, 2, 9–10
  historical and history, 147
Science, 7, 227
Science Citation Index, 7
scientific paper, reading order, 229
scope, *see* generality
searching cost, 209
selection, 61–2
selective comparison, 248
self-correlation, 41–6
self-delusion, 248
self-thinning, 45, 75
semelparity-iteroparity dichotomy, 87
seral stage, 40, 201
  and typology, 100
shifting ground in debate, 242–3
significance level, 243, 249
simplicity, 17, 34–8, 265, 275, 298–9
simplification, 118, 274
  food web analyses, 199
simulation, 16, 146
  calibration, 116
  economy, 119–21, 217
  examples, 115–16
  lake, 115, 127
  limitations, 116, 270
  possible fad, 230
  purpose, 126–8
  qualitative analysis, 210
  resource competition, 206–8
  top-down control, 143
Simuliidae, *see* black flies

singularist causes, 131, 172–3
Sitta, see rock nuthatch
size efficiency hypothesis, 186, 226
SLOSS, 188
sociobiology, 161
sociology of science, 23, 186
solubility, 285
sorites, see heaper
sorption, 285
specialization, 48, 87
  tunnel vision, 113, 135
special pleading, 247
species
  -area relations, 75, 246, 289, 294
  classification, 80–1
  diversity, 24, 183, 205, 274
  most sensitive, 215
  number and connectance, 47
  number and stability, 96, 204, 289
  packing, 75, 97
specification of complex models, 119–22, 207
specificity, *see* generality
spurious analysis, *see* self-correlation
statistics and causal attribution, 130
stability
  concept, 292, 297
  components, 92–7
  cyclic, 93
  definitions, 93–4
  diversity, 161
  global, 94
  intractability, 79–80
  neighbourhood, 94
  neutral, 94
  structural, 94
  trajectory, 94
standard
  procedures, 229
  role in operationalization, 77
standing stock of grasslands, 276
statistical models, 108, 238, 301
  rules, 231–3
steady-state, 94–5
stochasticity, 275, 302
  and evolution, 15, 63, 75
stock recruitment, 191–3, 204
strategy, 167, 245
stress, 94–5, 202, 289, 297
strong inference, 295
style, 218–19, 221
subjectivity

concept, 96
  evaluating ecology, 9, 17–18
  megafaunal dispersal, 248
subject/motive shift, 244–5, 268
succession, 24, 201–2, 209
  classification, 87
  competition, 267
  models, 154, 214
  phytoplankton, 152
  tree-by-tree, 154
  theory, 153
sufficiency and causal attribution, 129,
    181–4
  competition, 260
  predictive ecology, 304
superorganism, 5, 80
  conflation, 91
supervenience, 69
survivorship and fitness, 62–4
syllogism, 39, 225
symposia, 5–6
synthesis, 21–5, 161
systems analysis, *see* model, simulation

tabulations and colligative explanations,
    162
tautology, 25, 38–73
  behavioral goals, 133
  characteristics, 58–60
  community assembly rules, 197–9
  competition, 263–6
  definition, 38–9
  explanation, 265
  Lotka-Volterra, 160
  natural selection, 60–73
  r-, *K*-selection, 202–3
  succession, 201–2
  theory, 73
  verbal models, 201
Taylor's power law, 49
team research, 16, 112
technological solutions, 229–30
teleology and teleonomy, 153
temperature and plant form, 283
terminal science, 144, 185–6
terminology, *see* jargon
territoriality, 75, 165
testability of hypothesis, 14, 17, 275
tests, 26–7, 222
  superficial and partial, 182, 246–7
textbooks
  intractable questions in, 13–15

theorems, 38–9, 59
theory
  of animal abundance, 19–20
  Cartesian model of, 26–8
  data, 30–1
  definition, 18–20, 36–7
  empirical, 108, 303
  example, 19–20
  explanation, 105, 303
  law and, 20, 106
  logic and, 38–9, 53
  non-theory, 21–4
  operationalization and, 76–7
  prediction and, 23, 105–6
  provisional nature of, 293
  risk, 212
  tautology and, 73
theoretical ecology, 15, 38, 51–60
  English, 169
  models, 105
  specificity of its critics, 212
thermal balance, as a purpose, 112
thermodynamics
  as a hard core, 71
  laws, 151
  trophic dynamics and, 83
top-down approach and causality, *see*
    hierarchy
top-down control, 143, 226
toxicity, 284
trade-offs, 191
transfer functions, 115, 119
transparency, water, 102, 110, 235
trend analysis, 196, 201–3, 248
  competition, 271
trivia, 197, 201
  and novelty, 227
trophic cascade, 143–4, 186
  possible fad, 226
trophic dynamics and levels, 75, 83–6,
    103
  purpose, 12
trophic links, 46
trout, thermal niche, 92
truisms
  environmental, 98–100
  first maxim for Balliol men, 244
  normic explanation, 155, 167
truth, 105, 293
truth-is-always-in-the-middle damper,
    243–4
tunnel vision, 113, 135, 250–1

Tyler prize, 24
typology, *see* classification

unAmerican fallacy, 241–2
uncertainty, 29–30, 33, 122–6, 151–2, 216, 265
    statistics, 231–5
understanding, 14, 28–9, 147–77
    art and natural history, 176
    definitions, 148, 156
    prediction, 170
    undefined research, 225
undistributed middle, 242
unicorns, 26
uniformitarianism, 157
unique systems 168, 172–5, 212
    singularist causation, 172–3
    medical analogy, 174–5
unit phenomena, *see* unit questions
universality, 150
universal laws, 149–50, 167
    empirical rules, 150–2
universal statements, 149
unusual events, 158, 247
utility, 29, 199

vacuous contrast, 249–50, 302
vagueness, 4, 181
    competition, 271
    limiting similarity, 266
    qualifiers, 262–3
vandalism, 139
variables *see also* predictor and response variables
    potential, 102–4
    sound, 274, 299–300
    state, 115, 119
    uncontrolled, 138, 268
    winnowing, 216
variance, 49, 122–6
    explained, 114

reduction of, 136
temporal-spatial, 231
variation, 61–2, 232, 268
verbal models, 196–201
    competition, 263
verification, *see* confirmation
Viburnum whitefly
    all subset regression, 239
    pseudoreplication, 233
vicarious re-enactment, *see* rational reconstruction
visionaries, 23
vitalism, 107, 110

water balance, 282
water mixing, 207
water quality indices, 85
water residence time, 280
watershed characteristics, 182
    phosphorus loading, 277–81
Whatever-follows-must-be-the-consequence fallacy, 244
Why questions, 141, 25, 301
wisdom and ingenuity, 226, 254
wombat, 215
woodpecker, 196
writing, *see* style

yellow fever, 178

zebra, 67–8
zero growth isoclines, 57–8, 208
zero-one rule, 159, 170
zooplankton
    biomass and chlorophyll a, 235
    excretion, 117, 121–2, 248
    feeding, 110, 186, 207–8, 230
    general models of feeding and excretion, 213–14
    herbivory, 143